MW01096951

The Three Sisters

Scrivener Publishing
100 Cummings Center, Suite 541J
Beverly, MA 01915-6106

Publishers at Scrivener
Martin Scrivener (martin@scrivenerpublishing.com)
Phillip Carmical (pcarmical@scrivenerpublishing.com)

The Three Sisters

Acid Gas Injection, Carbon Capture and Sequestration, and Enhanced Oil Recovery

Edited by
Ying Wu,
John J. Carroll and Yongle Hu

WILEY

This edition first published 2019 by John Wiley & Sons, Inc., 111 River Street, Hoboken, NJ 07030, USA and Scrivener Publishing LLC, 100 Cummings Center, Suite 541J, Beverly, MA 01915, USA

For more information about Scrivener publications please visit www.scrivenerpublishing.com.

Wiley Global Headquarters
111 River Street, Hoboken, NJ 07030, USA

For details of our global editorial offices, customer services, and more information about Wiley products visit us at www.wiley.com.

Library of Congress Cataloging-in-Publication Data

ISBN 978-1-119-51006-2

Cover image: Natural Gas Plant, Sompong Rattanakunchon | Dreamstime.com
Cover design Kris Hackerott

Set in size of 11pt and Minion Pro by Exeter Premedia Services Private Ltd., Chennai, India

10 9 8 7 6 5 4 3 2 1

Contents

Preface

The seventh edition of the International Acid Gas Injection Symposium (AGIS VII) was held in Calgary, Canada in May 2018. The Symposium covers topics related to acid gas injection (AGI), carbon capture and sequestration (CCS), and the use of CO_2 for enhanced oil recovery (EOR). This volume is a collection of select works presented at the Symposium.

The Keynote address was presented by Jim Maddocks, the CEO of Gas Liquids Engineering in Calgary, Canada and this is the first chapter of the book.

For the first time there was an award for the Best Student Paper, which was presented to Hanmin Tu. Miss Tu is doing a joint PhD at Southwest Petroleum University in Chengdu, China and the University of Regina in Regina Saskatchewan. Her paper is Chapter 6.

Another paper worth noting was our first contribution from Japan. Yoshihiro Sawada presented a review of the Tomakomai CCS Demonstration Project in the Hokkaido Prefecture in northern Japan. This is Chapter 12 in this volume.

Alice Wu wrote a history of AGIS up to and including this book. In her paper, Chapter 2, you will discover why we chose the title for this volume.

YW, JJC, & YH
June 2018

1

Acid Gas Injection: Engineering Steady State in a Dynamic World

Jim Maddocks

Gas Liquids Engineering Ltd. Calgary, Alberta, Canada

Abstract

Acid gas injection (AGI), while widely used throughout the energy industry, still has significant learning and development opportunities, particularly as flows and pressures increase, and design engineers begin to stretch the limits of conventional AGI design thinking.

This paper and presentation is intended to provide some background and thoughts on the nature of acid gas injection, the uncertainties, the possibilities, and the myriad of issues that can be present during a typical design cycle. With a significant number of rapidly changing inter-related process variables, the engineering/design team needs to consider acid gas injection as a very transient and dynamic process. Adapting the design process and altering the way we think about, approach, design, fabricate and finally operate acid gas injection systems requires some innovative critical thinking skills, some learning on the custom nature of the systems involved, and finally the realization that the AGI system will ultimately have to adapt to the environment.

Keywords: Acid gas injection, compression, water content

1.1 Introduction

Acid gas is composed of a mixture of H_2S and/or CO_2 and usually water vapour. Acid gas, a byproduct of gas treating systems, is often considered to be a simplistic binary mixture of H_2S and CO_2. There are frequently other contaminants including methane, BTEXs, amine, and other hydrocarbon

Corresponding author: jmaddocks@gasliquids.com

Ying Wu, John J. Carroll and Yongle Hu (eds.) The Three Sisters, (1–22) © 2019 Scrivener Publishing LLC

components. Carbon capture streams are typically pure CO_2 although there are other contaminants co-captured with the carbon dioxide.

The very nature of acid gas compression and injection means that these AGI "systems" are nothing more than the glorified "garbage trucks" of gas processing. The design team has a very limited ability to manage the inputs to the process, a very limited ability to control the outputs, and the system barely even gets to talk to the upstream "garbage generator". Basically, it's desirable that this "garbage truck" simply do its job, quietly, efficiently, and as cheaply/painlessly as possible with minimal operator intervention and low maintenance.

1.2 Steady-State Processes

As engineers, we're trained to understand, develop, and then control steady-state processes. We do this hundreds of thousands of times with pumps, compressors, turbines, heaters, coolers, and a multitude of other processes. This is one way of simplifying a design problem. In many cases, we don't know anything other than steady-state and we proceed blindly assuming that the process will operate in "steady-state". Nothing could be further from the truth.

Every process, every moment, and every single second of our day is filled with processes in constant transition. Everything from our furnace, to our cars, to the internet, is filled with constant change. Nothing is steady-state. The temperature of your house is constantly changing even with the best "smart" thermostat. The speed of your car is always changing. Your heart rate goes up and down according to external and internal stimuli, energy and forces. Even a simple system like cruise control on your vehicle is constantly experiencing micro-changes in inputs (like hills, curves, vehicle mass, and wind) and the system control output is expected to compensate for all these changes. It does this (usually seamlessly) and the user seldom notices the minor variations in speed.

Every part of our world is constantly in a state of dynamic transition. The weather is never constant; our ambient is always warming or cooling slightly, moisture content is changing, and the system is adapting. The belief in steady-state performance is a misnomer.

In order to understand and evaluate the dynamic nature of this process, it's important to establish a different way of understanding this process.

Part of this learning process involves the use of "critical thinking skills". These skills are defined as:

> Critical thinking is that mode of thinking — about any subject, content, or problem — in which the thinker improves the quality of his or her thinking by skillfully analyzing, assessing, and reconstructing it. Critical thinking is self-directed, self-disciplined, self-monitored, and self-corrective thinking. It presupposes assent to rigorous standards of excellence and mindful command of their use. It entails effective communication and problem-solving abilities, as well as a commitment to overcome our native egocentrism and sociocentrism [1].

This style of thinking forces us to reconsider our basic assumptions in a design process – it's an essential tool in the design of an acid gas system.

As engineers and designers, we're expected to make steady-state approximations because they allow us to essentially draw a design line in the sand. Obviously, if a client stated that the compressor would see a flow varying from 30 to 60 e^3m^3/day of acid gas, we'd likely establish a "steady-state" design flow condition of 60 e^3m^3/day and then find a way to adapt to the low flow case. However, even recognizing that this high flow case is unsteady and dynamic, there are dozens of other non-steady-state variables in the system. Many are unnoticeable, some actually cancel each other out, and finally, some compound themselves into potentially serious process issues. Many of these changes take place over seconds, hours, days, or even months as the machines and processes begin to experience wear. Even the frequency of the change itself is changing. In order to fully understand, design, and control a process, we have to know that our understanding of steady-state is at best, a semi-educated guess.

1.3 Basic Process Requirements

1.3.1 Process Needs

The acid gas streams are often captured at low pressure (40–80 kPa[g]) from either a gas treating facility or a carbon capture system. Carbon dioxide (or CO_2) gathered from EOR systems may be captured at moderately higher pressures (170 kPa[g]) and pure makeup CO_2 supply pressures are often higher. As the AGI equipment and injection process is often downstream of many other larger process units, the acid gas system is expected to handle everything extracted in the amine unit or recovered from the reservoir. This means that flows, composition, temperature, and often pressures are highly variable and can change quickly without notice (and often without apparent reason). In order to prevent process upsets, shutdowns, and potentially regulatory non-compliance, it's important that the acid gas injection system be able to adapt quickly (and with stability) to the changes.

1.4 Process Input Variabilities

1.4.1 Compositional Variances

While an acid gas injection system is usually designed as a steady-state process, it is far from a steady-state operation. Typically, the EPC engineering team requests a primary plant feedstock analysis from a different business unit within the owner's company. In many cases, this feedstock is a known parameter and the owner company has a high degree of confidence in the analysis. However, in many newer developments, particularly in shale gas and tight gas developments, the field or reservoir has insufficient flowing history. In some cases, the Owner development engineering teams "take their best shot" at the anticipated composition. In order to provide for maximum regulatory and design flexibility, these teams often inflate or exaggerate the H_2S fraction in the feed gas to reduce the capital risk of under-design. In many cases, they aren't aware of the risk of having a large design allowance. The end result is that the composition of feed gas to the facility is often dramatically different than anticipated. As the field development matures, the composition can also evolve – in some cases, this may mean:

- A changing hydrocarbon content with altered gas dew points and heavy components like C_{5+};
- H_2S and CO_2 can vary in either direction – sometimes rapidly;
- New and unexpected components may appear including oxygen, elemental sulphur, mercaptans, toluene, benzenes, and COS.

In some cases, the feedstock may contain other unexpected contaminants. Various wellhead treating chemicals and production chemicals like wax dispersants, sulphur solvents, triazine, hydrate inhibitors including methanol, and asphaltenes solvents can be troublesome for amine plants and can trigger operational upsets and foaming events. These foaming events are often random and unpredictable and can:

- Alter the pickup of H_2S and CO_2 by shifting system kinetics and mass transfer

- Potentially alter the co-adsorption of hydrocarbons
- Result in significant (and rapid) flow variances as the foam breaks and re-forms

Often, these changes are simultaneous and rapid, making the prediction of acid gas compressor steady-state feedstock a challenging issue.

Redeployment of facilities often means that the industry is installing equipment in a service that may not be an ideal fit for the system. In other cases, the owner companies may need to significantly customize or even swap the solvent in the process. These seemingly minor process changes can have a dramatic effect on how the system performs.

In the natural gas midstreaming industry, it's common for the midstream clients to vary (intentionally or otherwise) their production rates, compositions, water content, and even contaminants. While midstream operators usually specify maximum contaminant levels (wax/asphaltenes/solids), it's difficult to manage or control to the degree required and some producers use the midstreamer as a receiver of all things. Producers can (and often do) inject unusual chemicals and experimental well treating technologies that can cause havoc inside an operating amine plant.

1.4.2 Flow Variances

Flow variances are often the most frequent, and most obvious (and thought to be the most challenging) to respond to in the design and operation of an acid gas injection system. As noted above, it is not uncommon for amine plant regeneration systems to swing and allow for a significant (and rapid) variation in the flow of Stage 1 acid gas. Designers expect a very flexible control system as the requirements can go from design flow to less than 25% (4:1 turndown) in a matter of seconds if the amine system and/or control system is unstable. Since the majority of acid gas compression systems are reciprocating positive displacement compressors, this significant decrease in flow is a dramatic change in 1st stage suction volumetric flow. Even minor changes in volumetric capacity must be managed quickly otherwise suction pressure will rise and fall rapidly. While the compressor is designed often to a single design point, the amine regeneration system will deliver the acid gas depending on the regen system performance. This amine regeneration performance is dependent on a number of system variables:

- System cleanliness and "normal" contaminants
- Ability of the system to generate stripping steam

- Control system performance, valve performance, and system dynamics
- Size of the system
- Amine Regeneration Condenser performance
- Time and thermal lags
- Amine reboiler thermal momentum, driving force and performance

Most design teams are able to manage the flow changes with process control, typically a combination of speed and auto-bypassing; however, these systems need to have a rapid response without generating their own inherent process instability. Previous papers have examined the need for acid gas compressor capacity management, but it should be noted that the bypassing arrangements may not be identical as the composition shifts. Depending on the shifting composition, pressures, and phase behaviour, the bypass valves may not perform as required and the system will fail. Engineering the bypass assembly is a critically important step in acid gas system design.

1.4.3 Initial Suction Temperature Variances

Temperature variances are somewhat less disruptive as long as the system has the mechanical limits to manage the temperature swings. The primary issue with feed acid gas temperature is the changing water content. Clearly, lower inlet gas temperatures will suppress water content and will lower the amount of water rejected by the AGI compressor. It should be noted that, while this does lower overall water recovery rates, lower initial suction does not alter the final hydrate in the injection fluid.

As well, lower temperatures will lower the 1st stage discharge temperature and will also lower the volumetric requirement. For example, decreasing suction temperature from 30 to 10 °C will increase available compressor capacity by just over 4% with only a 2% increase in required power. Water content will decrease considerably but this usually has very little impact if the remaining stages operate as before. In addition to the water content, rapidly swinging inlet temperatures will trigger intercooler instability. If the system is operating well, the 1st stage after cooler will manage this bouncing temperature and will dampen out the variances. However, in some cases, oscillating cooler inlet temperatures can put the process cooling system into a poor response resulting in even bigger cooler outlet temperature swings leading to an eventual shutdown.

Even moderate ambient temperature swings can have a significant effect on cooling performance (both in discharge cooling as well as piping systems) and corresponding compression system consistent performance. This means that part of the success of any capacity management system is predictable temperature stability. This is even more important in an acid gas compressor where temperature, composition, and water content are closely connected. As well, long suction headers can introduce a time lag or system delay into the response of the compressors; long piping runs can also provide for condensation, fluid buildup, slugging, and carryover.

Poor or incorrect designed recycle systems within the compressor can have an effect on system Stage 1 suction temperatures. Excessive use of the cold recycle arrangement can begin to swing the suction temperature. It is exceedingly difficult for a compressor aftercooler to manage this as the temperature instability will be pushed through the machine. This, in turn, will likely generate more system instability.

1.4.4 Initial Suction Pressure Variances

First-stage suction pressure has the most direct impact on system capacity. At the lower suction pressures common in most AGI systems, even a 10 kPa decrease in suction pressure will have a major effect on available system capacity. Available compressor capacity drops by almost 8% with a similar power drop with only a 10 kPa drop – this reinforces the need for good suction pressure control as well as consistent system performance. Given the low system pressures, the system must control to a very precise pressure. While 50–100 kPa[g] swings in a normal natural gas compressor application suction pressure are not uncommon (and are not usually considered a problem), this system is expected to control AGI compressor suction pressure within a 3–7 kPa range or tighter. Large swings in control will cause feedback to the amine unit and cause process instability resulting in sour amine, acid gas and main gas flaring, off-spec products, and increased operating costs. In some facilities, the acid gas compressor is used to directly manage the pressure in the regeneration system. This can be quite successful and avoids any additional pressure drops on the suction of the acid gas compressor system or pressure increases in the amine regeneration system; however, it takes careful tuning of the compressor control loops. Suction pressure loop stability is one of the most important criteria in compressor design planning. Equally important is the need to perform a full range of system sensitivity runs to establish capacity vs. suction pressures.

1.4.5 Amine Plant Generated Instability

Amine processing systems can display a high degree of inherent instability. This is generated by a number of factors including poor designs, insufficient residence time, foaming, control instability, off-design operation, ambient sensitivity, amine chemistry issues, and many others. As such, it's not uncommon for an amine process package to experience some process swings. Some of these are due to control system bounces, while others are due to ambient temperature variations, thermal momentum, and system dynamics. In many cases, this instability generates rapid changes in acid gas flows and compositions. The acid gas compression system is expected to manage this seamlessly.

Many amine plant designs utilize an MDEA-based solvent (either generic or proprietary) to manage the H_2S pickup and CO_2 slip. While the H_2S reaction is typically an instantaneous one (as long as the amine is relatively lean), the degree of CO_2 pickup is based on system kinetics and fluid chemistry. These kinetic factors are triggered by mass transfer, thermodynamics; factors which are influenced by system physical parameters like residence times, weir heights, tray counts, level control response, and of course, temperature. While reboiler temperature will be relatively stable, many of the other system temperatures including lean amine cooling and reflux condensing are variable. Thus, many of the factors that affect CO_2 slip are also driven by system factors that are also changing. In many cases, the temperature, pressure, flow and composition are changing simultaneously. It will be almost impossible to have a steady or near-steady-state operation with multiple integrated changes happening simultaneously.

Amine plant regeneration systems often have a considerable thermal time lag. This is due to reboiler kinetics and response to increased heat requirements, large thermal mass of process fluid and steel, and delays in temperature response to changes in inputs. In some cases, this thermal time lag results in system instability and temperature swings. It should be noted that increased heat input will not change the reboiler temperature but will drive increasing amounts of steam into the system until the overhead temperature begins to register the increase. In some cases, the control system will overheat the system until a series of thermal swings becomes evident and the system stability is impaired. These thermal lags usually result in system swings in acid gas rate, composition and conditions.

In addition to this, amine system contaminants, like heat stable salts, can alter amine solution chemistry resulting in erratic performance, poor absorption, off-spec products, increased corrosion, and operating upsets. The use of anti-foams (silicone or otherwise), corrosion inhibitors, physical solvent enhancers, and other "process chemicals can also alter the amine plant's performance. Many facilities use variants of MDEA which uses a mass transfer based kinetic based CO_2 pickup rather than a conventional primary or secondary amine. This means that system turndown can potentially affect the pickup of CO_2 and consequently the composition and phase behaviour of the acid gas.

Certain amine plant designs utilize amine solution chemistry that allows for the pickup of mercaptans like Shell Sulfinol™ or Huntsman DGA™. In many cases, these physical solvent additives or hybrid solutions allow for the increased pickup of heavy hydrocarbons as a co-adsorbent. While these hydrocarbons do not typically make up a significant % of the acid gas flow, they can have other deleterious impacts including:

1. Altered phase envelopes with shifting hydrocarbon dew points.
2. Altered water equilibrium data
3. Poor wellbore density

Finally, the discovery of new reserves can force the amine unit into an operational regime that was not conceived during the initial design. The redeployment of used equipment has a similar challenge in that seldom is this equipment a perfect fit for the process.

Depending on the level and type of contamination, the acid gas system has considerable potential for instability simply generated from the upstream amine system.

1.5 Process Output Variables

1.5.1 Discharge Pressure Variances

In most cases, compressor discharge is held steady either by a compressor discharge block valve or a wellhead PCV. In reality, this PCV really just sets a minimum discharge pressure that will keep the compressor in some sort of steady-state condition. The actual discharge pressure can and will vary depending on the injectability of the fluid into the formation. The operator, for the most part, has very little control over this pressure. Should the formation experience damage, become plugged with debris or contaminants, or become inoperable in some way, then this compressor discharge pressure will rise to try and push the acid gas into the formation. The nature of the injection process is that the wellbore fluid head +the discharge pressure, must overcome the friction losses, and the reservoir pressure. Thus, the success of the injections scheme is directly related to the density of the wellbore fluid, which in turn, is directly related to temperature, pressure, and composition.

The introduction of non-condensable gas into the wellbore will drive down the fluid level as the gas breakout sits in the top of the fluid level. The resulting loss of fluid head will require a corresponding increase in compressor discharge pressure. Similar to this issue, the density of the fluid in the wellbore is a function of fluid temperature – acid gas arriving hotter or colder than anticipated will alter the fluid density. Luckily in most AGI applications, the fluid pipeline velocity is so slow that the acid gas often shows up at the wellsite at ground temperature regardless of the temperature entering the pipeline. However, in shorter injection schemes, the varying temperature of the fluid can have dramatic effects on both the fluid density as well as the injection pressure.

Varying discharge pressure will certainly make itself known through interstage pressure variances. This, in turn, will alter the phase equilibrium of the acid gas, will alter the water capacity of the fluid, and will alter the cooling and water extraction needs of the system. Varying discharge pressure may move the system into a different part of the phase envelope and may impair system performance – for example, on an interstage dehydration system that needs to operate at a certain point in the phase envelope for optimal water removal. In an acid gas pumping application, a pressure move that results in fluid at pump suction at the critical point, will almost certainly result in unpredictable pump performance. As well, varying discharge pressure can have an effect on recycle valve performance.

Depending on the type of acid gas injection scheme, reservoir life and capacity can also have an effect on system performance. As the reservoir begins to fill and achieves ultimate capacity, we can expect bottom hole pressure to rise. This is seldom a rapid change but it is not controllable and will obviously result in an increase in tubing head injection pressure and ultimately a loss of injectability.

An additional variable, not often considered, is the effect of pushing acid gas into the formation. Depending on the permeability, the fluid, and the flow, it may take significant pressure to overcome the sandface differential. Varying flows and to a lesser degree, composition, can make wellhead pressure variable. The presence of wellbore issues, while seldom predictable, can also result in flow based pressure issues. Downhole check valves, improperly performing tubular equipment, undersized tubing and contaminants like solid sulphur can make the system very dynamic, and very unpredictable.

1.6 AGI Process Associated Variables

1.6.1 Variable Response Mechanisms

In many process systems, process response to a single input can be approximated by a sinusoidal wave. While process tuning attempts to reduce this offset, it is not uncommon to see this type of response in a cooler. In many cases, this sinusoidal response is small enough to be insignificant (see Figure 1.1).

In some cases, the instability of the multiple loops can actually cancel each other out resulting in a system that "appears" to have a steady-state response. As an example, this could represent an increase in suction gas flow combined with a decrease in suction temperature thus negating the swings in volumetric flow. See Figure 1.2 for a dual variable system.

Alternately, in some cases, the variable frequency is identical and in a similar fashion and the resulting overall system response can be additive. This can be represented by the response to a cooler control issue combined with a capacity control issue. The overall response is dramatically worse as can be seen in Figure 1.3 (light grey line):

However, in other cases, depending on a number of factors and process variables, the instability is propagated through the system until the final process has departed dramatically from its original setpoint (light grey line), see Figure 1.4. This is a combination of multiple variables like compressor capacity control (speed plus recycle), scrubber level control, and cooler control.

Finally, Figure 1.5 shows a series of multiple variables showing the potential overall system response on the light grey line. This is the combined

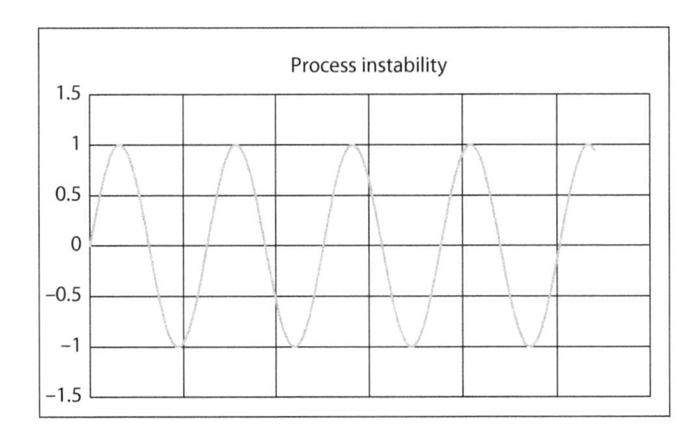

Figure 1.1 Process Instability - Single Variable.

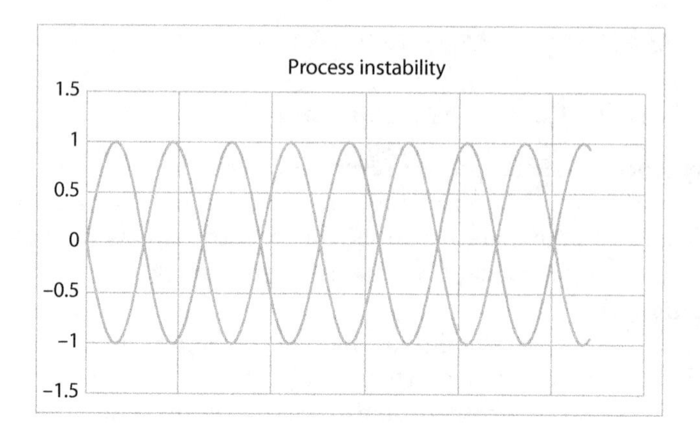

Figure 1.2 Process Instability - Dual Variable – Destructive Offset.

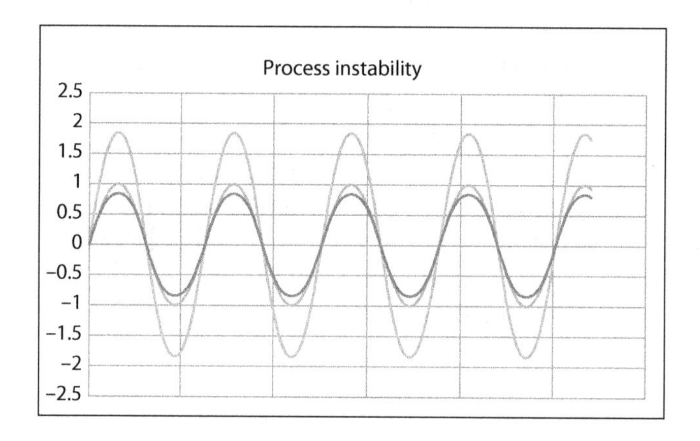

Figure 1.3 Process Instability - Dual Variable - Constructive Offset.

response of many system variables, some self-cancelling (good) and some additive. This system is unpredictable as the frequency of change is itself changing. While we've exaggerated the response to show the uncertainty, this exists in most AGI compression systems to some degree.

As shown above, each variable contributes in some way to process stability or instability. The key point to remember is that the frequency and amplitude of the process response is often different as the variables are often co-dependent. Simple, and often small, process changes can have major impacts on how the system responds. Seemingly unrelated variables can have complex interaction resulting in unexpected process changes and system instability.

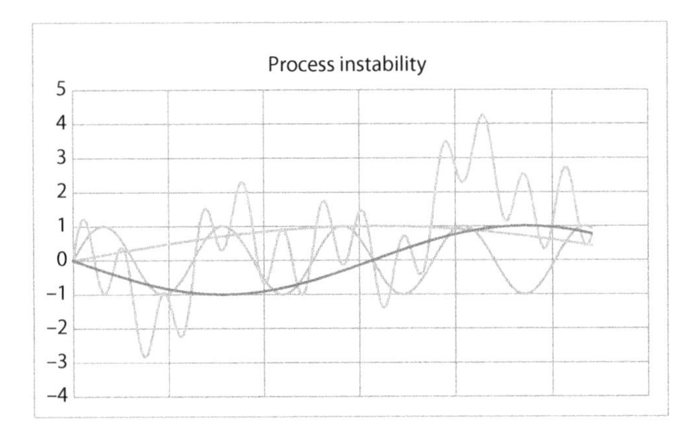

Figure 1.4 Process Instability - 3 Variables.

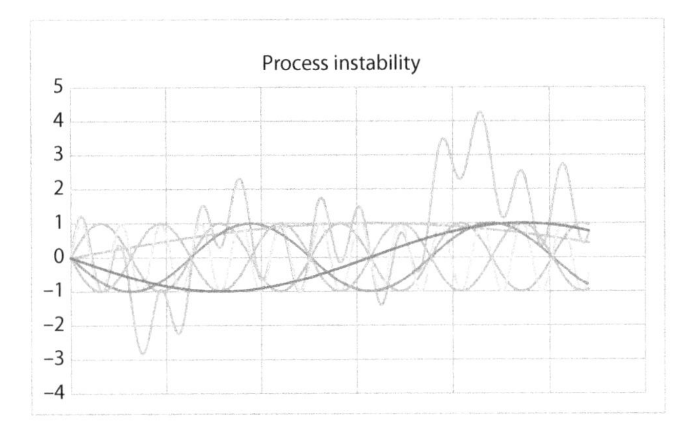

Figure 1.5 Process Instability – Multi Variable Offset.

1.6.2 Cooling Variances

Seasonal ambient temperature variations can have a considerable effect on the acid gas system. Higher seasonal ambient temperatures have a number of effects:

- High aftercooler temperatures resulting in higher residual water content.
- Correspondingly higher suction temperatures resulting in higher discharge temperatures and shorter valve life.
- Lower density of the final injection fluid resulting in higher tubing pressure at the wellhead. While this may seem negligible, it's surprisingly common to see acid gas wellhead

injection pressures peak in the late afternoon/early evening and then decline through the evening – especially during summertime.

In a few cases, this high ambient temperature (and high process fluid temperature) does provide assistance in keeping the fluid further away from its hydrate point; however, in most cases, the hot ambient also results in a substantially higher water content and a correspondingly higher fluid hydrate temperature. Given that the pipeline ditch temperature doesn't vary a great deal at burial depth and the fluid usually reaches ground temperature relatively quickly in the length of the pipeline, the higher starting temperature has little benefit and many downsides. Allowing the formation of an aqueous phase in the final injection fluid can have dramatic consequences on corrosion and material integrity. The management of the water content remains the dominant issue in acid gas injection system design.

The situation is reversed during winter months. If the acid gas cooler is designed for a 35 °C (95 °F) ambient, the typical single bundle LMTD (Log Mean Temperature Difference) is approximately 33 °C. Once the winter ambient temperature has decreased to −31 °C, the resulting LMTD jumps to 99 °C. This means that the cooler is now 3X larger than originally designed. Without the aid of warm air recirculation, this cooler will overcool the acid gas and will almost certainly condense the acid gas returning to the compressor scrubbers. In order to prevent this, design engineers often specify warm air recirculation as a way of maintaining a false ambient in the cooler. A typical warm air recirculation temperature setpoint of 25 °C (77 °F), still results in a cooler that is potentially 1.5X larger than required for the cooler ambient. Any turndown in flow or duty will exacerbate this situation resulting in a cooler that is much too large for the process conditions. This situation can be forced through the cooler as a low suction temperature on Stage 1 (due to overcooling of the amine reflux stream) now results in a low 1st stage discharge temperature. This now results in a low outlet temperature moving on to the next stage. This phenomenon moves through the system until the cooler is struggling to maintain bundle temperatures.

Cooler manufacturers are typically focused on achieving sufficient bundle tube area and adequate fan air movement during the high ambient cases. However, it is equally important (if not more so) for the design team to look at cooler performance, controls and air recirculation during cold ambient conditions. In many cases, a simplistic heating coil is installed with the intention of adding heat to the system to prevent the bundle from freezing. However, the dewpoint temperatures and even the hydrate temperature is so high that even moderately warm air will result in the possible formation of hydrates in the cooler tubes–particularly on the air side passes. This can happen during turndown operation and this may not be obvious when the tubeside pressure drop is very low. As well, system air leakage can result in the loss of much of the retained heat before it has a chance to provide warmth to the tube bundle.

An interesting effect is the dynamic nature of the air cooling system. In many acid gas coolers, it is common to gang multiple services into a single frame. In small systems, or early designs, it is not unusual to see 5–6 separate bundles installed into a single frame. Hopefully, each of these bundles is attempting to achieve the desired outlet temperature based on adjusting louver position and fan speed. As well, the cooler chamber may also be adjusting the recirculation system, fan speed (depending on configuration), and potentially even fan pitch. However, thermal systems can have a significant thermal mass that results in erratic control.

In many cases, these unstable loops do not take into account the thermal momentum of the system and they are often put into manual due to perceived instability. These temperature variations often occur during compressor throughput changes when more or less acid gas is received from the amine plant. The quicker the change in flow, the quicker and more dramatic the response.

Each stage, and each bundle/louver group, goes through a series of moves to achieve the setpoint; however, because each loops responds differently, it can be very difficult for the cooler to find its "happy place". Depending on the stability of the amine plant, the flow of acid gas can vary almost constantly.

The result of cooler temperatures variances continues. The amount of water condensing (and potentially acid gas during a phase envelope incursion), is proportional to the aftercooling temperature. Thus, the compressor suction scrubber level buildup is not a steady-state process either. In fact, a 15 °C drop in aftercooler temperature, in addition to potentially encroaching on the hydrate line, can result in significantly more liquids condensing.

Cooler control can be challenging and represents one of the most significant challenges in the design and operation of an acid gas system; particularly in cold weather locations or locations experiencing major continental climates. The use of bundle louvers to slow air flow can be difficult as the louvers are not a linear response to a changing process variable. Even a cooler with the fan turned off can have significant induced air flow simply from the thermal differential. Finally, the tuning of louver controllers can be very challenging as the ambient temperature can alter cooler design/overdesign very quickly. Even fully closed louvers will allow a small amount of inlet air flow and it can be difficult to achieve acceptable process temperatures during very cold ambient temperatures, especially using simplistic non-linear louvers. The use of derivative control can potentially assist this by using a rate-of-change control methodology however, this can lead to massive loop instability if mismanaged.

1.6.3 Speed Management Variances

In an ideal situation, the compressor will ramp linearly (with a rapid but stable response) in speed to adjust to the changing flows. This response is usually quite quick in an electric motor and it's not unusual to see a 3:1 or even 5:1 turndown allowing for a large range in system speeds and response before other capacity modifications become necessary. While compressor power is nearly linear with speed, the use of speed management can result in operating cost savings although the acid gas compressor is seldom the largest consumer of power; saving power is seldom a consideration. Utilizing speed management in an engine-based system is more challenging. The available turndown is usually quite small, and depending on the engine configuration, the available power/torque is not a linear function of the speed. In order to access engine power while remaining in the sweet spot of the power band, auto-bypassing is often utilized earlier and more often in an engine drive arrangement. It is common for the aerial cooler fans to be electric and variable speed drive; this means that the system avoids the more common engine/cooler direct coupling of bigger natural gas compressors.

1.6.4 Scrubber Level Management Variances

Most acid gas compressors are set up with a cascading scrubber dump system. The liquids in the Stage 5 suction scrubber dump to the Stage 4 suction scrubber, and then to Stage 3 and finally on to Stage 2 where they are usually level controlled off-skid to a water management/disposal system. These systems can be set up with either continuous or batch configuration depending on the system, the owner, and selected instrumentation. The advantages are numerous in that the water from the higher stage pressures is allowed to degas in an upstream stage allowing the release and capture of dissolved acid gas. As well, the lower control valve pressure drops means that the water dump valves will operate further from a hydrate point. Finally, the use of the cascade dumps allows for safer water handling in the associated downstream equipment. Dumping high pressure, sour water, from a high pressure scrubber to an atmospheric tank, can result in tank over-pressuring and a loss of containment with the subsequent release of sour gas. Dumping lower pressure, degassed water is much safer, and can be more predictable. Should the system inadvertently condense acid gas or hydrocarbons through poor cooling control, the cascade system allows for a method of dealing with these liquids. They can potentially build up in the system, and they will impair acid gas compressor capacity, but they won't be sent to an atmospheric system where they will expand uncontrollably creating a hazardous situation.

Depending on system response and valve sizing, it is possible for the higher stage dump valves, to suddenly overwhelm the compressor with vapour from a higher pressure stage. This can be noticeable when the valves ports are larger than required, when the system response is poor or when liquid is dumped from Stage 4 or 5 to a lower stage. The rapid flow of water, which degasses almost instantly, can cause suction pressure bumps in certain stages, however, unless they lead to a major increase in Stage 1 suction pressure, will appear to be fairly insignificant within the compressor and will dissipate quickly.

In many cases, the level control valves will operate in either an =% mode, or a linear mode. Depending on the system performance, the valves may need periodic tuning depending on other system variables.

1.6.5 Contaminant Based Variances

The presence of certain acid gas contaminants can cause system instability and process deterioration. In some amine systems, the use of DGA or a physical solvent like Sulfolane can dramatically increase the pickup of hydrocarbons. This manifests in increased C5 +pickup in the amine as well as significantly increased pickup of aromatic components like benzene and toluene. Because of their low flash point and low volatility, they are not released in the amine flash tank but are usually evolved during the amine regeneration cycle.

These hydrocarbons are destroyed in the sulphur plant reaction furnace and their presence is considered unimportant. However, in the acid gas compression cycle, these components can condense, flash and re-condense within the compression cycle due to the cascade dumping system. This doesn't harm the system, but the presence of these hydrocarbons will consume compressor capacity and will alter the water holding capacity and hydrate point of the injection fluid depending on the type and concentration. Depending on the type and density of the fluids, they may not be picked up by the scrubber level detection systems and may in fact, find their way into the compressor without a high level shutdown. It's critical that the design team be observant of this phenomenon and ensure that the system allow for this. This usually means breaking the cascade cycle and installing a 3-phase flash tank to allow for separation and degassing of the hydrocarbons.

Additional contaminants like mercaptans typically have very little impact. The presence of light hydrocarbons like methane and ethane will alter the fluid hydrate temperature as well as impair the injection fluid density. The presence of hydrocarbons is seldom sufficient to generate any other process instability. Periodic dumping of system scrubbers to an off-skid drain system can help to break the loop and allow the system to start over with no trapped hydrocarbons.

The use of methanol within acid gas compression is well known. In many cases, this methanol is used to suppress hydrate formation in the final discharge fluid or to prevent hydrate formation in the interstage coolers. This injection of methanol into the final compressor discharge will have no effect on the acid gas compressor "pseudo-steady-state" performance. However, methanol injected upstream of the scrubbers will "pull down" more water and will also lower the hydrate point of the final fluid. This type of injection system will increase water extraction and will increase dumping rates of the compressor scrubbers. Depending on conditions, this methanol can become trapped in the system. Eventually it will reach a level where it will leave the system but it can still consume small amounts of system capacity. As well, methanol can be visually difficult to witness and difficult to manage as a fluid as the density is very similar to that of acid gas.

1.6.6 Recycle Control Based Variances

Acid gas compression systems utilize a number of capacity control methods. As stated previously, the primary method is via the driver speed. The secondary method is often a bypass assembly to allow the system to false-load the compressor with recycle gas. The key point of a recycle system is not to generate instability and to allow for a smooth transition between real process gases and recycle gas.

Ideally, the recycle gas is introduced into the system such that the compressor doesn't know that it is dealing with recycle gas. This means that the recycle gas should be the same temperature and composition as the main gas so as to not upset the compressor steady-state. This also means that

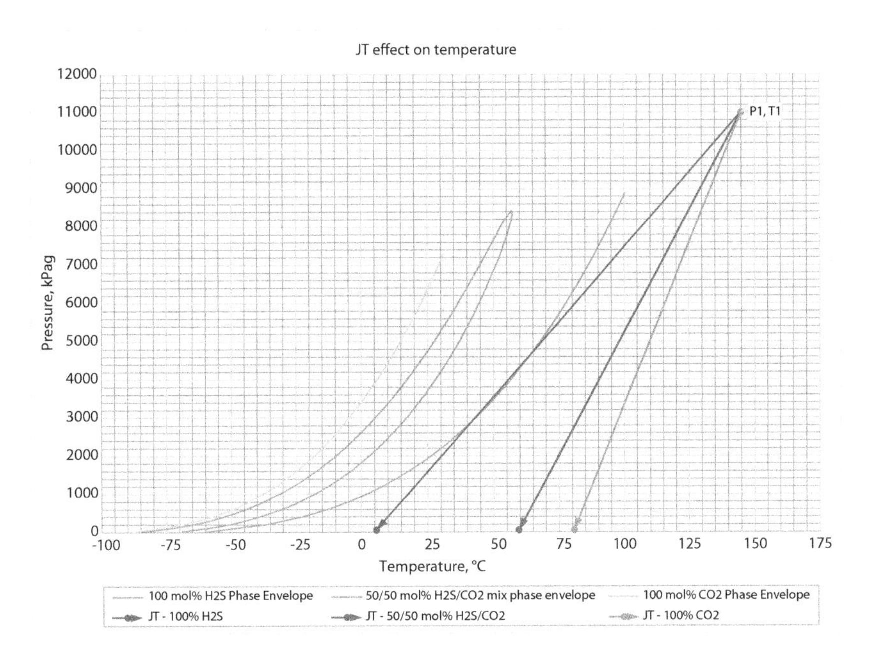

Figure 1.6 JT Effect of Varying Composition.

the recycle system should be smooth, fast responding, and not generate overshoot inside the compressor control system.

Initially, many acid gas compressors utilized a single hot recycle arrangement. This worked well as long as the final pressure was constant and the fluid composition remain unchanged. However, the single recycle valve was often sensitive and required considerable tuning to prevent instability due to such a large pressure drop and rapid changes in flow with a small valve change. Dual recycle valve assemblies began to make an appearance–these often used a 3–1 cool gas recycle with a 5–3 hot recycle. While this made for a smoother system, it required tuning of two (2) control loops (often simultaneously) and required some ramping considerations and pressure cutoffs, particularly during startup and shutdown. Recently, a number of projects have been using a temperature controlled bypass system. Varying composition and pressure often means that a system that works for a CO_2 dominant case will not work for an H_2S dominant case with differing JT performance through the valves.

This type of recycle valve assembly uses a hot valve, a cold valve, and a mixed pressure letdown valve to manage the mixed temperature of the recycle fluid. These systems work by directly managing the recycle temperature but can generate some process instability if they begin to hunt looking for the correct mixing ratio of hot and cold gas. The key to successful recycle management is tuning that is sensitive enough to work at dropping suction pressure while not bumping the process with an excess amount of recycle gas in several seconds. In an ideal situation, the recycle valve would not engage until the compressor speed was at its minimum while not providing feedback to the amine plant regeneration system. This must be avoided at all costs. Having the compressor generate amine plant instability backwards through the acid gas system is not acceptable. It is preferable to tolerate some acid gas compressor suction crossover to prevent any amine plant pressure bounces in the regeneration system as these types of scenarios can cause operational havoc in an amine unit which in turn, will generate forward dynamic instability until the entire system is swinging in an uncontrollable manner. The operator comes to the simple conclusion that "acid gas injection doesn't work".

1.7 Conclusion

There are a large number of system variables within the acid gas compression system. While many appear independent, they can have a defining effect on the overall performance of the system. Some of the variables like composition, remain unchangeable, while other variables, like the speed control and cooler controls tuning can be managed via engineering and operations. Each compressor and each amine system will operate slightly differently due to system characteristics, VFD responses, control loop performance, composition, and system instability generated outside of operational control. In many cases, a minor system instability can lead to other loops becoming unstable and the system performance can appear unstable. As an example, a rapid jump in acid gas flow due to the addition of amine plant anti-foam, can quickly result in a dramatic increase in suction pressure as well as altered composition including hydrocarbon carryover. The system responds quickly to increase compressor speed, but the cooler performance remains unchanged until system temperatures have changed. The increased flow demands a higher duty from the cooler and the cooler must either speed up or open louvers to match this demand for increased cooling. At the same time, the rate of liquid condensing (water and/or hydrocarbons) will increase, and compressor scrubbers will begin to dump more frequently to get rid of the increased liquid accumulation.

The opposite happens during a sudden decrease in acid gas flow, which is often the case in a post-foaming event when things start to return to normal. This may also have occurred due to the loss (or poor response) of the amine reboiler or plant heat medium system. The sudden flow reduction means the compressor speed must decrease rapidly to try and match the new, suddenly lower flow requirements. Depending on how quickly this system responds, the recycle may or may not engage to keep the compressor operating. The quick drop in system throughput will mean that the compressor aftercoolers will be over-cooling until the louvers can close, and the cooler fan speed can decrease to match the lowered duty loads. Each bundle in the cooler (depending on mass and temperature) will have a thermal momentum meaning the system will not respond immediately until the bundle outlet temperature begins to drop. Hopefully, the cooler will respond quickly enough to prevent the system from hitting a hydrate temperature or condensing acid gas. Cooler temperature cycling is not unusual.

This really means that the acid gas compressor is really never operating in a steady-state mode. In fact, operations and design teams can be assured that constantly changing conditions are considered a perfectly normal mode of operation for this process. This assumption of steady-state operation is incorrect. The acid gas compression and cooling system is one of the most dynamic services in a sour gas plant or EOR system.

Assuming steady-state operation will almost certainly result in poor system performance. In order to manage this system instability, it's important to know how the system will respond, it's important to tune the controllers regularly as process variables change, and it's important to monitor compressor performance over time. The system must have sufficient inherent flexibility to manage varying inputs, and the system sensitivity on flow, pressure, and composition must be understood.

The engineering design teams as well as operations should attempt to make use of critical thinking skills in developing and managing acid gas injection systems. It's important to challenge conventional assumptions of steady-state operation, and to critically examine established thought patterns in the design process. This must be done on every component including compression, cooling, water/level management, and finally the injection wellbore/reservoir.

It must also be realized that the acid gas compression system is a keystone in both plant on-line time and overall plant regulatory compliance. While the purpose of this paper is to focus on the misnomer of steady-state acid gas performance, the performance of the entire upstream facilities from separation through compression and treating will have an effect on the entire acid gas system. For this reason, the plant acid gas management system must be examined in its entirety.

Acid gas compressors warrant extra attention to ensure strong, robust, and rapid responding (yet stable) control systems. Finally, the engineering, design, and operations teams must understand that dynamic acid gas system performance is normal, expected, and should be anticipated.

Reference

1. Thinking, T.F. (n.d.). Available from: http://www.criticalthinking.org/pages/our-concept-of-critical-thinking/411www.criticalthinking.org.

2

A History of AGIS

Ying (Alice) Wu

Sphere Technology Connection, Calgary, Alberta, Canada

Abstract
The International Acid Gas Injection Symposium (AGIS) is a semi-regular event organized by Sphere Technology Connection (STC) of Calgary, Alberta. The first Symposium was a two-day event and included two keynote speeches, paper presentations, and poster displays. Authors came from Canada, USA, and China. Over the years AGIS has had attendees from North America, Europe and Asia. The foundation of AGIS continues to be the paper presentations anchored by a well-known keynote speaker. However, AGIS has expanded to include workshops and a roundtable.

Keywords: Acid gas injection, enhanced oil recovery, carbon capture and storage, symposium

2.1 Introduction

Although it references "acid gas" in the title, the International Acid Gas Injection Symposium covers what we call "The Three Sisters": 1. Acid gas injection (AGI), 2. Carbon dioxide for enhanced oil recovery (EOR), and 3. Carbon capture and storage (CCS) or as some prefer, carbon capture, utilization and storage (CCUS). The analogy was inspired by the mountain near Canmore, AB – one mountain with three peaks (see Figure 2.1). If you are spiritual, you would refer to the peaks as Faith, Hope, and Charity, and if you were more analytical you would simply refer to them as Big Sister, Middle Sister, and Little Sister, but we refer to them as AGI, CCS, and EOR.

These three technologies are used to produce and exploit natural resources in an environmentally friendly manner. They all involve an acid

Corresponding author: alicewu@spheretechconnect.com

Ying Wu, John J. Carroll and Yongle Hu (eds.) The Three Sisters, (23–28) © 2019 Scrivener Publishing LLC

Figure 2.1 The Three Sisters, near Canmore Alberta.

gas (CO_2, H_2S, or a mixture of the two) and include compression, transportation, injection, and reservoir interactions.

The main part of the Symposium is the paper presentation, led by a keynote speaker. However, over the years, AGIS has expanded to include workshops, a roundtable, and an evening reception. At AGIS we attempt to promote technology transfer through both technical presentations and networking. You can listen to our keynote speaker and then have a chat with them over a beverage!

For all seven of the Symposia, John Carroll of Gas Liquids Engineering has served as the Chair of the Technical Committee. With the help of the committee members, who have changed over the years, he is responsible for the technical content of the Symposium.

Over the years, AGIS has attracted participants from Canada, USA, The Netherlands, England, Scotland, France, Germany, Switzerland, Italy, Russia, Australia, China, and for the first time in 2018, Japan–it truly is an international event.

2.2 Venues

The first AGIS was held in Calgary in October 2009 and the seventh was again in Calgary. Based on the success of the first AGIS, an expanded version of the symposium was held in 2010.

For the third AGIS, we moved to the beautiful Rocky Mountains west of Calgary and the resort town of Banff, Alberta, in the world-famous Banff National Park. The third AGIS saw the introduction of the GLE Reception,

Table 2.1 Location and venues for the first seven AGIS.

	Date	Location	Venue
AGIS I	5–6 October 2009	Calgary, Canada	Coast Plaza Conference Centre
AGIS II	27–30 September 2010	Calgary, Canada	Radisson Airport Hotel
AGIS III	29 May - 1 June 2012	Banff, Alberta	Banff Park Lodge
AGIS IV	24–27 September 2013	Calgary, Canada	Coast Plaza Conference Centre
AGIS V	19–22 May 2015	Banff, Alberta	Banff Park Lodge
AGIS VI	25–28 October 2016	Houston, Texas	Omni Galleria Hotel
AGIS VII	22–25 May 2018	Calgary, Alberta	Sandman Hotel City Centre

held the evening before the event begins in earnest. This provides the opportunity for participants to mingle in a social setting before the actual hard work begins. Since then, the reception has been sponsored by Gas Liquids Engineering at every AGIS.

Over the years, the location, venue, and timing has changed a bit but an overview is presented in Table 2.1. On average, there are roughly 18 months between events. Most of the Symposia have been held in Alberta but in October 2016 we ventured outside of Canada for the first time and AGIS VI was held in Houston, Texas.

2.3 Keynote Speaker

We believe we have had an outstanding list of keynote speakers. In order to appeal to both surface and subsurface aspect, the first AGIS had two keynotes speakers, John Carroll, a Director at Gas Liquids Engineering, and Mehran Pooladi-Darvish, Vice President of Fekete and Professor at

Table 2.2 AGIS Keynote speakers and their titles.

AGIS I	Mehran Pooladi-Darvish, Fekete and The University of Calgary, Canada *Acid Gas Injection into Aquifers – Reservoir Engineering Considerations*
	John J. Carroll, Gas Liquids Engineering, Canada *Acid Gas Injection: The Past, Present and Future*
AGIS II	Stefan Bachu, Alberta Innovates – Technology Futures, Canada *Subsurface Characteristics of Acid Gas Disposal Operations in Western Canada: A Precursor to CO_2 Storage*
AGIS III	Abbas Firoozabadi, Yale University and RERI, USA *Carbon Dioxide Injection in the Subsurface and Considerations on Physics and Numerical Modelling*
AGIS IV	Tim Wiwchar, Shell Canada, Canada *Shell's Quest Carbon Capture and Storage Project*
AGIS V	Geert Versteeg, Procede Gas Treating, The Netherlands *Rate-Based Simulation of Absorption Processes: Fata Morgana or Panacea?*
AGIS VI	Susan Hovorka, The University of Texas at Austin, USA *Progress Toward Commercialization of Carbon Capture, Use, and Storage for Greenhouse Gas Management*
AGIS VII	Jim Maddocks, Gas Liquids Engineering, Calgary, Canada *Acid Gas Injection: A Very Dynamic Process*

The University of Calgary. The list of keynote speakers for all of the AGIS is given in Table 2.2.

2.4 Workshops

AGIS II included two workshops, one presented by Ray Tomcej, President of Tomcej Engineering and the other by Gas Liquid Engineering's John Carroll. A workshop has been included in each of the subsequent Symposia. Workshop topics have included: acid gas injection, natural gas sweetening, process simulation, corrosion, compression, and wells and the subsurface.

2.5 Roundtable

New for AGIS IV was the Roundtable discussion. A blue-ribbon panel of experts was collected to answer questions from the participants. The first Roundtable was moderated by Johnny Johnson, who at the time was with URS in Denver, Colorado, a good friend of AGIS. Although not held at every AGIS, the Roundtable is an important part of the Symposium.

2.6 Sponsors

We are proud of the list of sponsors over the years. These include Gas Liquids Engineering (Calgary, Alberta); Virtual Materials Group, now a part of Schlumberger (Calgary, Alberta); GEOLEX Inc. (Albuquerque, New Mexico); WSP, formerly WSP | Parsons Brinckerhoff (Houston, Texas); Corrosion Resistant Alloys, CRA (Houston, Texas); Fekete (Calgary); LEWA (Leonberg, Germany); GE Oil & Gas (Florence, Italy); and Alberta Sulphur Research, ASRL (Calgary, Alberta). The sponsors not only provide financial support, they often contribute presentations and workshops. Without the sponsors it is probably fair to say that there would be no AGIS.

2.7 AGIS Books

After the first AGIS we saw some value in publishing the papers and an arrangement was made with Scrivener Publishers to publish the proceeding for all of the Symposium. The first book was appropriately called *Acid Gas injection and Related Technologies* (ISBN: 978-1-118-01664-0). Select papers presented at the Symposium, both oral presentations and posters, have been included in the books. The covers of the first six volumes are shown in Figure 2.2.

2.8 Conclusion

Since 2009, AGIS has been held on a semi-regular basis approximately every 18 months. AGIS has a proud past and a bold future.

Figure 2.2 Covers of the First Six AGIS Books Published by Scrivener Publishing.

Table 2.3 AGIS Workshop titles and presenters.

AGIS I	None
AGIS II	*Sour Gas Treatment* Ray Tomcej, Tomcej Engineering
	Acid Gas Injection John Carroll, Gas Liquids Engineering
AGIS III	*Acid Gas Injection – Design by Example* John Carroll, Gas Liquids Engineering
AGIS IV	*Process Simulation for Acid Gas Injection* James van der Lee, Virtual Materials Group
AGIS V	*Corrosion Considerations for Acid Gas Injection* Sandy Williamson, Ammonite Corrosion Engineering
AGIS VI	*Acid Gas Injection – From Well Design to Reservoir Evaluations* Liaqat Ali and Russell Bentley, WSP|Parsons Brinckerhoff
AGIS VII	*Reservoir Engineering Aspects of Acid Gas Injection* Liaqat Ali, XHorizons
	Acid Gas Compressors and Compression Randy Franiel, Compass Compression

3

Acid Gas Injection: Days of Future Passed

John J Carroll

Gas Liquids Engineering, Calgary, Alberta, Canada

Abstract
At the First International Acid Gas Injection Symposium (AGIS), I presented a paper on the past, the present, and the future of AGIS [1]. The purpose of this paper is to look back at some of those predictions and see which ones have come to fruition and which did not, and also to examine what new things were not predicted but have started to emerge as potential new technologies. The focus of this paper will largely be on things presented at the six AGIS leading up to this paper but a couple of other things will also be discussed.

For example, during this time we saw the first offshore project to inject H_2S on the East Coast of Canada. On the other hand, we have yet to see the first mega project–one injecting more than 100 MMSCFD (2.8×10^6 Sm/d).

Several new processes have emerged to enhance the overall injection process. These include DexPro™ and the Controlled Freeze Zone (CFZ™), which have been discussed at AGIS. And these will be reviewed in this paper. Other new technologies will also be discussed.

Keywords: Acid gas injection, new technologies

3.1 Introduction

Acid gas injection began as an environmentally friendly way to monetize small, sour gas resources using small processing facilities. In the past, such plants probably would have been permitted to flare the acid gas, which is a by-production of the upgrading of raw natural gas.

This paper is a review of acid gas injection and in particular material presented at the International Acid Gas Injection Symposia. This paper

Corresponding author: jcarroll@gasliquids.com

Ying Wu, John J. Carroll and Yongle Hu (eds.) The Three Sisters, (29–38) © 2019 Scrivener Publishing LLC

is an attempt to highlight some of the important work reported at the Symposia. Unfortunately not all of the papers are mentioned, but they are important contributions nonetheless.

3.2 State of the Art

The state of the art for acid gas injection remains basically unchanged. It is a process where acid gas, a mixture of carbon dioxide and hydrogen sulfide, is removed from a sour natural gas stream. The acid gas is produced at low pressure (less than about 2 bar [30 pisa]) and is saturated with water. The gas is compressed, transported to an injection well via pipeline, and injected into a subsurface formation. In many cases, the acid gas can be sufficiently dehydrated using compression and cooling alone. Depending upon the nature of the gas and the conditions experienced or the specification of the operating company, this stream may have to be dehydrated. Figure 3.1 shows the standard block diagram for AGI developed by the author and used in many publications.

In the 2010 paper, three generations of AGI were identified. These are summarized in Table 3.1. The basic distinction between the three generations were the sizes, that is, the volume of gas injected. Most of the acid gas injection projects are of the First Generation type and to date there have been no Third Generation projects.

The "granddaddy" of all AGI projects is at Acheson, near Edmonton, Canada, which started up in 1989 [2], and [3]. This was the first acid gas injection project and like many of the First Generation projects, the

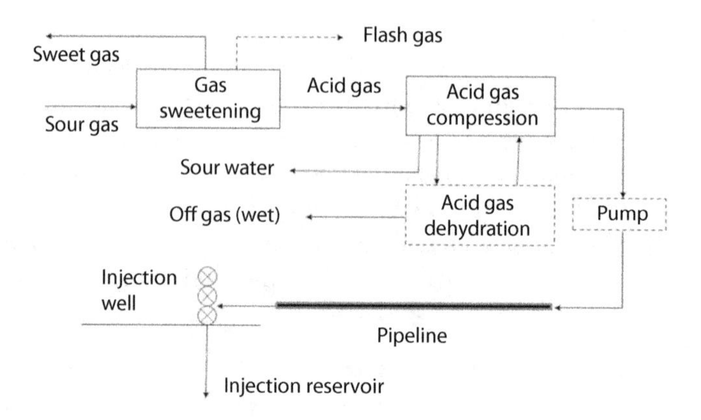

Figure 3.1 Simplified Block Diagram for Acid Gas Injection Process.

Table 3.1 Three generations of acid gas injection projects.

	First generation	Second generation	Third generation
Rate	Less than 10 MMSCFD (300 × 10^3 Sm^3/d), many less than 1 MMSCFD	10 to 100 MMSCFD (0.3 to 3 × 10^6 Sm/d)	Greater than 100 MMSCFD (3 × 10^6 Sm/d)
Compressor	Single reciprocating machine	Multiple reciprocating or single centrifugal	Multiple centrifugal machines (including redundancy)
Well	Single, dedicated injection well Usually a new well, but could be old well re-completed	Often a single well, but perhaps two	Multiple injection wells (possibly including redundant wells)
Pipeline	Typically a small-diameter, short pipeline		Pipeline network to distribute injection fluids to the wells Are you willing to have higher pressure drop in order to reduce pipeline inventory?

injection was proposed in order to meet new legislation on the reduction of sulfur emissions.

3.2.1 Third Generation

In 2009, it was predicted that possibly a third generation injection project would be constructed. However, to date, no such large projects have been built. Several factors may drive producers/processors to consider AGI over the production of elemental sulfur:

1. The unpredictability of the sulfur markets, particularly as several more large projects come on-stream.
2. The significant space required to stockpile elemental sulfur from a large project.
3. Poor quality acid gas (less than around 25%), which cannot be directly sent to a Claus-type sulfur plant.
4. Potential carbon credits or tax saving from the injection of CO_2. In a conventional Clause-type sulfur plant, the CO_2 is emitted to the atmosphere.

As an example of a third generation project consider the Shah Gas Development (now Al Hosn) in Abu Dhabi, U.A.E. Al Hosn was a huge sour gas development processing 28×10^6 Sm^3/d (1 BCSFD) of high sour gas. Some of the data for this project are given in Table 3.2. The feed is taken from Schulte *et al.* [4] and the acid gas is merely the H_2S and CO_2 from the stream. In reality, it would contain a small amount of hydrocarbons, but the point is to get the magnitude of such a project.

A project of this size would require a huge amount of compression, multiple injection wells, and a network of pipelines (sort of a gathering system in reverse) to deliver the acid gas. Suitable disposal reservoirs are not available in the Gulf region so undoubtedly the acid gas would have to be re-injected into the producing formation. There are clearly advantages to injecting into the producing formation: (1) the volume injected is less than the volume removed so there should be sufficient pore volume to contain the injected fluid. (2) The cap rock is sufficient to seal the original fluid for geological time; therefore it should contain the injected fluid. (3) The original fluid in the reservoir is highly sour and thus there should be no negative interactions between the injected acid gas and the fluid and rock of the original reservoir.

Table 3.2 Data from the al hosn (shah sour gas development) as an example of a third generation acid gas injection project.

	Feed	Acid gas
Rate (Sm^3/d)	28.3×10^6	9.15×10^6
(BSCFD)	1.0	0.323
Composition (mol%)		
hydrogen sulfide	22.8	70.6
carbon dioxide	9.5	29.4
methane	56.3	–
ethane	4.3	–
propane plus	7.1	–
Sulfur Equivalent ton/d	10,240	10,240
tonne/d	9290	9290
CO_2 Emissions ton/d	5510	5510
tonne/d	5000	5000

3.2.2 Offshore

Even in 2009, there was a significant acid gas injection project offshore: Sleipner in the Norwegian North Sea [5]. This is a large project injecting approximately 1.4×10^6 Sm^3/d (50 MMSCFD) of carbon dioxide (there is no hydrogen sulfide at Sleipner).

In 2013 an offshore project started on the East Coast of Canada: Encana's Deep Panuke in the coastal waters of Nova Scotia [6]. The raw gas contains a small amount of H_2S and the acid gas is disposed of using acid gas injection. The sweetening unit and acid gas compressor are on the platform and the injection well is on the seafloor below the platform. This project remains the only offshore project that injects a mixture of H_2S and CO_2.

This project faced the usual problems of estimating injection pressure, water content optimization, hydrate avoidance, etc., but was faced with a new challenge – estimating the flow rate and consequences from a blow out of the injection well situated on the seafloor.

3.3 New Processes

In the typical natural gas plant, the desired products are saleable natural gas and natural gas liquids (ethane, liquefied petroleum gas (LPG), butanes, condensate, and others). However, the raw gas often contains H_2S and CO_2, the so-called acid gases, which must be removed in order to make saleable products.

Current technologies for the removal of the acid gases are based on aqueous solutions of alkanolamines, such as methdiethanolamine (MDEA). These are energy intensive and produce a low-pressure acid gas by-product. Much research is currently underway to find new solvents and process to reduce the cost and also on new processes capable of producing a high-pressure acid gas (and thus reduce compression costs).

3.3.1 New Solvents

This may be driven more by the carbon capture world than the natural gas world, but there continues to be a search for new solvents to remove the acid gas components from the raw gas stream. At AGIS we have had papers on two new classes of solvents in particular: (1) de-mixing amines and (2) ionic liquids.

The groups from Université Clermont Auvergne (formerly Université Blaise Pascal) and the University of Guelph presented a series of papers on de-mixing amines [7, 8]; and [9].

We have had a few papers discussing the properties of ionic liquids, such as Zoubeik, *et al.*, [10]. These may result in new solvents for improved recovery at lower cost.

We also had a presentation on Vitrisol® a 100% selective process for H_2S removal in the presence of CO_2", which has potential for acid gas injection applications [11].

3.3.2 DexPro

Water remains a concern in the design of an injection scheme. The acid gas entering the disposal unit is saturated with water, which raises concerns regarding corrosion and hydrate formation. In many cases, dehydration can be achieved using compression and cooling alone. However in certain cases additional dehydration is required. The conventional methods such as a regenerative glycol system can be used to dehydrate acid gas but there are problems associated with them.

A new dehydration process, developed at Gas Liquids Engineering, called *DexPro* has emerged. *DexPro* has been the subject of several presentations at AGIS including Maddocks *et al.*, and Anwar, *et al.*, [12].

3.3.3 CFZ™

The major cost for an acid gas injection scheme is the compressor. However, if the acid gas could be removed under pressure the amount of compressor required could be reduced or even eliminated [13].

3.4 Modelling

Early in the development of acid gas injection, the software tools were not sufficiently accurate to precisely model the behaviour of the gas and two areas in particular: (1) the water content of acid gas and (2) the density and other transport properties. *AQUAlibrium*, a software package developed by the author of this paper, stood out for its accuracy. However, software developers have developed new packages that better do such predictions.

We have had several papers from Virtual Materials Group on the application of their software to acid gas injection processes [14], and [14], and from AspenTech [15] Optimized Gas Treating [16].

3.5 More Data

AGIS has also given a forum for researchers to present new data related to AGI. At the first Symposium researchers from the University of Calgary presented new data for the water content of acid gas and described a valuable new method for accurately measuring the water content [17, 18]; and [19].

3.6 The New Future

The future will be mostly like the past – first generation projects: small gas plant monetizing reserves in an environmentally friendly way. This will probably continue to be the most common application of AGI.

In addition we will probably see more second generation plants, especially those converting their sulfur plants to AGI in order to avoid stockpiling of sulfur or to deal with poor quality acid gas (low H_2S content).

The biggest step as we head toward the future is a Third Generation project. Such a project will be based in the Middle East or the Caspian Sea region. Time will tell if we see such a project.

However, moving forward with these projects will need new and higher quality data, models based on these data that more accurately predict the behavior and properties of these mixtures, and new processes (or continued development of newer ones). These will always be important topics in future Symposia!

References

1. Carroll, J.J., Acid gas injection: past, present, and future. In: Wu Y, Carroll J. J, eds. *Acid Gas Injection and Related Technologies*. Salem, MA, Scrivener Publishing, 2010.
2. Lock, B. Acid gas disposal. A field perspective.*Seventy-Sixth GPA Annual Convention*,San Antonio, TX, pp. 161-170, 1997.
3. Bosch, N. Acid gas injection – a decade of operating history in canada. 14th operations maintenance conference,Calgary, AB.
4. Schulte, D., Graham, C., Nielsen, D., Almuhari, A.H., Kassamali, N., The Shah Gas Development (SGD) Project – A New Benchmark. *SOGAT* 31 March – 1 April. pp. 9–35, 2009.
5. Baklid, A., Korbd, R., Owren, G.Sleipner vest CO_2 disposal, CO_2 injection into a shallow underground aquifer. SPE annual technical conference and exhibition.
6. Skrypnek, T.Acid gas injection at deep panuke. Energy and research development forum.
7. Fandiño, O., Yacyshyn, M., Cox, J.S., Tremaine, P.R., Speciation in Liquid-Liquid Phase-Separating Solutions of Aqueous Amines for Carbon Capture Applications by Raman Spectroscopy. In: Wu Y, Carroll J. J, Zhu W, eds. *Acid Gas Extraction for Disposal and Related Topics*. Salem, MA, Scrivener Publishing. pp. 81–94, 2016.
8. Ballerat-Busserolles, K., Lowe, A.R., Coulier, Y., Coxam, J.-Y., Calorimetry in Aqueous Solutions of Demixing Amines for Processes in CO_2 Capture. In: Wu Y, Carroll J. J, Zhu W, eds. *Acid Gas Extraction for Disposal and Related Topics*. Salem, MA, Scrivener Publishing. pp. 69–80, 2016.
9. Coulier, Y., El Ahmar, E., Coxam, J.-Y., Provost, E., Dalmazzone, D., Paricaud, P., *et al*., New Amine Basecd Solvents for Acid Gas Removal. In: Wu Y, Carroll J. J, Zhu W, eds. *Carbon Dioxide Capture and Acid Gas Injection*. Salem, MA, Scrivener Publishing. pp. 128–145, 2017.
10. Zoubeik, M., Mohamedali, M., Henni, A., Carbon Capture Performance of Seven Novel Immidazolium and Pyridinium Based Ionic Liquids. In: Wu

Y, Carroll J. J, Zhu W, eds. *Carbon Dioxide Capture and Acid Gas Injection.* Salem, MA, Scrivener Publishing. pp. 71–90, 2017.

11. Wirmink, W.N., Ramachandran, N., Versteeg, G.F., Vitrisol® a 100% Selective Process for H_2S Removal in the Presence of CO_2. In: Wu Y, Carroll J. J, Zhu W, eds. *Carbon Dioxide Capture and Acid Gas Injection.* Salem, MA, Scrivener Publishing. pp. 91–126, 2017.
12. Anwar, M.R., Wayne McKay, N., Maddocks, J.R., Enhanced Dehydration using Methanol Injection in an Acid Gas Compression System. In: Wu Y, Carroll J. J, Li Q, eds. *Gas Injection for Disposal and Enhanced Recovery.* Salem, MA, Scrivener Publishing. pp. 129–152, 2014.
13. Oelfke, R.H., Denton, R.D., Valencia, J.A., Controlled Freeze Zone™ Commercial Demonstration Plant Advances Technology for Commercialization of North American Sour Gas Resources. In: Wu Y, Carroll J. J, Zhu W, eds. *Sour Gas and Related Technologies.* Salem, MA, Scrivener Publishing. pp. 79–90, 2012.
14. Van der Lee, J., Carroll, J.J., Satyro, M., A Look at Solid CO_2 Formation in Several High CO2 concentration Depresurizing Scenarios. In: Wu Y, Carroll J. J, Zhu W, eds. *Sour Gas and Related Technologies.* Salem, MA, Scrivener Publishing. pp. 117–128, 2012.
15. Said-Ibrahim, S., Rumyantseva, I., Garg, M., Employing Simulation Software for Optimized Carbon Capture Process. In: Wu Y, Carroll J. J, Zhu W, eds. *Acid Gas Extraction for Disposal and Related Topics.* Salem, MA, Scrivener Publishing. pp. 39–46, 2016.
16. Alvis, R.S., Hatcher, N.A., Weiland, R.H., Expectations from Simulation. In: Wu Y, Carroll J. J, Zhu W, eds. *Acid Gas Extraction for Disposal and Related Topics.* Salem, MA, Scrivener Publishing. pp. 47–68, 2016.
17. Bernard, F., Marriott, R.A., Giri, B.R., Equilibrium Water Content Measurements for Acid Gas at High Pressures and Temperatures. In: Wu Y, Carroll J. J, Zhu W, eds. *Sour Gas and Related Technologies.* Salem, MA, Scrivener Publishing. pp. 3–20, 2012.
18. Marriott, R.A., Fitzpatrick, E., Bernard, F., Wan, H.H., Lesage, K.L., Davis, P.M., *et al.*, Equilibrium Water Content measurements for Acid Gas Mixtures. In: Wu Y, Carroll J. J, eds. *Acid Gas Injection and Related Technologies.* Salem, MA, Scrivener Publishing. pp. 3–20, 2010.
19. Ward, Z.T., Marriott, R.A., Koh, C.A., Phase Equilibrium Investigations of Acid gas Hydrates: Experimental and Modelling. In: Wu Y, Carroll J. J, Zhu W, eds. *Acid Gas Extraction for Disposal and Related Topics.* Salem, MA, Scrivener Publishing. pp. 107–113, 2016.
20. Maddocks, J., McKay, W., Hansen, V., Acid Gas Dehydration – A DexPro™ Technology Update. In: Wu Y, Carroll J. J, Zhu W, eds. *Sour Gas and Related Technologies.* Salem, MA, Scrivener Publishing. pp. 91–115, 2012.
21. Van der Lee, J., Wichert, E., Investigation of the Use of Choke Valves in Acid Gas Compresion. In: Wu Y, Carroll J. J, Zhu W, eds. *Sour Gas and Related Technologies.* Salem, MA, Scrivener Publishing. pp. 165–181, 2012.

4

Calorimetric and Densimetric Data to Help the Simulation of the Impact of Annex Gases Co-Injected with CO_2 During Its Geological Storage

F De los Mozos, K Ballerat-Busserolles*, B Liborio, N Nénot, J-Y Coxam and Y Coulier

Institut de Chimie de Clermont-Ferrand (ICCF), Université Clermont Auvergne Clermont-Ferrand, France

Abstract

Most of the industrialized countries have started to significantly reduce their greenhouse gases emissions. CO_2 capture and geological storage processes entered an operational phase with the emergence of several pilot-sites where different technologies are tested. The gas mixture composition to be stored can vary considerably both qualitatively and quantitatively, based on the origin of CO_2, the chosen process of capture and the concerned industrial sector. Indeed, in addition to CO_2, several components can be present at various concentration levels, including O_2, N_2, SO_x, H_2S, N_yO_x, that are a concern to the energy industry. These impurities can have an impact on the chemical reactivity with water, reservoir or cap-rock forming minerals, and the materials constituting injection or monitoring wells installed in a repository site. Moreover, some impurities (SO_x, H_2S, N_yO_x, CO) are toxic for human health and the environment, even at low concentrations.

This work is part of the objectives of the French ANR program SIGARRR: Simulation of the Impact of Annex Gases (SO_x, N_yO_x, O_2) co-injected with CO_2 during its geological storage on the Reservoir-Rocks Reactivity. Based on the technical experience of the consortium, it was appropriate to propose a project to study the behavior of co-injected gases mixtures under storage conditions.

This chapter will focus on the determination of two key properties for the mixtures containing pure gases: Enthalpies of mixing, solubility, and Molar volumes at

**Corresponding author:* karine.ballerat@uca.fr

Ying Wu, John J. Carroll and Yongle Hu (eds.) The Three Sisters, (39–54) © 2019 Scrivener Publishing LLC

infinite dilution of aqueous solutions of salts, loaded with different gases, namely CO_2 and SO_2.

Keywords: Density, enthalpy, solution, electrolytes, carbon dioxide, sulfur dioxide, water

4.1 Introduction

Most of the industrialized countries have now started to significantly reduce their greenhouse gases emissions. Capture and geological storage processes entered an operational phase with the emergence of several pilot-sites, notably in France, where several technologies are currently being tested.

During the storage of CO_2 in geological formations, four main mechanisms have to be considered: (i) Structural trapping, which is the presence of an impermeable cap-rock which prevents CO_2 from escaping from the outset, (ii) Residual CO_2 trapping, where CO_2 is trapped by capillary forces in the interstices of the rock formation, (iii) Solubility trapping, where the CO_2 dissolves in the water found in the geological formation and sinks and (iv) Mineral trapping happening when dissolved CO_2 chemically reacts with the rock formation to produce minerals.

Considering the whole treatment network of CO_2 storage in geological repositories, the major problem in the separation step is the elimination of some recalcitrant impurities during the treatment of industrial effluents. The gas mixture composition can vary considerably both qualitatively and quantitatively (varying from 80% to 99% of CO_2) [1], based on the origin of CO_2 and the chosen process of capture (oxycombustion, post-combustion, pre-combustion). The presence of annex gases and the purity level of CO_2 during its separation are essential aspects of capture and are well documented [2]. In addition to CO_2, several components are of concern to the energy industry and can be present at various concentration levels, including O_2, N_2, SO_x, H_2S, N_yO_x, H_2, CO, and Ar. Some of these impurities (SO_x, H_2S, N_yO_x and CO) are toxic for human health and the environment, even at low concentrations [3].

The presence of impurities in the CO_2 stream may have an effect on all types of geological storage scenarios, especially in terms of changes in storage capacity and injectivity due to changes in phase behavior with respect with pure CO_2. In addition, impurities could have a significant effect on injectivity through geochemical reactions in the vicinity of injection wells. Geochemical effects such as dissolution of CO_2 and reactions with minerals may determine the long-term fate of injected CO_2, especially in deep saline formations; the impurities may affect the risk profile of storage sites, as geochemical reactions are widely seen as a key mechanism for the stabilization of pressure and brine displacement. Geochemical reactions can also affect the integrity of caprock sequences above storage complexes.

Until recent years, the role of gas impurities on the geochemical behavior of storage sites was not well documented. Knauss *et al.*, [4] evaluated effects of H_2S and SO_2 on the evolution of a mineralogical assemblage that is characteristic of a geological reservoir at the Frio pilot site (USA). They observed that the effects of H_2S are negligible, while SO_2, in oxidizing conditions, maintains a low pH that is not compatible with carbon sequestration by carbonates precipitation. The authors then concluded that laboratory data are lacking and it is consequently hard to validate the numerical approaches even if simulations tend to improve in those contexts [5, 6].

Determining equilibrium conditions between various phases is crucial when investigating a fluid with several components. Dissolution, transport, and precipitation of chemical species will occur differently depending on the state of the fluid phase (monophasic liquid, biphasic liquid or liquid +vapor). The well-known systems of gas mixtures or water/salt/gas mixtures are highly simplified because of the limited number of species that are considered (binary or ternary mixtures [7–9]; Moreover, data on solubility of gas mixtures used to construct phase equilibrium diagrams are limited to conditions of temperature and pressure much lower than those typical of the geological storage sites. In addition, the diagrams generally do not take into account high ionic strength aqueous fluids present within geological formations considered for CO_2 storage.

Concerning risk assessment, the impacts of impurities on vulnerable assets (human health, aquifers, ecosystems, etc.) in case of leakage are poorly known. One of the main reasons for this lack of knowledge is the absence of data concerning the expected composition of leakage, first studies consequently focused on pure CO_2 leakages. To overcome this issue, usual approaches for impact assessment [10] assume that the leaking gas has the same composition as the injected gas, which is a very strong hypothesis. Furthermore, the toxicity of impurities is strongly dependent on the chemical form and thus on the aqueous speciation of pollutants in the reservoir waters.

To answer this question, the use of models for simulation of the system has to be improved. Different levels are considered, starting with geological models to simulate the rock-aquifer interactions [11] and the influence of injected gases on those interactions. It is then obvious that the understanding of the rock-aquifer behavior is driven by the physicochemical, and moreover by the thermodynamic properties of the brine loaded with gas. More specifically, the acidity of the solutions, the solubility of the gases and the energetic properties during mixing are key data in the simulations.

In order to start with thermodynamic modeling, the study of dissolution of pure gas in aqueous solutions of electrolytes present in the brine has to be carried out. These simplified ternary systems {gas-salt-water} allow determining interaction parameters, hardly obtained in more complex mixtures.

Gas dissolution can be represented by a thermodynamic model that takes into account physical and chemical equilibria. For physical dissolution, the equation for gas-liquid equilibrium is usually based on a dissymmetrical approach $\gamma-\varphi$ [12, 13]. The general equation of the model is reported in Eqn. (4.1), below.

$$H_{i,W}\left(T,p^{SAT}\right) exp\left(\frac{V_{i,W}^{\infty}\left(p-p^{SAT}\right)}{RT}\right)\gamma_i^* m_i = p y_i^g \varphi_i^g \quad (4.1)$$

where m_i is the molality of the dissolved gaseous component i, γ_i is the activity coefficient, $H_{i,W}\left(T,p^{SAT}\right)$ is Henry's law constant at water saturation pressure, $v_{i,W}^{\infty}$ is the molar volume at infinite dilution, y_i^g is mole fraction in the gas phase and φ_i^g is fugacity coefficient.

Non ideality in liquid phase is represented by an activity coefficient, taking into account interactions between the gas molecules, water and ions. The activity coefficient γ_i^* is estimated using thermodynamic electrolyte models, such as e-NRTL or Pitzer modified models. These models include interaction parameters which need to be adjusted using experimental solubility data and enthalpy data; the enthalpy of solution can be derived from Eqn. (4.2) using van't Hoff equation. Henry's constant and volume at infinite dilution are determined from solubility and experimental volume data obtained for low gas dilute solution.

Non ideality in gas phase is represented by the fugacity coefficient, derived from an equation of state such as Viriel or Peng-Robinson. The interaction parameters in gas phase are adjusted on vapor-liquid equilibrium data [13].

In order to obtain the experimental data required for optimizing the thermodynamic models, different experimental methods have been developed in our laboratory. This paper focuses on the determination of volumes at infinite dilution of gas, namely CO_2, using a new experimental methodology for measurements of densities of gas loaded solutions in aqueous solutions. Enthalpies of solution and gas solubility of CO_2 and SO_2 will be also presented, and a comparison of the effect of the two gases on energetic properties and solubility will be described.

4.2 Material and Methods

4.2.1 Densimetric Measurements

The original methodology of the vibrating tube densimeter was modified to allow density measurements for aqueous solutions loaded with controlled amount of gas. Measurements can be performed between 293.15 K and 423 K at pressure up to 15 MPa. Different vibrating tube densimeters have been used, depending on the domain of temperature investigated. At low temperature, the measurements can be carried out using an Anton Paar MPDS five densimeter, while a home-made densimeter is used for the highest temperatures [14].

The density of the solution is derived from the period of vibration of the vibrating tube following Eq. (4.2).

$$\Delta\rho = \rho - \rho_0 = K\left(\tau^2 - \tau_0^2\right) \tag{4.2}$$

ρ and τ are the density and the period of vibration related to the solution; ρ_0 and τ_0 are the density and the period of vibration related to a reference fluid (water) and K the calibration constant of the densimeter, experimentally determined as function of temperature and pressure.

An original set-up has been designed in order to obtain the densities of gas loaded solutions with well-known composition. The schematic view is shown in Figure 4.1.

The densimeter worked in flowing mode. This method makes it possible to prepare the mixture flowing through the vibrating tube, with a controlled gas loading charge. This preparation of solution in flowing mode prevents bubble formations and avoids corrosion problems.

Firstly, a well-known quantity of gas is carefully introduced in a first high pressure syringe pump. The quantity of gas injected in the pump is calculated from the volume indicated by the pump, previously calibrated and the pressure inside the pump, read with a pressure transducer (precision 0.025% full scale). The temperature of the pump is maintained constant using a julabo thermostatic bath (precision 0.1 K). Then, a precise quantity of liquid is added using a second high pressure pump. The injection pressure of liquid is slightly higher than the initial pressure of the

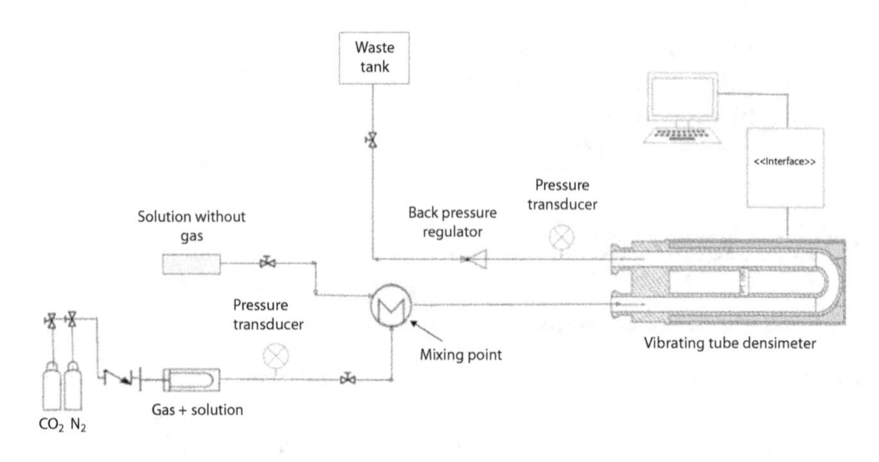

Figure 4.1 Schematic View of the Densimetric Device for Measurements on Gas Loaded Solutions.

first pump filled with the gas. The quantity of liquid is obtained from the volume variation, the temperature and pressure of the second pump.

In this experiment the main issue is to prepare homogeneous solutions loaded with gas but avoiding any gas phase. It is then necessary to prepare a solution with a composition below the limit of gas solubility, estimated at temperature and pressure of the experiments, to insure no degassing.

The experimental procedure consists in measuring the period of vibration of a reference fluid (in our case water) and of the gas loaded solution. Different decreasing gas loading charges are obtained by dilution of the initial gas loaded solution, prepared as described before, with a solution which does not contained gas. The dilution occurs at a mixing point located on the flow lines, as shown in Figure 4.1. The new gas loading charge is calculated from pump flow rates and solution densities.

The apparent molar volume $V_{\varphi,CO2}$ is calculated from the density, using the Eqn. (4.3).

$$V_{\varnothing,CO2} = \frac{M_S}{\rho_0 + \Delta\rho} - \frac{\Delta\rho}{m_{CO2}\left(\rho_0 + \Delta\rho\right) \cdot \rho_0} \tag{4.3}$$

where M_s is the molar mass of the solute (the gas in our case) and m_{CO2} the molality of the solution. The subscript 0 refers to water.

In order to validate the procedure of preparation of the gas loaded solution, the experiments are repeated in same conditions of temperature and pressure, but starting with two different initial solutions (different CO_2 compositions). Experimental results for densities and apparent molar volume for {CO_2-water} system are given in Figure 4.2.

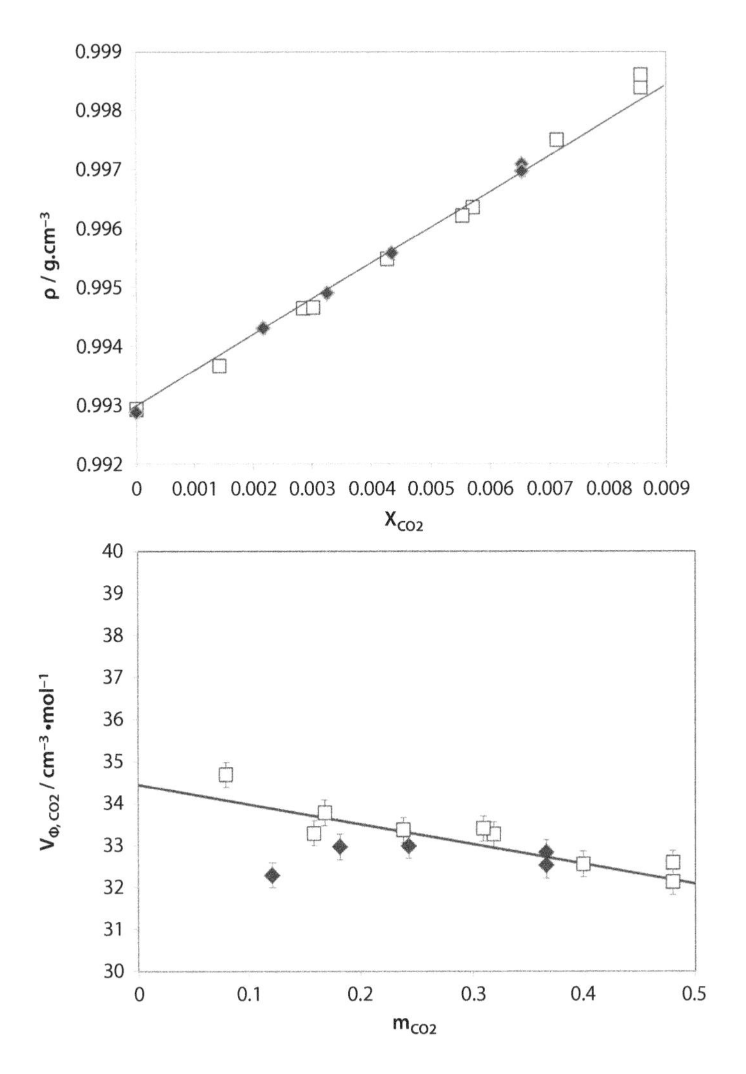

Figure 4.2 Density (left) and Apparent Molar Volumes (right) of Aqueous Solutions of CO_2 at 314 K and 2.6 MPa. The Molar Compositions of the Initial Gas Loaded Solutions Are: ◆,x_{CO2} = 0.0065,□ ;x_{CO2} = 0.0086.

Figure 4.2 shows a good consistency between the measurements obtained from with two different initial solutions of CO_2 mole fractions 0.065 and 0.086. This technique makes it possible to obtain molar volume at infinite dilution $V_{CO_2}^{\infty}$ required in equation (1), by extrapolation of $V_{\varnothing,CO2}$ at $m_{CO2} = 0$.

Density measurements as function of CO_2 loading have been performed for the binary mixture water + CO_2 at temperatures from 293

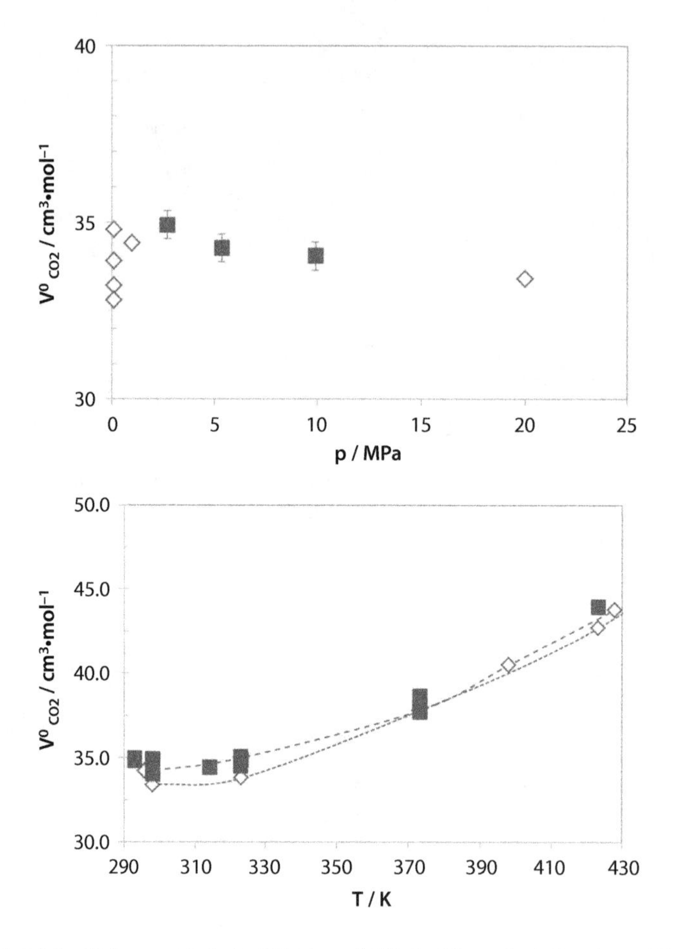

Figure 4.3 Molar Volume at Infinite Dilution of CO_2 in Water. Left: Results at 298.15 K, Versus Pressure; Right: Results Versus Temperature. ▪: our data, ◇: literature [15–17].

K to 428 K and pressures up to 10 MPa. Results for molar volumes at infinite dilution versus pressure and temperature are reported on the Figure 4.3.

The results are in good agreement with the literature values [15–17]. Figure 4.3 shows that the pressure has a very small impact on the volume. The molar volume at infinite dilution of CO_2 in water is increasing with temperature, as expected for non-electrolyte molecule in water [18] and will diverge when reaching the critical temperature.

This study will now be extended to other gases in water and aqueous solutions of electrolytes, to give robust data for adjusting models representative of dissolution of annex gases in brines.

4.2.2 Calorimetric Measurements

Enthalpy of solutions and solubility of gas in aqueous solutions have been determined using a method largely used in the group. The enthalpies are obtained as a function of gas loading charge, α (moles CO_2/mole amine), from the experimental heat flux measured by a Setaram C-80 calorimeter during absorption of CO_2. The method is detailed elsewhere [19–21].

A schematic overview of the experimental set up is given in Figure 4.4. The experiments are performed at constant temperature (±0.01 K) and

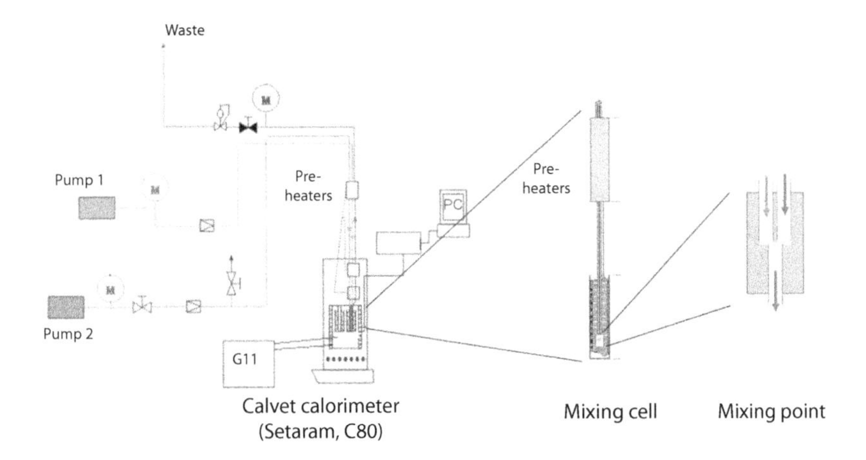

Figure 4.4 Schematic Representation of the Calorimetric Measurements.

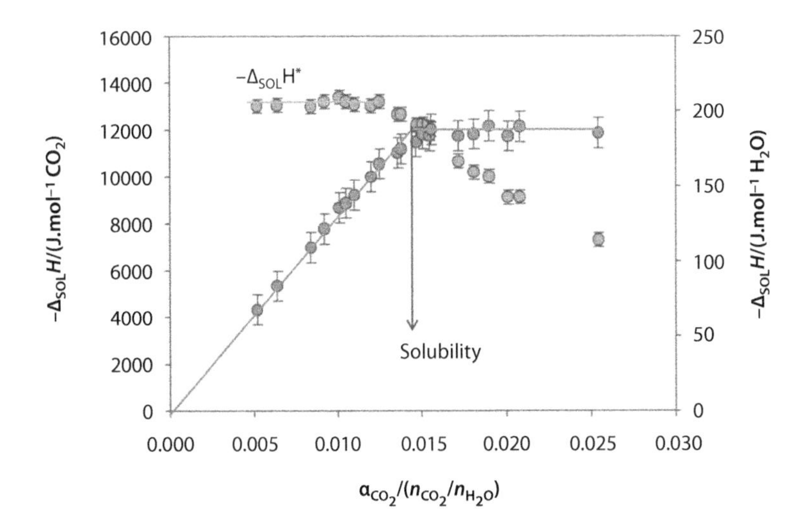

Figure 4.5 Enthalpy of Mixing of CO_2 in Water, at 323 K and 5.1 MPa.

pressure (±0.025 MPa) in a flow mode. The aqueous solution and the carbon dioxide are injected into a mixing cell located inside the calorimeter. The two fluids flow at controlled volume flow rate using two high-pressure syringe pumps (Isco Model 260 DM). These two pumps are temperature regulated using a thermostatic bath in order to maintain a constant molar flow rate. The gas loading charge is determined from volume flow rates and fluids densities.

The enthalpy of solution is calculated from the heat flux signal measured during gas absorption following Eqn. (4.4).

$$\Delta_{SOL}H = \frac{F - F_{BL}}{K.\acute{n}_i} \tag{4.4}$$

Where $\Delta_{SOL}H$ is the enthalpy of solution, F the heat flux signal during gas absorption, F_{BL} the heat flux signal of the baseline recorded when only one fluid is flowing in the mixing cell, K the calibration constant and $\acute{n}_i$ the molar flow rate of compound i, namely gas or water. The enthalpy of solution [Eqn. (4.4)] can be expressed in joule per mole of gas or per mole of solvent. Figure 4.5 shows an example of enthalpy of solution per mole of CO_2 or per mole of water as function of gas loading charge.

when the enthalpy of solution is expressed per mol of gas, a constant $\Delta_{SOL}H$ value is observed at low loading charges. This behavior makes it possible to determine an enthalpy of solution at low loading charges that could be compared to enthalpy value derived from Henry's constant. At highest loading charge, where the liquid is gas saturated, the calculated $\Delta_{SOL}H$ is decreasing because CO_2 considered in Eqn. (4.4) is not fully absorbed CO_2. The absolute uncertainty on the enthalpy of solution is estimated to be about 5%.

When expressed per mol of water, the enthalpy of solution increases linearly with CO_2 loading charge. The slope corresponds to the enthalpy value on the plateau previously obtained. For the highest loading charges, when the liquid is gas saturated, the calculated $\Delta_{SOL}H$ is more or less constant. The first point of this plateau corresponds to the gas solubility. The gas solubility can then be graphically determined from calorimetric measurement. The main difficulty with this method is to determine accurately low gas solubilities. The experiments have to be carried out at low gas loading charges that mean to work with small gas flow rate. The global uncertainty on the determination of gas solubility will then increase with the relative uncertainty on volume flow rate. In this work, the relative uncertainty on the limit of gas solubility is estimated to be ±10%.

4.2.3 Absorption of CO_2 in Aqueous Solutions

The results obtained for dissolution of CO_2 in water and in aqueous solutions of electrolytes at 323 K and 5 MPa are reported in Figure 4.6.

The enthalpy of solution in pure water, expressed per mole of CO_2 (Figure 4.6, left) is exothermic and the value of the plateau is about −13 kJ.mol^{-1}. This value is in good agreement with reported literature data [21]. When adding a salt, the enthalpy slightly decreases. $\Delta_{SOL}H$ in $CaCl_2$ aqueous solution is close to −12.5 kJ.mol^{-1}, and reaches −11 kJ.mol^{-1} in Na_2SO_4 aqueous solutions. These enthalpy values are consistent with a mechanism of physical dissolution, and with no chemical absorption. The models used

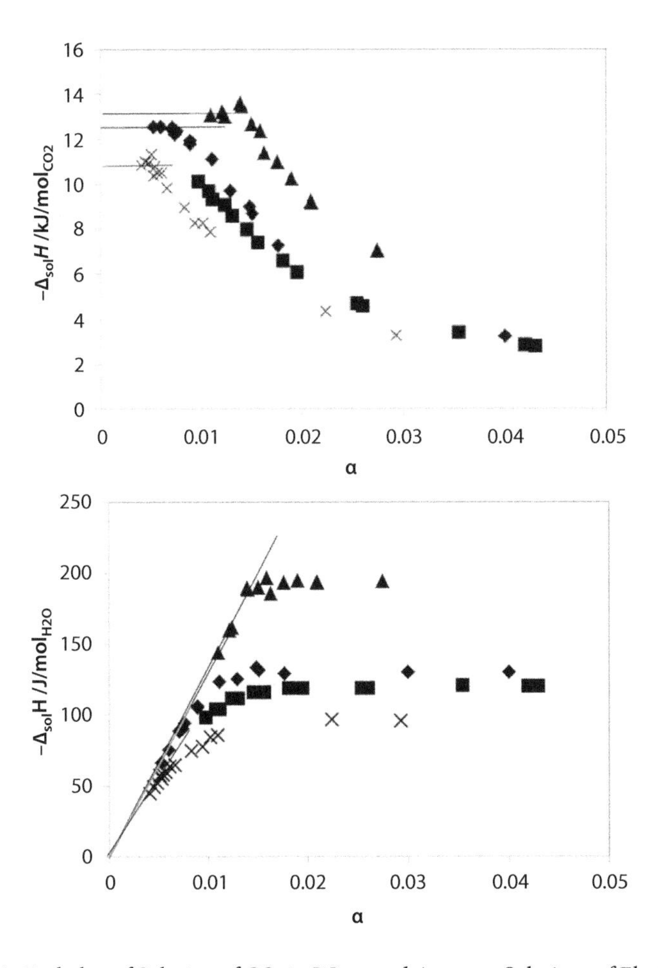

Figure 4.6 Enthalpy of Solution of CO_2 in Water and Aqueous Solutions of Electrolytes, at 323 K and 5 MPa. Left: Enthalpy per Mole of CO_2; Right : Enthalpy per Mole of Water. ▲ : pure water; X: water + Na_2SO_4; ▪: water +NaCl; ♦: water + $CaCl_2$.

to describe these mixtures will then consider only the physical absorption term.

The solubility of CO_2 is determined graphically from the plot of the enthalpy per mole of water (Figure 4.6, right). The value obtained in pure water, expressed in CO_2 molar fraction, is close to 0.014. This result is in good agreement with literature data [21–23]. This solubility decreases when adding salts, as it can be seen on Figure 4.6 (right figure). This behavior corresponds to the salting-out effect, which can be quantified using the Eqn. (4.5).

$$S.O.\left(\%\right) = \frac{x_0 - x}{x_0} \cdot 100 \tag{4.5}$$

where x_0 and x are the solubilities in pure water and salted solution, respectively. The salting out effect calculated at 323 K is about 35% for NaCl; this value does not depend on pressure. The salting out effect is due to the structuration of water in presence of ions. In salted solutions, water is less available for the gas solvation because of electrostatic interactions with the ions. The salting out effect slightly increases in Na_2SO_4 solutions, with however a difference compared to water within the experimental uncertainty. This value decreases more significantly for $CaCl_2$ (25%). This may indicate that the charge of the cation has a larger influence on the salting out effect than the anion.

4.2.4 Absorption of SO_2 in Aqueous Solutions

The same experimental work has been performed with sulfur dioxide instead of carbon dioxide. The experiment were carried out at lower pressure than for CO_2 (namely 0.3 MPa) for safety and experimental constraints. The results are reported in Figure 4.7, for dissolution in water and in aqueous solutions of NaCl and Na_2SO_4.

The enthalpy of solution of SO_2 in water per mole of gas is close to −28 $kJ.mol^{-1}$. This value is consistent with literature value reported at 298.15 K [24]. The mechanism of gas dissolution includes here chemical reactions such as ionizations of SO_2 to form HSO_3^- and SO_3^{2-}. The addition of salt decreases significantly the enthalpy of solution (Figure 4.7, left).

When plotting the enthalpy per mole of water (Figure 4.7, right) for the calculation of gas solubility, it can be observed that this graphical determination of a first point on the plateau is more difficult than for CO_2. The solubility of SO_2 in water reported in literature [25] is about 0.022 in molar fraction. Our calorimetric data shows no clear plateau around this value. However, close to this value, it can be observed a change in the slope of

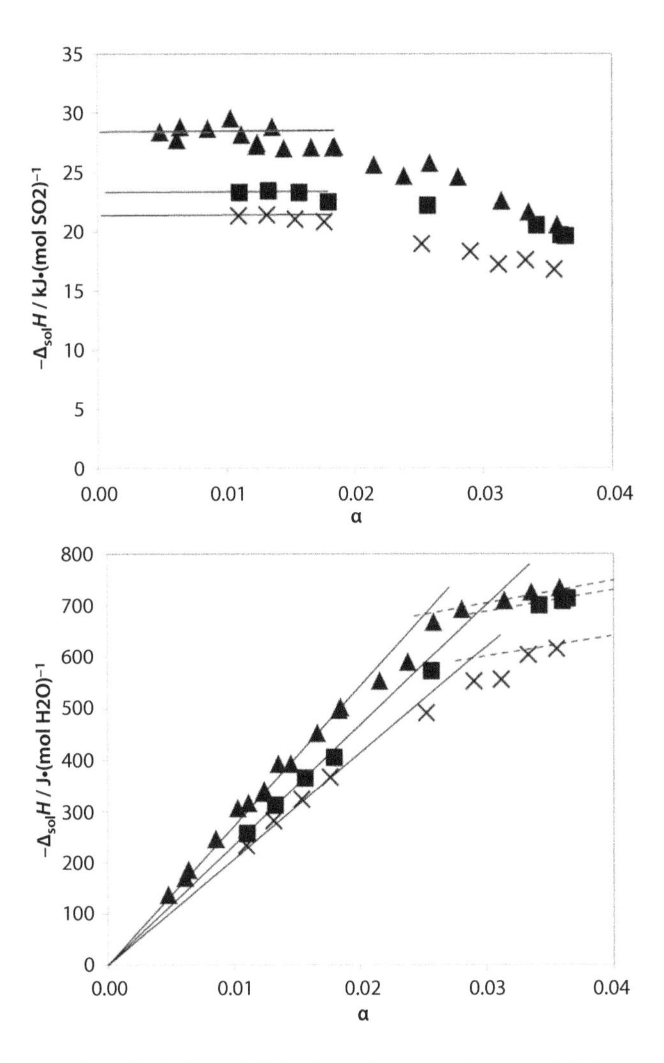

Figure 4.7 Enthalpy of Solution of SO_2 in Pure Water and Aqueous Solutions of Electrolytes, at 323 K and 0.3 MPa. ▲: Pure Water; ×: Water + Na_2SO_4; ▪: Water +NaCl.

the $\Delta_{SOL}H$, probably linked to this solubility. The residual slope has to be attributed to intermolecular reorganization in the solution in presence of ions. When adding salt in the solution, it can be observed that the change in the slope related to the gas solubility increases exhibiting an opposite behavior compared to CO_2; the presence of salt leads to a salting-in effect of the gas. A deeper study must be carried out to explain how interactions with ionic species can affect the chemical reactions during gas dissolution.

4.2.5 Comparison Between the Two Gases

The experimental enthalpies of solution of gas in water and aqueous solutions of NaCl and Na_2SO_4 has been reported in Figure 4.8. Comparison between the different salted solutions is performed at a same ionic strength. This work shows that for CO_2 dissolution, the enthalpy of solution does not vary when adding electrolytes; the physical dissolution of gas is not so much sensible to the nature and to the electrical charge of ions.

For SO_2 dissolution, $\Delta_{SOL}H$ decreases significantly when adding NaCl. This change is larger when adding Na_2SO_4, probably because of the presence of divalent anion. The chemical reactions in water leading to the formation of hydrogen sulfite increases both the enthalpy of solution and gas solubility compared to physical dissolution of CO_2. The presence of salt decreases the enthalpy. A thermodynamic representation of SO_2 dissolution, including chemical reactions, will permit to explain the influence of the presence of ions, in term of ion-pair formation with the sulfite or structuration of the solution due to electrostatic forces. The thermodynamic modeling could also help to explain the salting-in effect observed with SO_2, while the presence of salts generates a salting-out effect in presence of CO_2.

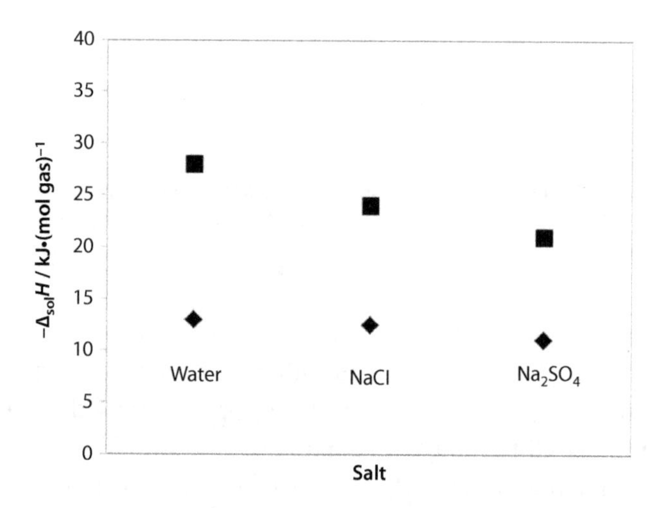

Figure 4.8 Enthalpy of Solution of Gases as Function of the Composition of the Solution; ▪: SO_2; ◆: CO_2.

Acknowledgement

This work is part of the French ANR program SIGARRR (ANR-13-SEED-0006): Simulation of the Impact of Annex Gases (SO_x, NyO_x, O_2) co-injected with CO_2 during its geological storage on the Reservoir-Rocks Reactivity.

References

1. Wang, J., *et al.*, Effects of impurities on geological storage of carbon dioxide. *in IEAGHG*, 2011.
2. Broutin Paul, L.E., LECOMTE fabrice, *CO_2 Capture. Technologies to Reduce Greenhouse Gas Emissions*, 2010.
3. Toxicology, N.R.C.U.Co., Emergency and continuous exposure limits for selected airborne contaminants, 2, 1984.
4. Knauss, K.G., Johnson, J.W., Steefel, C.I., Evaluation of the impact of Co_2, co-contaminant gas, aqueous fluid and reservoir rock interactions on the geologic sequestration of CO2. *Chem. Geol.*, 217(3), 339–350, 2005.
5. Xiao, Y., Xu, T., Pruess, K., The effects of gas-fluid-rock interactions on Co_2 injection and storage: Insights from reactive transport modeling. *Energy Procedia*, 1(1), 1783–1790, 2009.
6. Xu, T., Apps, J.A., Pruess, K., Yamamoto, H., Numerical modeling of injection and mineral trapping of Co_2 with H_2S and So_2 in a sandstone formation. *Chem. Geol.*, 242(3), 319–346, 2007.
7. Carroll, J.J., Phase diagrams reveal acid-gas injection subtleties. *Oil and Gas Journal*, 92–96, 1998.
8. Carroll, J.J., Acid-gas injection encounters diverse H_2S, water phase changes. *Oil & Gas Journal*, 96(10), 57–59, 1998.
9. Carroll, J.J., Lui, D.W., Density, phase behavior keys to acid gas injectionp. Medium: X; Size. *Oil and Gas Journal*, 63–72, 1997.
10. Loschetter, A., Modular toolbox for assessing the impact of Co_2 leaking from a geological storage reservoir into a building. *Tenth Annual conference on Carbon capture and sequestration*. United States, Pittsburgh, 2011.
11. Karay, G., Hajnal, G., Modelling of groundwater flow in fractured rocks. *Procedia Environmental Sciences*, 25, 142–149, 2015.
12. Corvisier, J., Ahmar, E.E., Coquelet, C., Sterpenich, J., Privat, R., Jaubert, J.-N., *et al.*, Simulations of the impact of co-injected gases on Co_2 storage, the SIGARRR project: first results on water-gas interactions modeling. *Energy Procedia*, 63, 3160–3171, 2014.
13. Hajiw, M., Corvisier, J., Ahmar, E.E., Coquelet, C., Impact of impurities on Co_2 storage in saline aquifers: Modelling of gases solubility in water. *International Journal of Greenhouse Gas Control*, 68, 247–255, 2018.

14. Perfetti, E., Pokrovski, G.S., Ballerat-Busserolles, K., Majer, V., Gibert, F., Densities and heat capacities of aqueous arsenious and arsenic acid solutions to 350°C and 300bar, and revised thermodynamic properties of , and iron sulfarsenide minerals. *Geochim. Cosmochim. Acta*, 72(3), 713–731, 2008.
15. Crovetto, R., Wood, R.H., Solubility of Co_2 in water and density of aqueous Co_2 near the solvent critical temperature. *Fluid Phase Equilib.*, 74, 271–288, 1992.
16. Hn edkovsk ý, L., Wood, R.H., Majer, V., Volumes of aqueous solutions of CH_4, Co_2, H_2S and NH_4 at temperatures from 298.15 K to 705 K and pressures to 35 MPa. *J. Chem. Thermodyn.*, 28(2), 125–142, 1996.
17. McBride-Wright, M., Maitland, G.C., Trusler, J.P.M., Viscosity and density of aqueous solutions of carbon dioxide at temperatures from (274 to 449) K and at pressures up to 100 MPa. *J. Chem. Eng. Data*, 60(1), 171–180, 2015.
18. Slavík, M., Šedlbauer, J., Ballerat-Busserolles, K., Majer, V., Heat capacities of aqueous solutions of acetone; 2,5-hexanedione; diethyl ether; 1,2-dimethoxyethane; benzyl alcohol; and cyclohexanol at temperatures to 523 K. *J. Solution Chem.*, 36(1), 107–134, 2007.
19. Arcis, H., Rodier, L., Ballerat-Busserolles, K., Coxam, J.-Y., Enthalpy of solution of Co_2 in aqueous solutions of methyldiethanolamine at T=322.5K and pressure up to 5MPa. *J. Chem. Thermodyn.*, 40(6), 1022–1029, 2008.
20. Coulier, Y., Lowe, A.R., Coxam, J.-Y., Ballerat-Busserolles, K., Thermodynamic modeling and experimental study of Co_2 dissolution in new absorbents for post-combustion Co_2 capture processes. *ACS Sustainable Chem. Eng.*, 6(1), 918–926, 2018.
21. Koschel, D., Coxam, J.-Y., Rodier, L., Majer, V., Enthalpy and solubility data of Co_2 in water and NaCl(aq) at conditions of interest for geological sequestration. *Fluid Phase Equilib.*, 247(1-2), 107–120, 2006.
22. Takenouchi, S., Kennedy, G.C., The binary system H_2O - Co_2 at high temperatures and pressures. *Am. J. Sci.*, 262(9), 1055–1074, 1964.
23. Takenouchi, S., Kennedy, G.C., The solubility of carbon dioxide in nacl solutions at high temperatures and pressures. *Am. J. Sci.*, 263(5), 445–454, 1965.
24. Goldberg, R., Parker, V.B., Thermodynamics of solution of So_2(g) in water and of aqueous sulfur dioxide solutions, 90, 1985.
25. Hales, J.M., Slitter, S.L., Solubility of sulfur dioxide in water at low concentrations. *Atmospheric Environment*, 7(10), 997–1001, 1973.

5

Densities and Phase Behavior Involving Dense-Phase Propane Impurities

JA Commodore, CE Deering and RA Marriott*

Department of Chemistry, University of Calgary, Calgary, Alberta, Canada

Abstract

Produced liquid petroleum can contain varying amounts of carbonyl sulfide (COS) which tends to concentrate within commercial propane. This is because propane and COS have relatively similar normal boiling temperatures, – 42.1 and – 50.2 °C, respectively. In the presence of water, COS can hydrolyze to produce hydrogen sulfide (H_2S) and carbon dioxide (CO_2) which can change the phase behavior, density or reaction chemistry. The COS can also react with itself to produce carbon disulfide (CS_2) and CO_2. While COS itself is not corrosive, the hydrolysis product H_2S is, especially in the presence of water. Numerous studies have shown that the presence of H_2S in the propane stream frequently causes corrosion on copper and brass tubing used in propane-fed appliances. This is made evident in the failure of ASTM copper strip corrosion tests. Due to this reason, many applications such as alkanolamine absorption have been applied in industrial processes for decades to separate COS from the propane rich stream. Information on thermophysical properties such as fugacities and vapor-liquid phase behavior of propane mixtures would contribute to the design of facilities for separating non-hydrocarbon impurities from the rich hydrocarbon streams. One way to obtain such thermodynamic information is through density measurement.

In this work, the volumetric influence of minor amounts of impurities in a propane rich stream were studied through high-precision density measurements at conditions up to p = 35 MPa and temperatures ranging from T = 50 to 125 °C. The measured density data for propane and CS_2 were used to calculate the apparent molar volumes which were then utilized in fitting the adjustable parameters for a fluctuation solution theory model. The densities and apparent molar volumes for propane + CO_2 mixture in this work were used to validate the accuracy of the mixing coefficients from Kunz and Wagner. Also, the mixing coefficients were

*Corresponding author: rob.marriott@ucalgary.ca

Ying Wu, John J. Carroll and Yongle Hu (eds.) The Three Sisters, (55–62) © 2019 Scrivener Publishing LLC

validated against the literature vapor-liquid-equilibrium (VLE) data. In both comparisons, the calculation performed with the mixing coefficients showed a good agreement to experimental data.

Keywords: Density, vapor-liquid equilibrium, propane, carbonyl sulfide

5.1 Introduction

Raw natural gas is a multi-component mixture of widely varying composition with methane as the main constituent and further economically valuable components which include ethane, propane, butane and some other heavier hydrocarbons [1]. In addition, some undesirable impurities such as water, carbon dioxide (CO_2), hydrogen sulfide (H_2S) and carbonyl sulfide (COS) among others can be present in the gas stream. For instance, the presence of water within the gas stream can cause flow assurance issues (by formation of hydrates) in pipelines. Therefore, the raw natural gas is treated to a particular sale gas specification in order to avoid and/or minimize such problems. The treated gas stream (dried hydrocarbon) which is mostly methane is then sent *via* pipelines to the end user. Liquefied natural gas (LNG) is another fuel product [2]. Another important reason for LNG removal is to help with continual supply of petrochemical feedstocks to meet the ever-increasing demand for olefin production. To separate the LNG from the dried hydrocarbon stream, it is sent into a distillation tower to separate mostly propane and butane from the rest of the hydrocarbon stream.

LNG can contain varying amounts of sulfur impurities such as COS. The COS impurity tends to concentrate within the commercial propane fractions because of the similarity in their vapor-liquid properties, which makes separation by traditional methods difficult [3]. The COS within the propane stream can hydrolyze in the presence of water to produce toxic H_2S and CO_2 which can alter the phase behavior, density and reaction chemistry. It can also react with itself to produce carbon disulfide (CS_2). In this context, one way to understand the reaction chemistry and/or thermo-physical properties such as phase behavior of these impurities within the propane stream is via accurate density data. While many equations-of-state are robust for calculating the vapour-liquid-equilibria (VLE) the high-pressure pressure-volume-temperature (pVT) properties in the dense phase often need to be corrected using additional terms. Thus, density derived fugacity information needs to be accurate at both the phase boundaries and at higher pressures for robust chemical equilibria calculations. Although a substantial amount of literature density data are available for propane and CO_2 mixtures, [4–7] there are no previous literature density data for CS_2 as far as we can ascertain.

In this work, we report the binary density measurements of minor amounts of impurities (CS_2 and CO_2) dissolved in a propane rich stream using a high-precision vibrating tube densimeter at pressures up to, $p = 35$ MPa and temperatures ranging from $T = 50$ to 125 °C. The measured volumetric data for CS_2 and propane were used to obtain new fitting parameters for the fluctuation solution theory model. These same parameters can be used in a fugacity model to calculate fugacity for CS_2, which can then be applied in the Gibbs energy minimization equation to investigate reaction equilibria in real fluids (ultimate goal for this research).

5.2 Experimental Section

The densities of the propane rich binary systems were obtained using a high precision vibrating tube densimeter, which was fabricated by Deering *et al.*,[8, 9] Details of the experimental procedure i.e., mixture preparation, calibration, and measurement procedure used to obtain the densities in this work have been reported elsewhere [10]. This device has been shown in previous works to measure density to an average uncertainty of ±0.07 kg·m^{-3} at T = 50–125 °C and up to p = 35 MPa. [11], Temperature measurements were completed with four-wire 100 Ω platinum resistance thermometers (PRT) which were calibrated according to the International Temperature Scale of 1990 (ITS-90) [12]. The overall estimated standard uncertainty in the temperature measurements was found to be ±0.005 °C. Pressure measurements were completed with a Paroscientific 410KR-HT-101 Digiquartz pressure transducer which was re-calibrated with a deadweight tester yielding an accuracy of 0.0052% [12]. With this device, densities were obtained from the time period measurements of the oscillating tube. The period of oscillation, τ, of the tube was recorded with a Berkeley Nucleonics Corporation (BNC) model 1105 universal frequency counter with a resolution of 40 ps. The relationship between the time period of oscillation of the tube and the density is given by Eqn. (5.1).

$$\rho \quad \rho_0 = K_T(\tau^2 \quad \tau_0^2), \tag{5.1}$$

where ρ and ρ_0 are the fluid of interest and reference densities, respectively; τ and τ_0 are the time period of fluid of interest and reference fluid, respectively; and K_T, is the pressure-dependent calibration constant. In this work, the calibration constant was determined with two fluids: water as the calibration fluid and nitrogen as the reference fluid. These fluids were chosen because their densities have been well studied and are readily calculable for the experimental conditions. The Wagner and Pruss [13] equation-of-state (EOS) was used to obtain the density for water and Span *et al.*, [14] EOS was used for the densities of nitrogen.

5.3 Results and Discussion

The measured volumetric data for CO_2 and propane mixture were used to verify the accuracy of the optimized mixing coefficients from Kunz and Wagner [15]. As shown in Figure 5.1 and Figure 5.2, the calculations performed with the mixing coefficients were found to be in good agreement with our measured data. However, because chemical equilibria calculations require an EOS that can perform both pVT and VLE calculations with high accuracy, the Kunz and Wagner mixing coefficients were also validated against available literature VLE data. The calculation showed a good agreement with available literature data (Figure 5.3). This highlights the fact that Kunz and Wagner [15]. mixing coefficients can be used to accurately calculate fugacities for equilibria studies in dense fluid phase.

There are no high-accuracy equations for CS_2; therefore, the measured density data for the CS_2 and propane mixture were used to calculate the

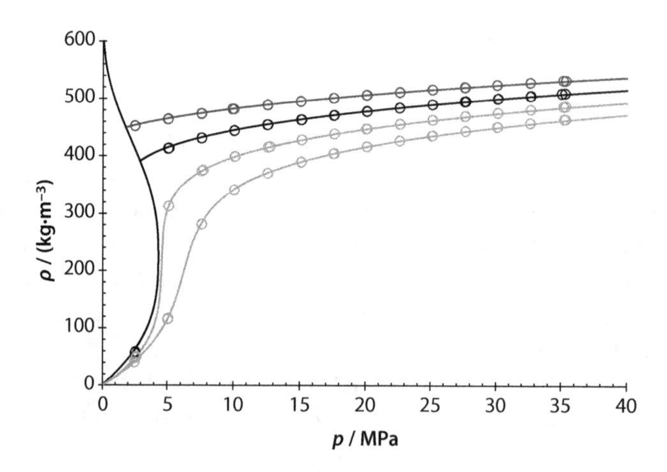

Figure 5.1 The density measurement of CS_2 dissolved in a dense propane phase at $T = 50–125$ °C and up to $p = 35$ MPa. (○), denotes experimental data from this work; —, denotes calculated densities using the high quality reference equations of state combined with the optimized mixing coefficients from Kunz and Wagner.[15] Black curve denotes the two phase region.

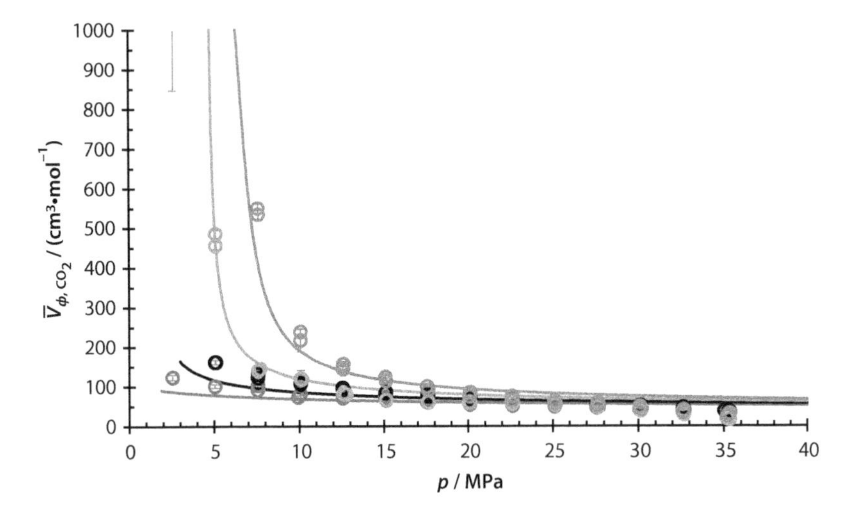

Figure 5.2 The apparent molar volumes of CO_2 dissolved in a dense propane phase at $T = 50–125$ °C and up to $p = 35$ MPa. (○), denotes experimental data from this work; —, denotes calculated molar volumes using the high quality reference equations of state combined with the optimized mixing coefficients from Kunz and Wagner [15].

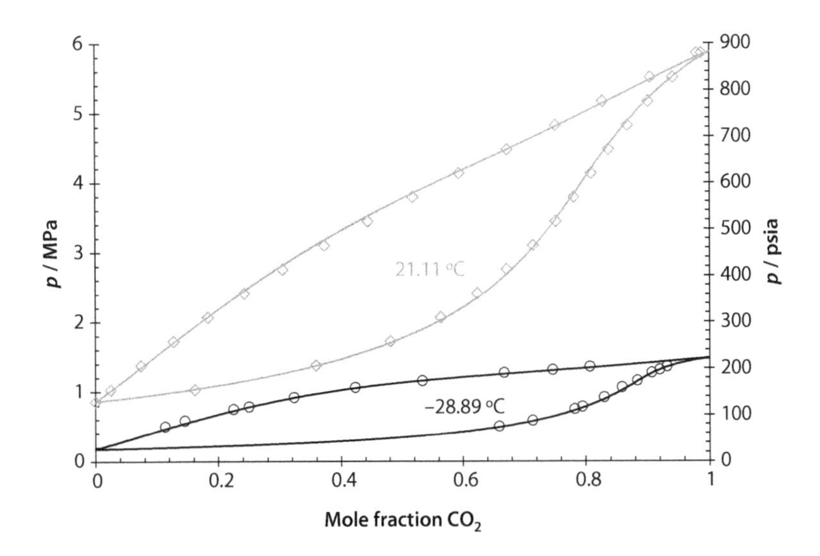

Figure 5.3 The p-x phase diagram for a system of CO_2 and propane at different temperatures. Experimental data from; (o), Reamer and Sage [16], (◊), Hamam and Lu [17]. (—) and (---), were calculated using the reduced Helmholtz EOS with the optimized mixing coefficients from Kunz and Wagner [15] .

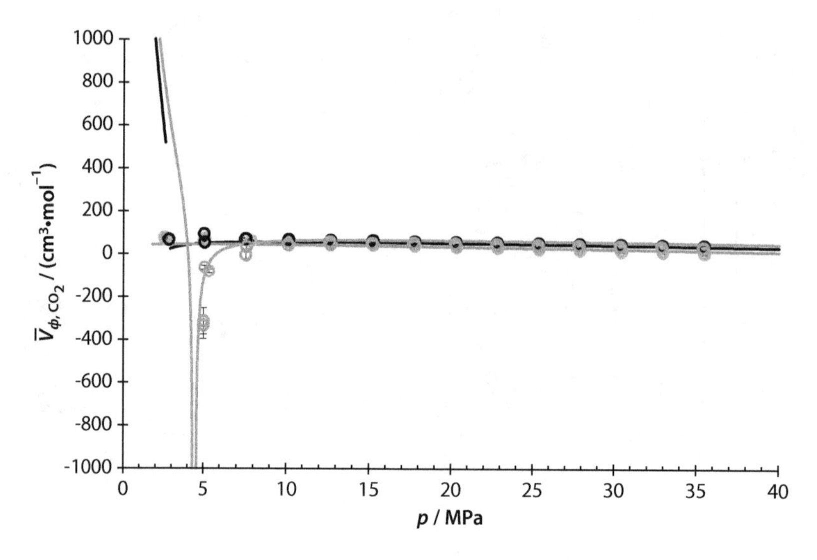

Figure 5.4 The apparent molar volumes of CS_2 dissolved in a dense propane phase at T = 50–125 °C and up to p = 35 MPa. (○), denotes experimental data from this work; —, denotes calculated molar volumes using the optimized mixing coefficients.

apparent molar volumes which were further used to optimize the adjustable parameters in the fluctuation solution theory model proposed by O'Connell *et al.*, [18]

$$\bar{V}_i^{\infty} / \left(\text{cm}^3\text{mol}^{-1}\right) = \bar{V}_1^o + \kappa_{T,1}^o \text{RT} \left\{ a_{\text{ij}} + b_{\text{ij}} \left[\exp\left(c_{\text{ij}} / \bar{V}_1^o \right) - 1 \right] \right\} / \bar{V}_1^o \quad (5.2)$$

where a_{ij}, b_{ij}, c_{ij} are adjustable semi empirical coefficients, $\bar{V}_i^{\infty}$ is the partial molar volume of solutes at infinite dilution, $\bar{V}_1^o$ is the molar volume of the pure solvent (pure CO_2), R is the ideal gas constant and $\kappa_{T,1}^o$ is the isothermal compressibility of the pure solvent. Figure 5.1 shows the calculated volumes performed with the adjustable parameters found in this work for the CS_2 and propane binary mixture. A substantial drop towards a negative infinite apparent molar volume can be observed as the critical point of the pure solvent is approached due to the large compressibility. This effect can be attributed to the low vapor pressure of the CS_2.

5.4 Conclusion and Future Work

We have optimized new adjustable parameters for the mixture of CS_2+ propane which can be utilized in a fugacity model to calculate fugacities of CS_2 at infinite dilution. The calculated fugacities can then be applied in the Gibbs energy minimization equation to investigate equilibria reactions in real fluids. The mixing coefficients from Kunz and Wagner [15]. showed a good agreement to our volumetric data and the literature VLE data for CO_2+ propane; therefore, these mixing coefficients can be used with high accuracy to calculate the fugacity for CO_2+ propane in the dense single phase in addition to the VLE region. This work will be extended to impurities such as water, COS and H_2S dissolved in dense propane phase to explore the possible reactions within rich propane streams using a Gibbs Energy Minimization routine for real fluids.

References

1. Baker, R.W., Lokhandwala, K., Natural gas processing with membranes: an overview. *Ind. Eng. Chem. Res.*, 47(7), 2109–2121, 2008.
2. Althuluth, M., Mota-Martinez, M.T., Berrouk, A., Kroon, M.C., Peters, C.J., Removal of small hydrocarbons (ethane, propane, butane) from natural gas streams using the ionic liquid 1-ethyl-3 -methylimidazolium tris(pentafluoroethyl)trifluorophosphate. *J. Supercrit. Fluids*, 90, 65–72, 2014.
3. Andersen, W.C., Abdulagatov, A.I., Bruno, T.J., The ASTM copper strip corrosion test: Application to propane with carbonyl sulfide and hydrogen sulfide. *Energy Fuels*, 17(1), 120–126, 2003.
4. Niesen, V.G., Rainwater, J.C., Critical locus, (vapor + liquid) equilibria, and coexisting densities of (carbon dioxide + propane) at temperatures from 311 K to 361 K. *J. Chem. Thermodyn.*, 22(8), 777–795, 1990.
5. Blanco, S.T., Gil, L., García-Giménez, P., Artal, M., Otín, S., Velasco, I., Critical properties and high-pressure volumetric behavior of the carbon dioxide+propane system at T=308.15 k. Krichevskii function and related thermodynamic properties.. *J. Phys. Chem. B*, 113(20), 7243–7256, 2009.
6. Feng, X.J., Liu, Q., Zhou, M.X., Duan, Y.Y.. *J. Chem. Eng. Data*, 55 (9, 3400–3409, 2010.
7. de la Cruz de Dios, J., Bouchot, C., Galicia Luna, L.A., New p–ρ–T measurements up to 70 MPa for the system CO2 + propane between 298 and 343 K at near critical compositions. *Fluid Phase Equilib.*, 210(2), 175–197, 2003.
8. Deering, C.E., Cairns, E.C., McIsaac, J.D., Read, A.S., Marriott, R.A., The partial molar volumes for water dissolved in high-pressure carbon dioxide from T = (318.28 to 369.40) K and pressures to p = 35 MPa. *J. Chem. Thermodyn.*, 93, 337–346, 2016.
9. Deering, C.E., Design, construction, and calibration of a vibrating tube densimeter forvolumetric measurements of acid gas fluids, university of calgary, 2015.
10. Commodore, J.A., The volumetric properties for selected impurities in dense CO_2 destinedfor re-injection, university of calgary, 2017.

11. Deering, C.E., Saunders, M.J., Commodore, J.A., Marriott, R.A., The volumetric properties of carbonyl sulfide and carbon dioxide mixtures from $T = 322$ to 393 K and $p = 2.5$ to 35 MPa: Application to COS hydrolysis in subsurface injectate streams. *J. Chem. Eng. Data*, 61(3), 1341–1347, 2016.
12. Preston-Thomas, H., The international temperature scale of 1990 (ITS-90. *Metrologia*, 27(1), 3–10, 1990.
13. Wagner, W., Pruß, A., The IAPWS formulation 1995 for the thermodynamic properties of ordinary water substance for general and scientific use. *J. Phys. Chem. Ref. Data*, 31(2), 387–535, 2002.
14. Span, R., Lemmon, E.W., Jacobsen, R.T., Wagner, W., Yokozeki, A., A reference equation of state for the thermodynamic properties of nitrogen for temperatures from 63.151 to 1000 K and pressures to 2200 MPa. *J. Phys. Chem. Ref. Data*, 29(6), 1361–1433, 2000.
15. Kunz, O., Wagner, W., The GERG-2008 wide-range equation of state for natural gases and other mixtures: An expansion of GERG-2004. *J. Chem. Eng. Data*, 57(11), 3032–3091, 2012.
16. Reamer, H.H., Sage, B.H., Phase equilibria in hydrocarbon systems: Volumetric and phase behavior of the n-Decane-CO_2 system.. *J. Chem. Eng. Data*, 8(4), 508–513, 1963.
17. Hamam, S.E.M., Lu, B.C.Y., Isothermal vapor-liquid equilibriums in binary system propane-carbon dioxide. *J. Chem. Eng. Data*, 21(2), 200–204, 1976.
18. O'Connell, J.P., Sharygin, A.V., Wood, R.H., Infinite dilution partial molar volumes of aqueous solutes over wide ranges of conditions. *Ind. Eng. Chem. Res.*, 35(8), 2808–2812, 1996.

6

Phase Equilibrium Computation for Acid Gas Mixtures Containing H_2S Using the CPA Equation of State

Hanmin Tu[1], Ping Guo[1,*], Na Jia[2] and Zhouhua Wang[1]

[1]*State Key Laboratory of Oil and Gas Reservoir Geology and Exploitation, Southwest Petroleum University, Chengdu, China*
[2]*Program of Petroleum Systems Engineering, University of Regina, Regina, Saskatchewan, Canada*

Abstract
Phase equilibrium of mixtures which containing acid gases such as carbon dioxide (CO_2) and hydrogen sulfide (H_2S) are of great importance in the chemical engineering and petroleum industries, especially for these acid gases co-injection with water or alcohols, etc. The Cubic-Plus-Association (CPA) Equation of State (EoS) coupled with the conventional Peng-Robinson (PR) and Soave-Redlich-Kwong (SRK) EoS are applied to investigate the interaction mechanisms between acid gas (H_2S) and H_2O. The possibilities of self-association and cross association between H_2S (a self-association compound using 2B, 3B and 4C sites) and H_2O (using 2B, 3B and 4 C sites) are considered. The binary interaction parameters (BIPs) were optimized to match the experimental data from literature in a wide temperature range of 333 K to 453 K using a nonlinear least square method, and a better quadratic polynomial *T*- dependent correlation is developed. Then, a vapour-liquid equilibrium calculation at the temperature of 377 K, 410 K and 444 K are performed to match with the experimental data and the best association scheme for the H_2S-H_2O systems was determined. In general, the CPA provides more satisfactory results in comparison to the conventional EoS. More specifically, the best calculation results at 377 and 410 K are achieved if H_2S is treated as a non-self-association compound and the 2B site of H_2O is considered. The calculation results generated by 3B site and 4 C site of H_2O are followed by this best scenario. However, at 444 K or higher temperature, the best association scheme is the one if

Corresponding author: guopingswpi@vip.sina.com

Ying Wu, John J. Carroll and Yongle Hu (eds.) The Three Sisters, (63–90) © 2019 Scrivener Publishing LLC

2B or 3B sites of H_2S and 4 C site of H_2O are taken into account. The new model which requires the association parameters for multicomponent systems containing H_2S and H_2O at a wide temperature and pressure range needs the support of more experimental data; this will be considered in the future.

Keywords: Binary interaction parameter, phase behaviour, acid gas, CPA, association site, H_2S

6.1 Introduction

With the gradually increasing demand for energy and continuous decreasing of natural "sweet" gas reserves, new natural gas reservoirs with high acid gas contents are becoming one of the main resources for economical production. Acid gases like CO_2 and H_2S always cause some serious issues, for example, the toxicity of H_2S; both H_2S and CO_2 may cause the corrosion and crystallization in the pipelines during the production and transportation processes; both gases are able to form gas hydrates in the presence of water at high pressure and low temperature [1], etc. The aforementioned phenomena induce big challenges for the oil and gas industry. In order to overcome these problems and safely control the production process of "sour" natural gas reserves, the simulation of complex thermodynamic behaviour of acid gases with H_2O at different conditions is extremely necessary and important. However, for the H_2S related system, the experimental equilibrium data is rare due to the operation safety concern; only very limited literature information is available in the public domain. Thus, an urgent request is to develop the thermodynamic study handling H_2S containing mixtures, especially for H_2S-H_2O system.

Modelling of the production and processing of H_2S-H_2O mixtures requires accurate vapour-vapour, vapour-liquid, liquid-liquid or even vapour-liquid-liquid equilibrium data. At present, H_2S has always been considered as an inert (non-self-associating) compound. There is not enough experimental data to prove that specific strong interactions exist between H_2S molecules. Based on the structural comparison and ab initio calculations, some findings [2–4] demonstrated that H_2S can be considered as an associating compound although its associating capability is significantly less compared with H_2O, thus H_2S can cross-associate with other compounds containing hydrogen bonds. Sennikov *et al.*, [5] found that weak complexes of H_2S and H_2O molecules are formed in the liquid phase. Furthermore, they also demonstrated that a more stable hydrogen bond is observed when H_2O acts as a proton acceptor while H_2S acts as a proton

Table 6.1 Association energies for the H_2S-H_2O complex.

Interaction	Association energy (J·mol^{-1})	Technique	References
$H_2O...H_2S$	−13807	Ab initio calculation	[6]
$HOH...SH_2$	−12552		
$HOH...H_2S$	−10878	IR(experimental)	[5]
$H_2O...H_2S$	−6276	Ab initio calculation	[8]
$HOH...SH_2$	−5439		
$C_2H_5OH...SH_2$	−8786	IR(experimental)	[9]
$H_2S...H_2S$	−3766~−6276	Ab initio calculation	[10]
Typical hydrogen bond	−25000		[10]

donor [6–8]. A summary of association energies for H_2S-H_2O systems is presented in Table 6.1.

In petroleum and chemical engineering, conventional cubic equation of states (EoS) like PR and SRK [11] with the van der Waals one fluid mixing rules are always used to predict phase equilibrium behaviours. However, it should be noted that because of the association interaction (self-association and cross association) of hydrogen bonding forces between H_2S and H_2O molecules, the conventional EoS failed to simulate the system accurately. Nowadays, the CPA-like Equation of State has been successfully utilized for the thermodynamic calculation of associating mixtures [12–14]. This type of EoS can be converted back to the conventional EoS which is often used for the systems in the absence of hydrogen bonding compounds. There are several basic types of association schemes presented by Huang and Radosz [15] (see Table 6.2). Different molecules are characterized by using different association schemes.

At present, only few publications examined H_2S-H_2O mixtures. Some findings [10, 16] examined vapour-liquid equilibrium (VLE) by using CPA EoS with 4C and 3B sites for H_2S and 4C for H_2O. They concluded that good results are obtained when considering H_2S as a self-association compound using 3B site, or treat H_2S as a non-self-association compound but assuming it can cross associate with H_2O (solvation). It is also shown

Table 6.2 Association schemes for compounds.

Association type	Type symbols	Association sites
Non-self-association	Inert	No sites
	solvation	
Self-association	1A	One proton donor (1d)
	2B	One proton donor, and one proton acceptor (1d,1a)
	3B	Two proton donor and one proton acceptor (2d,1a)
		One proton donor and two proton acceptor (1d,2a)
	4C	Two proton donor and two proton acceptor (2d,2a)

Note: inert (non-self-association) compound means neither can self-associate nor cross associate; solvation means the compounds cannot self-associate but can cross associate with other compounds; self-association compound means it not only can self-associate with itself, but also can cross associate with other compound; "d" donates proton donor, and "a" donates proton acceptor.

that the CPA EoS would be failed with all binary interaction parameters set to zero. Other studies [1, 10, 12, 14, 17] demonstrated that if the solvation between H_2S and H_2O (4C) are accounted, a satisfactory mutual solubility data would be obtained. The best results of H_2S-H_2O system collected from literatures are presented in Table 6.3.

The purpose of this work is to provide a comprehensive investigation to accurately predict thermodynamic behaviour using the PR, SRK and CPA EoS and to determine the best association scheme for H_2S-H_2O system. The equilibrium calculation for all types of association schemes for H_2S-H_2O are performed. Moreover, a set of the binary interaction parameters K_{ij} are determined and a new correlation between the BIPs and temperature is proposed.

6.2 The Cubic-Plus-Association Equation of State

The CPA EoS [13] usually combines the physical term from the SRK EoS [11] with an association term derived from Wertheim perturbation theory [18]. The equation is proposed to account for the effect of hydrogen bonds and can be expressed as follows,

$$Z^{CPA} = Z^{SRK} + Z^{assoc} \tag{6.1}$$

Where the physical and association terms are expressed as,

$$Z^{SRK} = \frac{V_m}{V_m - b} - \frac{\alpha}{RT\left(V_m + b\right)} \tag{6.2}$$

$$Z^{assoc} = \rho \sum_i x_i \sum_{A_i} \left[\left(\frac{1}{X_{A_i}} - \frac{1}{2} \right) \frac{\partial X_{A_i}}{\partial \rho} \right] \tag{6.3}$$

Where V_m is the molar volume $(= 1/\rho)$,x_i is the mole fraction of molecule i,X_{Ai} is the mole fraction of A-site of molecule i that do not bond with other active sites. It can be calculated as,

$$X_{Ai} = \frac{1}{1 + \rho_m \sum_j x_j \sum_{B_j} X_{B_j} \Delta^{A_i B_j}} \tag{6.4}$$

$\Delta^{A_i B_j}$ is the strength association between A-sites on molecule i and B-sites on molecule j, which is a function of association energy and association volume, $\varepsilon^{A_i B_j}$ and $\beta^{A_i B_j}$.

Table 6.3 The summary of best results of H_2S-H_2O system.

Association type	**Association sites of H_2S**	**Association sites of H_2O**	**References**
Self-association	2B	-	-
	3B	4C	Ruffine *et al.*, [16] Tsivintzelis *et al.*, [10]
	4C	-	-
Non-self-association	Solvation	4C	Kontogeorgis *et al.*, [12] Tsivintzelis *et al.*, [10] ZareNezhad and Ziaee [14] Santos *et al.*, [1]
	inert	-	-

$$\Delta^{A_iB_j} = g\left(\rho\right)^{ref}\left[exp\left(\frac{\varepsilon^{A_iB_j}}{RT}\right) - 1\right] b_{ij}\beta^{A_iB_j} \tag{6.5}$$

A simpler expression of the radial distribution function [19],$g\left(\rho\right)^{ref}$, for the reference fluid system is defined as:

$$g(\rho)^{ref} = \frac{1}{1 - 1.9\eta}, \; \eta = \frac{1}{4}b\rho_m \tag{6.6}$$

The CR-1 combining rule [20] has been proved very successful in the applications of public work. The expressions of association energy and association volume for cross association are expressed as,

$$\varepsilon^{A_iB_j} = \frac{\varepsilon^{A_iB_i} + \varepsilon^{A_jB_j}}{2} \tag{6.7}$$

$$\beta^{A_iB_j} = \sqrt{\beta^{A_iB_i}\beta^{A_jB_j}} \tag{6.8}$$

In the case of solvation, the modified combing rule are:

$$\varepsilon^{A_iB_j} = \frac{\varepsilon_{association}}{2} \tag{6.9}$$

$$\beta^{A_iB_j} = \beta_{across} = \text{fitted to the experimental data} \tag{6.10}$$

The CPA model works very well for mixtures where hydrogen bonds of compounds are accounted for while the others compounds only have van der Waals bonds [21]. As the SRK and PR EoS are already well known to the public, they will not be presented here.

6.3 Association Schemes

The well-known CPA-like EoS combined with SRK EoS was used in this work to calculate the phase equilibrium of H_2S-H_2O. There are five pure compound parameters for the CPA-SRK EoS, two are used for associating compounds ($\varepsilon^{A_iB_j}$,$\beta^{A_iB_j}$) and three additional parameters$\left(a_0, c_1, b\right)$ are used for SRK term. They are all estimated by fitting the experimental data of vapour pressure and liquid density [22].

Table 6.4 summarizes CPA parameters for pure fluid of H_2S and H_2O published in recent works. [19, 23] An investigation was proposed to treat H_2S as non-self-associating as well as self-associating compounds using 2B, 3B, and 4C sites while H_2O as self-association using 2B, 3B and 4C sites.

Table 6.4 CPA parameters for H_2S and H_2O involved in this study.

Component	Association scheme	a_0 (bar·L2/mol²)	B(L/mol)	c_1	ε^{AB} (bar·L/mol)	β^{AB}	References
H_2S	n.a	4.4505	0.0285	0.60265	-	-	[10]
	2B	3.47972	0.0285	0.41107	80.8848	0.08581	
	3B	3.86049	0.0292	0.50222	54.3992	0.05832	
	4C	3.96977	0.0295	0.53703	37.2634	0.04745	
H_2O	2B	2.7311	0.0147	0.1752	269.26	0.0202	[19]
	3B	3.5746	0.0155	0.1528	233.49	0.0068	
	4C	1.2278	0.0145	0.6736	166.56	0.0692	[23]

n.a.: non-self-association

It is the purpose of this work to determine the best association scheme for H_2S and H_2O and to examine their interactions with each other. Huang and Radosz [15] show that the association geometric structure of H_2S molecule is similar to that of H_2O. Several detailed processes of cross association and self-association for H_2S (non-self-association, 2B, 3B and 4C) and H_2O (4C) system are presented in Table 6.5. As can be seen from these association processes, there are two kinds of hydrogen bonds designated by the number of "1" and "2", respectively. "1" denotes the hydrogen bond between S-atom of H_2S and H-atom of H_2O, and "2" represents the hydrogen bond between H-atom of H_2S and O-atom of H_2O. The arrow indicates the direction of proton transferred. That is to say, S-atom or O-atom acts as a proton donor while H-atom acts as a proton acceptor. And the dashed lines represent an interaction between two atoms. Actually, only few references were discussed about the application of 2B for H_2S and 2B or 3B for H_2O since some other interactions are assumed either weak or negative [5].

6.4 Results and Discussion

6.4.1 Binary Interaction Parameter

The binary interaction parameters were estimated by a nonlinear least square method. These calculations required experimental thermodynamic properties which are either solubilities or phase equilibrium data in the range of interest temperatures and pressures [24]. In this work, the experimental solubility data of H_2S in H_2O at temperature from 333 K to 453 K [25–29] were selected to fit these parameters, as the reported experimental data in vapour phase were scarce.

Due to the complexity of the mixture, a better quadratic polynomial T- dependent correlation is developed to fit the K_{ij} values for all types of association schemes of H_2S- H_2O systems.

$$K_{ij} = mT^2 + nT + k \tag{6.11}$$

Where, m, n and k are regressed parameters, and T is the temperature in Kelvin. The objective function expressed below,

$$F_{objective} = \sum_{i=1}^{N} \left(\frac{x_{exp} - x_{cal}}{x_{exp}} \right)^2 \tag{6.12}$$

Where x_{exp} is the measured solubility data in the unit of mole fraction, x_{cal} is the predicted value. N is the number of experimental data points.

The optimum values of the three parameters m, n and k are presented in Table 6.6. The average absolute deviations (AAD) of the predicted results

Table 6.5 Example cross association and self-association processes between H_2S and H_2O.

Association schemes	Association siteson H_2S	Association structures based on H_2S
Solvation	2d-0a	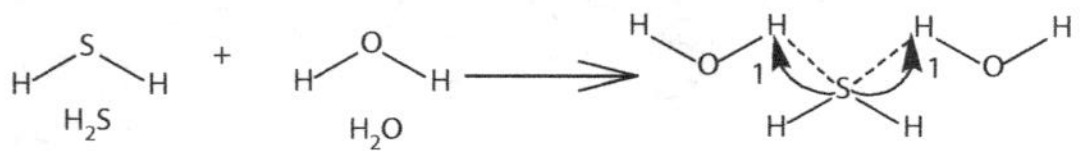 H_2S + H_2O →
H_2S-2B	1d-1a	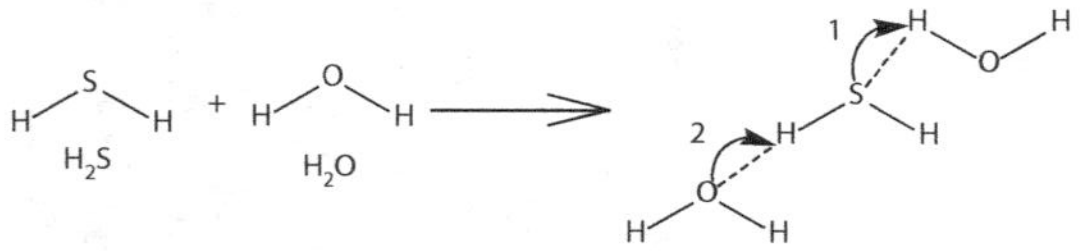 H_2S + H_2O →
H_2S-3B	2d-1a	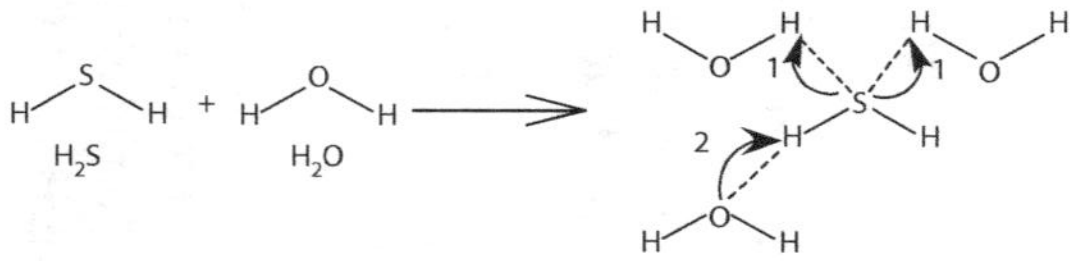 H_2S + H_2O →
H_2S-4C	2d-2a	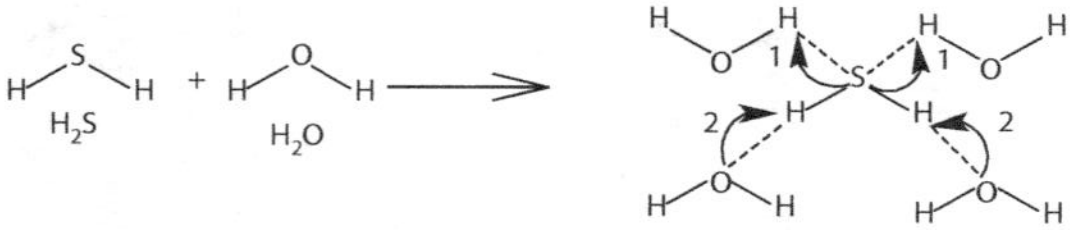 H_2S + H_2O →

(Continued)

Table 6.5 Cont.

Association schemes	Association siteson H_2S	Association structures based on H_2S
H_2O self -association	2d-2a	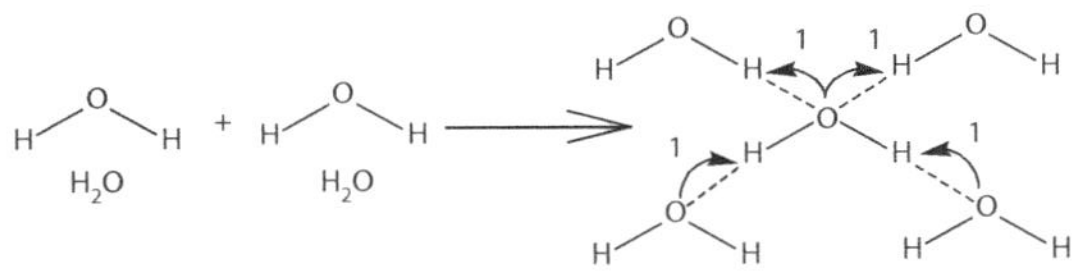 H_2O + H_2O →
H_2S self- association	2d-2a	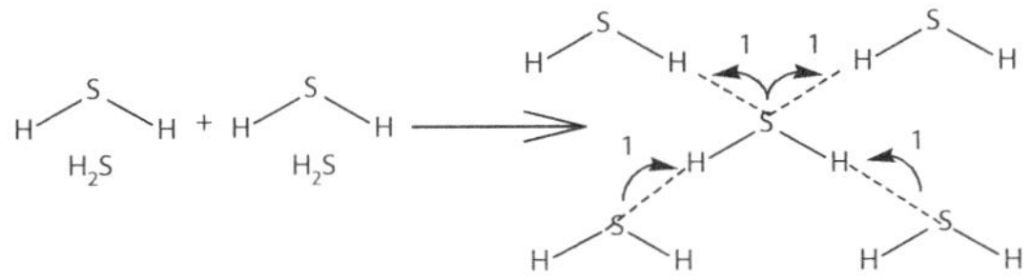 H_2S + H_2S →

Table 6.6 The coefficients of regression for different association schemes of H_2S and H_2O in the temperature range of 333 ~ 453 K.

H_2O		**2B**		
H_2S	m	n	k	$AAD\text{-}\Delta y$ (%)[a]
n.a.	-2×10^{-6}	0.0003	0.0594	1.01
Solvation	-2×10^{-6}	0.0008	0.1261	1.12
2B	-2×10^{-6}	0.0007	0.1275	1.15
3B	-2×10^{-6}	0.0006	0.18	1.18
4C	-3×10^{-6}	0.0001	0.158	1.16
H_2O		3B		$AAD\text{-}\Delta_y$ (%)[a]
H_2S	m	n	k	
n.a.	-1×10^{-6}	-4×10^{-5}	0.0142	0.89
Solvation	-1×10^{-6}	0.0005	0.0099	1.10
2B	5×10^{-6}	−0.0017	0.2527	1.23
3B	3×10^{-6}	−0.0011	0.2048	0.86
4C	4×10^{-6}	−0.0018	0.3626	1.12
H_2O		4C		$AAD\text{-}\Delta_y$ (%)[a]
H_2S	m	n	k	
n.a.	-2×10^{-6}	-4×10^{-5}	0.0034	1.54
Solvation	-2×10^{-6}	0.0003	0.0394	1.46
2B	-4×10^{-6}	0.0012	0.1667	1.23
3B	-4×10^{-6}	0.0013	0.1887	0.65
4C	-4×10^{-6}	0.0013	0.1938	0.66
SRK	-7×10^{-7}	0.0009	−0.0606	1.23
PR	-9×10^{-7}	0.0009	−0.0511	1.20

Note: $AAD\ (\%) = \frac{1}{N} \cdot \frac{|x_{exp} - x_{cal}|}{x_{exp}} \times 100$, y represents the mole fraction of H_2S in vapour phase.

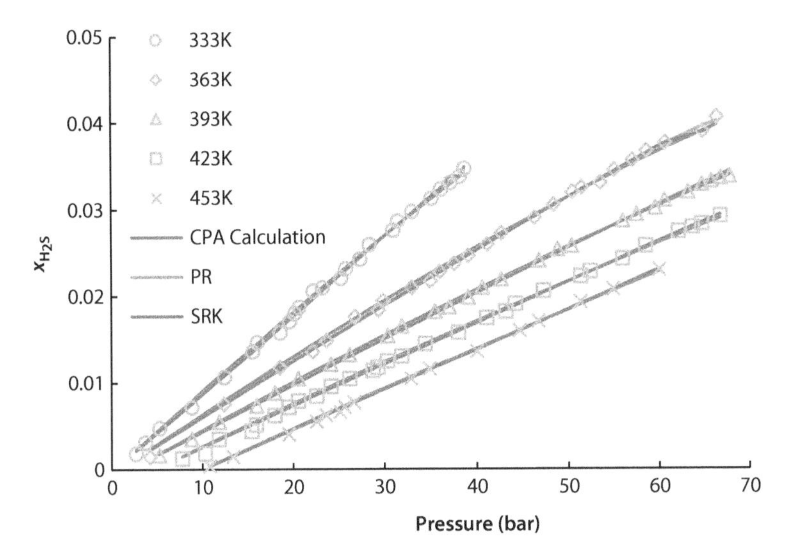

Figure 6.1 Comparison of calculation results against experimental data (CPA calculation of H_2S solubility in H_2O assuming that H_2S is non-self-association fluid and H_2O is self-association fluid using 4C association scheme with K_{ij} = −0.0041,−0.0122, -0.0234,−0.0352 and −0.0525 for temperature of 333K, 363K, 393K, 423K and 453K, respectively).

using the generalized expression for K_{ij} with the experimental data are presented in Table 6.6. The results show that almost all the deviations are below 1.5%. In particular, a relative minimum deviation was obtained if H_2O was considered as 4C association model while H_2S was considered as 3B or 4C (0.65 and 0.66%).

Typical calculation results for the solubility of H_2S in liquid water phase when considering H_2S as a non-self-association compound and H_2O as a self-association compound using 4C site are shown in Figure 6.1. There is a satisfactory agreement between the experimental data and the predicted values over wide temperatures. The *AAD*s of the three kinds of EoS are basically the same.

6.4.2 Solubility Calculation

The experimental solubility data for H_2S in H_2O rich phase and H_2O in H_2S rich phase have been collected from literature at three temperatures of 377 K, 410 K and 444 K and pressures up to 210 bar [30]. At the studied temperatures and pressures, the H_2S-H_2O system is in the states of gas and liquid.

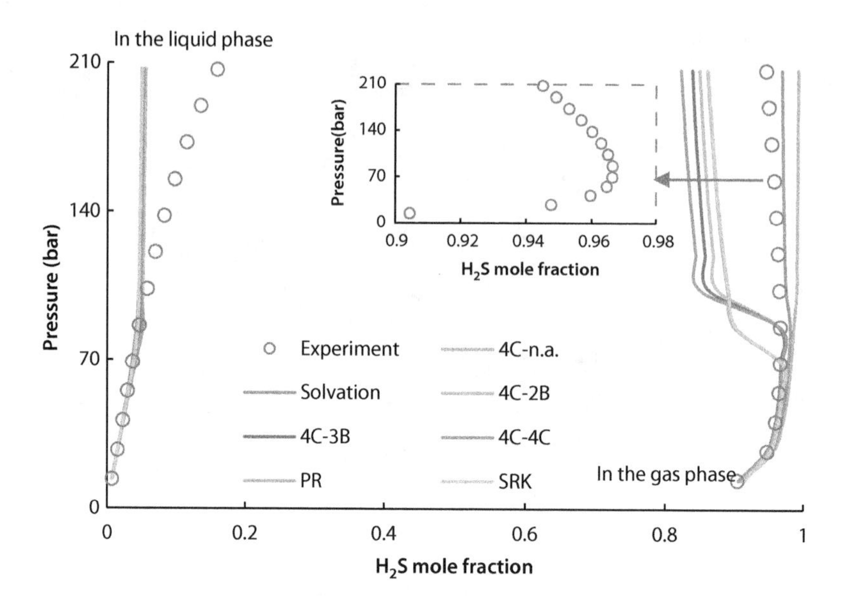

Figure 6.2 H_2S-H_2O Vapour-liquid Equilibria at 377 K (CPA Calculation Assuming that H_2O as Self-association Compound Using 4C Site and H_2S as Non-self-association and Self-association Using 2B, 3B and 4C).

The vapour-liquid equilibria data were modelled by using the CPA, PR and SRK EoS with the BIPs fitted in this work. Totally, 17 different models (15 CPA-EoS with different association schemes, one PR EoS plus one SRK EoS models) have been studied at each temperature. All equilibrium calculations used non-zero BIPs. The overall performance of CPA is better than conventional cubic equations of state, PR and SRK EoS. Figures 6.2–6.10 show the results between the simulation and experimental data. It can be seen that the H_2S content in a liquid phase increases quickly with the increasing pressure, while in the gas phase, initially the H_2S content increases quickly with increasing pressure, then after a certain pressure it decreases with increasing pressure. This phenomenon is very pronounced at low temperatures and can be seen from the insert plot (see Figure 6.2, Figure 6.5 and Figure 6.8). For instance, at the temperature of 377 K, the phenomenon changed pressure occurs at near 90 bar which is close to the critical pressure of H_2S. As temperature increases, the pressure increases to 120 bar and 170 bar at 410 K and 444 K, respectively.

As can be seen from the results of figures, different association schemes between H_2S and H_2O generate different results. The lowest deviations of solubilities are obtained if H_2S is considered as a non-self-association compound at the temperatures of 377 K and 410 K. From our calculation,

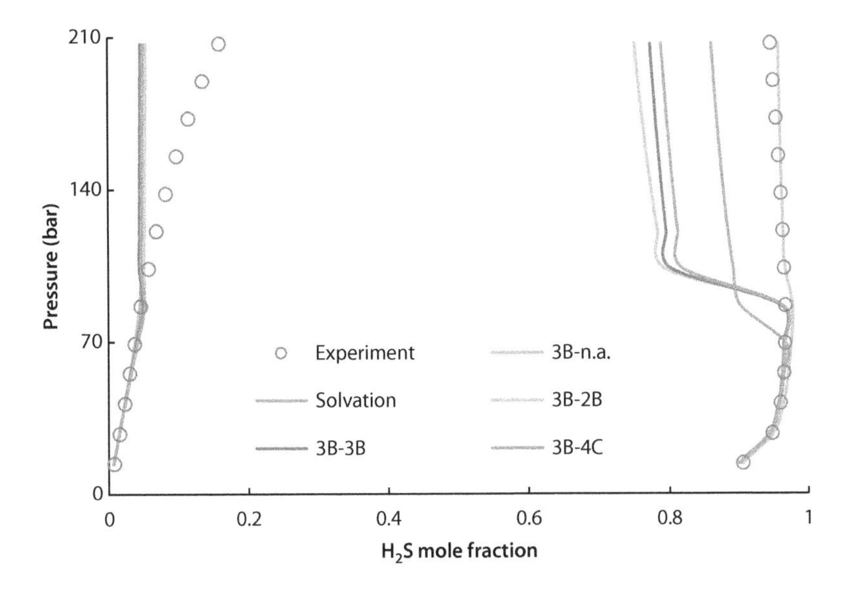

Figure 6.3 H_2S-H_2O Vapour-liquid Equilibria at 377 K (CPA Calculation Assuming that H_2O as Self-association Compound Using 3B Site and H_2S as Non-self-association and Self-association Using 2B, 3B and 4C).

the best results are not obtained as the literatures [31–34] suggested in which H_2O was considered as 4C association scheme; our results indicate that the best results are obtained when 2B association scheme of H_2O was accounted, was followed by 3B scheme, and finally was 4C scheme. When H_2O is treated as 4C scheme, the simulation results are basically the same whether H_2S can be able to cross associate with H_2O as the proton donor.

Relatively high deviations for experimental data are obtained if H_2S cross associate with H_2O but not self-associate (which is so-called the solvation). Very high deviations are obtained when H_2S is considered as self-association compound using 2B, 3B and 4C sites.

It is clearly shown that presently no association model or Equation of State could be able to completely describe the solubilities of H_2S in the liquid phase. CPA can accurately predict the H_2S solubility in the liquid phase to a certain pressure and above which the simulated results start to deviate from the experimental data as the rapid solubility increase tendency stop especially at the 377 and 410 K. The simulation results also indicate the best association scheme should be utilized for CPA calculation. For example, that CPA considering H_2S as a self-association compound failed to trace the curve shape of H_2S in the vapour phase. However, CPA which considering H_2S as a non-self-association compound and a proton donor

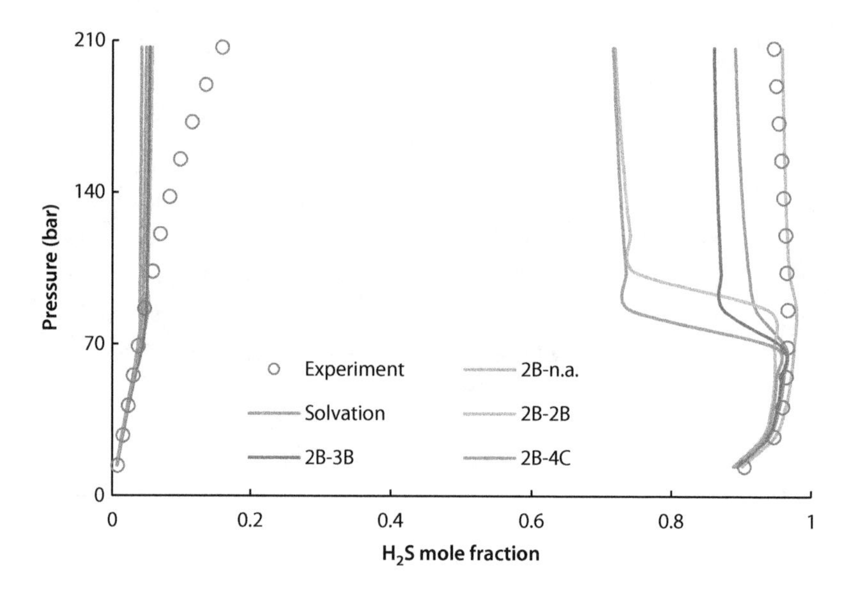

Figure 6.4 H_2S-H_2O Vapour-liquid Equilibria at 377 K (CPA Calculation Assuming that H_2O as Self-association Compound Using 2B Site and H_2S as Non-self-association and Self-association Using 2B, 3B and 4C).

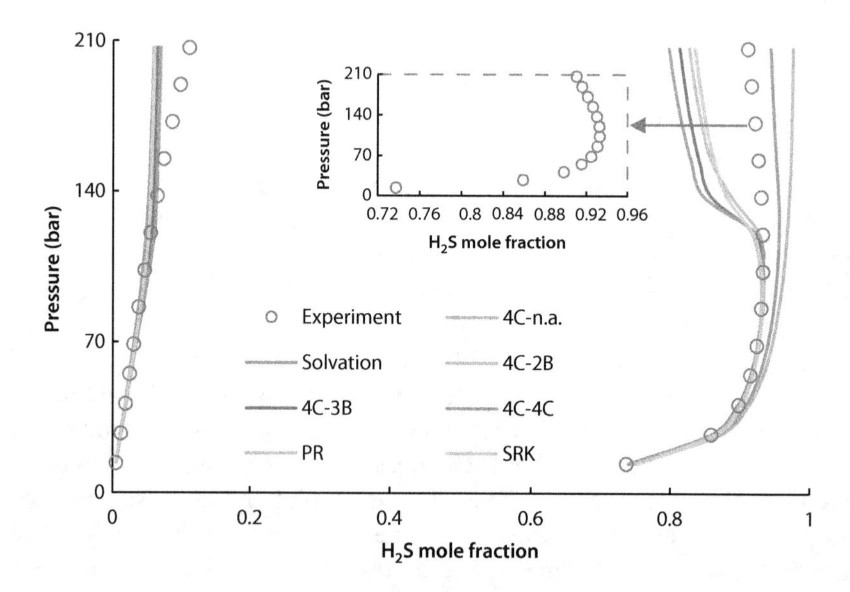

Figure 6.5 H_2S-H_2O Vapour-liquid Equilibria at 410 K (CPA Calculation Assuming that H_2O as Self-association Compound Using 4C Site and H_2S as Non-self-association and Self-association Using 2B, 3B and 4C).

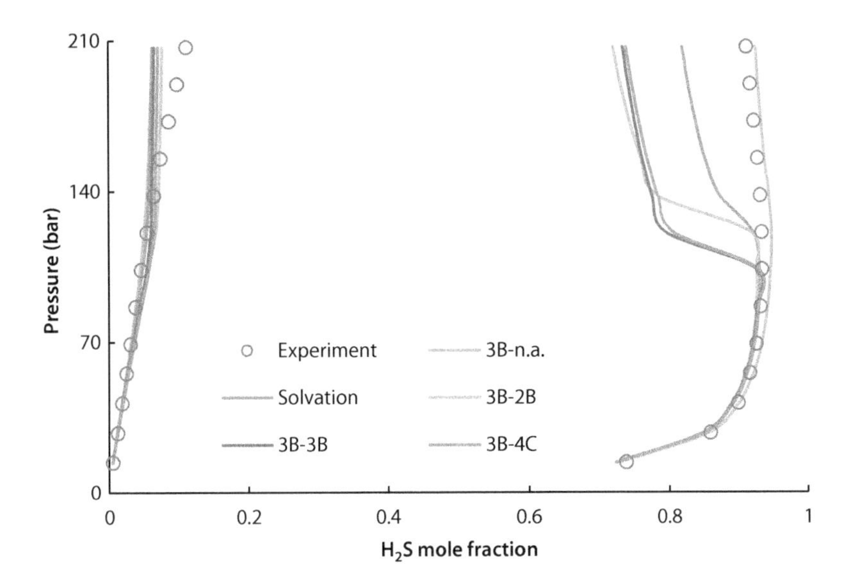

Figure 6.6 H_2S-H_2O Vapour-liquid Equilibria at 410 K (CPA Calculation Assuming that H_2O as Self-association Compound Using 3B Site and H_2S as Non-self-association and Self-association Using 2B, 3B and 4C).

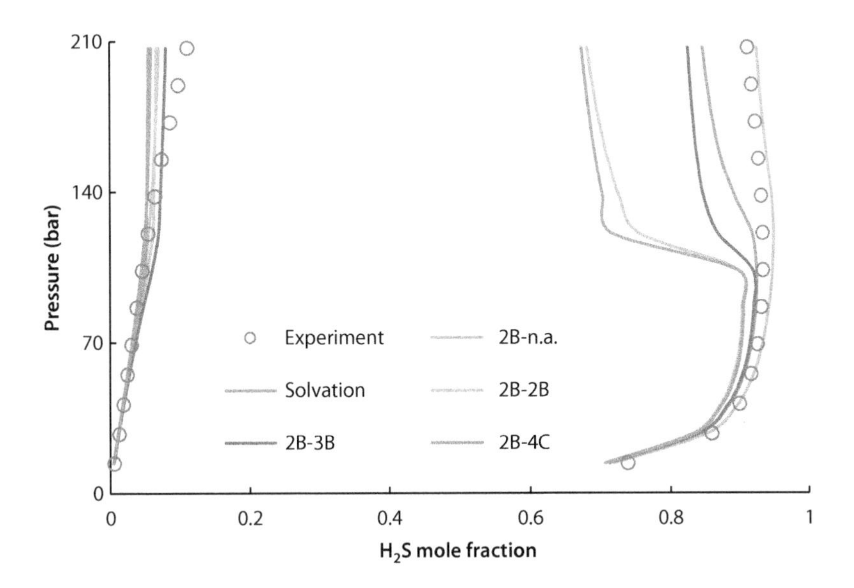

Figure 6.7 H_2S-H_2O Vapour-liquid Equilibria at 410 K (CPA Calculation Assuming that H_2O as Self-association Compound Using 2B Site and H_2S as Non-self-association and Self-association Using 2B, 3B and 4C).

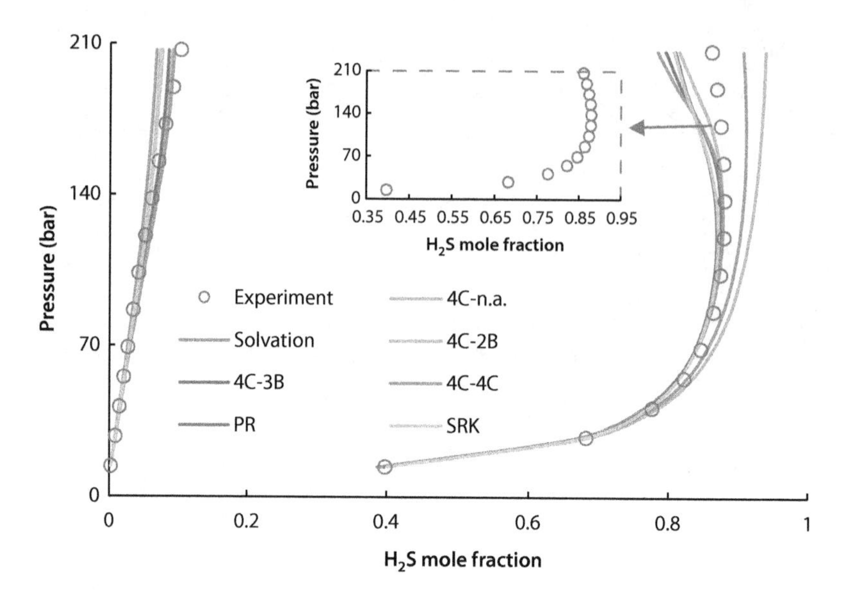

Figure 6.8 H_2S-H_2O Vapour-liquid Equilibria at 444 K (CPA Calculation Assuming that H_2O as Self-association Compound Using 4C Site and H_2S as Non-self-association and Self-association Using 2B, 3B and 4C).

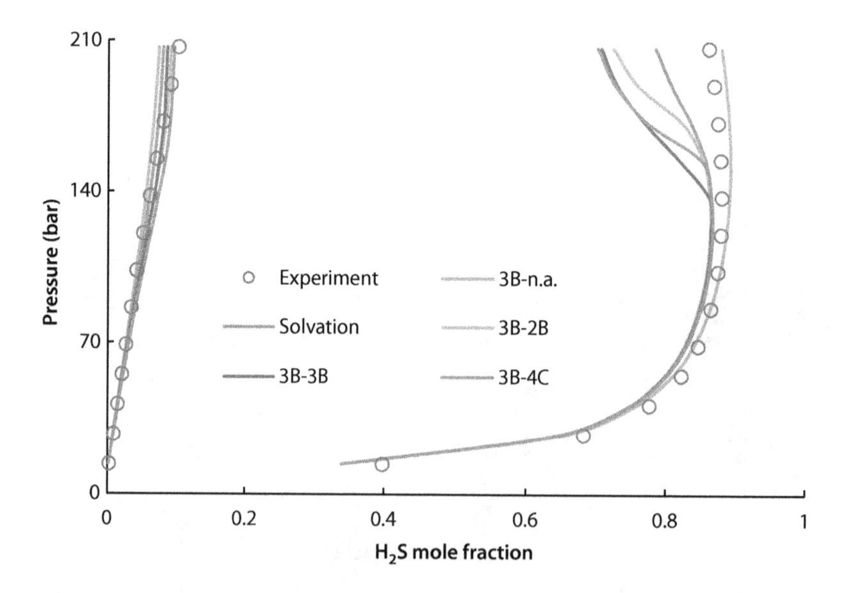

Figure 6.9 H_2S-H_2O Vapour-liquid Equilibria at 444 K (CPA Calculation Assuming that H_2O as Self-association Compound Using 3B Site and H_2S as Non-self-association and Self-association Using 2B, 3B and 4C).

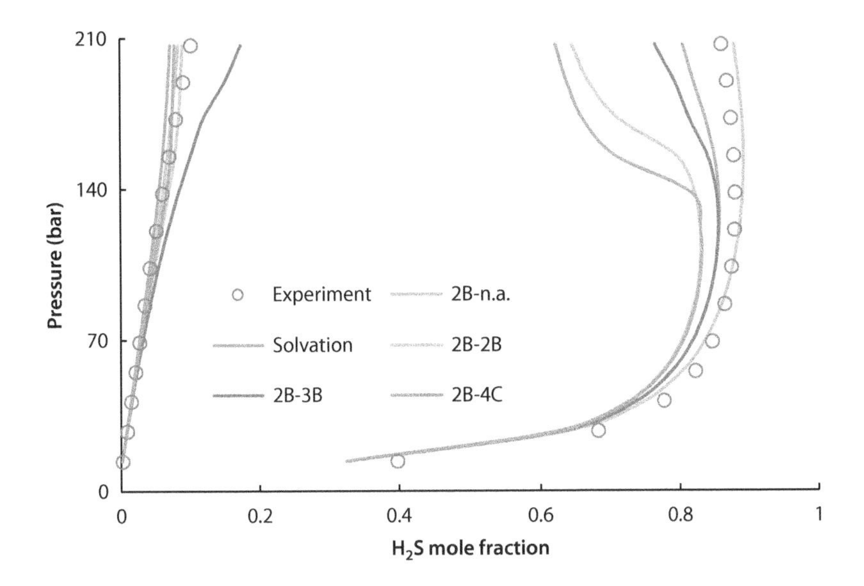

Figure 6.10 H_2S-H_2O Vapour-liquid Equilibria at 444 K (CPA Calculation Assuming that H_2O as Self-association Compound Using 2B Site and H_2S as Non-self-association and Self-association Using 2B, 3B and 4C).

can accurately depict the relationship between the mole fraction of H_2S in vapour phase and specified pressures.

With the increasing temperature, the simulation results are getting improved. Similar plots at temperature of 444 K are given in Figure 6.8–6.10 . Satisfactory results are achieved if H_2O using 4C site and H_2S using 2B or 3B site.

To easily understand the simulating trends and results, Tables 6.7–6.9 show the detailed comparison of absolute average deviations between the simulated results and experimental data. The BIPs used in this work are also presented. The average deviation between the experimental results and the calculated data almost decreased from 24.6% to 4.4% as temperature increases. The absolute deviations which are defined as $\Delta y = y_{exp} - y_{cal}$ and$\Delta x = x_{exp} - x_{cal}$ graphically shown in Figures 6.11–6.16 at 377 K, 410 K and 444 K, respectively.

Generally, the mole fraction of H_2S in vapour phase will be accurately predicted unless H_2S is treated as non-self-association compound. Meanwhile, at temperature of 444 K, both conventional equation of states of PR and SRK give underestimated results. It also can be found that as temperature increases, the pressure range for accurate prediction also increases.

Table 6.7 Simulation results of solubility for different association schemes at 377 K.

Association scheme of H_2O	Association scheme of H_2S	377 K			
		K_{ij}	*AAD -y* (%)	*AAD-x* (%)	Average (%)
4C	n.a.	–0.0224	2.5584	26.0443	14.3013
4C	n.a.(2d)	0.0490	1.2609	27.3713	14.3161
4C	2B	0.2482	5.4619	25.2524	15.3571
4C	3B	0.2806	6.0820	27.7146	16.8983
4C	4C	0.2857	6.9216	27.6567	17.2892
3B	n.a.	–0.0008	0.5496	25.6067	13.0781
3B	n.a. (2d)	0.0511	5.1896	26.5477	15.8686
3B	2B	0.1300	10.7062	26.0782	18.3922
3B	3B	0.1228	9.7367	29.9272	19.8319
3B	4C	0.2187	8.7579	28.8826	18.8203
2B	n.a.	0.0690	0.6348	23.9420	12.2884
2B	n.a. (2d)	0.1877	3.8240	27.9959	15.9100
2B	2B	0.1787	13.5850	27.8689	20.7270
2B	3B	0.2208	6.2251	27.4558	16.8405
2B	4C	0.2296	15.4544	33.7610	24.6077
SRK	-	0.0254	5.3598	27.5181	16.4390
PR	-	0.0328	5.3674	27.6677	16.5176

Calculated solubilities of H_2O in this work and the literature data are compared. The results of only H_2O being treated as 4C site are shown in Table 6.10. Although there is no significant improvement in the prediction of H_2O solubility in the liquid phase with CPA EoS combined with aforementioned quadratic *T*-dependence BIP correlations, relatively lower deviations were observed for H_2O in vapour phase against

Table 6.8 Simulation results of solubility for different association schemes at 410 K.

Association scheme of H_2O	Association scheme of H_2S	410 K			
		K_{ij}	$\Delta y(\%)$	Δx (%)	Average (%)
4C	n.a.	−0.0396	3.6499	13.9126	8.7812
4C	n.a.(2d)	0.0430	2.1915	16.3248	9.2581
4C	2B	0.2560	3.0732	12.2477	7.6604
4C	3B	0.2917	3.8855	13.9083	8.8969
4C	4C	0.2968	4.4617	13.7960	9.1288
3B	n.a.	−0.0100	1.0295	14.3534	7.6915
3B	n.a.(2d)	0.0596	3.8206	13.9642	8.8924
3B	2B	0.1136	7.9850	14.5292	11.2571
3B	3B	0.1104	8.6959	15.5705	12.1332
3B	4C	0.1911	8.2570	15.8912	12.0741
2B	n.a.	0.0630	1.2393	11.6512	6.4452
2B	n.a.(2d)	0.1982	3.3331	15.7031	9.5181
2B	2B	0.1859	12.4612	14.8079	13.6345
2B	3B	0.2247	5.1419	15.3209	10.2314
2B	4C	0.2387	13.2670	18.5663	15.9167
SRK	-	0.0496	3.2839	14.4938	8.8889
PR	-	0.0553	3.1370	14.7704	8.9537

Tsivintzelis *et al.*, [10]. who reported a deviation more than 100% when H_2S was considered as self-association compound. More experimental data and further mechanism studies are required in the future to improve the simulation capability for the H_2S-H_2O system.

Table 6.9 Simulation results of solubility for different association schemes at 444 K.

Association scheme of H_2O	Association scheme of H_2S	444 K			
		K_{ij}	Δy (%)	Δx (%)	Average (%)
4C	n.a.	−0.0619	3.9116	9.8423	6.8770
4C	n.a.(2d)	0.0322	2.4091	13.0002	7.7047
4C	2B	0.2549	1.6873	7.1445	4.4159
4C	3B	0.2940	2.2462	6.8382	4.5422
4C	4C	0.2991	2.2211	8.6460	5.4335
3B	n.a.	−0.0219	2.5666	11.2368	6.9017
3B	n.a.(2d)	0.0662	4.4330	8.0784	6.2557
3B	2B	0.1082	5.6839	8.2518	6.9678
3B	3B	0.1044	6.8061	8.9591	7.8826
3B	4C	0.1718	6.4278	12.5158	9.4718
2B	n.a.	0.0522	3.1321	7.2470	5.1895
2B	n.a.(2d)	0.2044	5.2700	12.0885	8.6793
2B	2B	0.1887	10.9767	9.8421	10.4094
2B	3B	0.2241	6.1382	30.3947	18.2665
2B	4C	0.2413	12.6027	10.1087	11.3557
SRK	-	0.072831	1.9625	8.5518	5.2572
PR	-	0.076483	2.0559	9.0832	5.5696

6.5 Conclusions

The CPA, PR and SRK EoSs are applied to H_2S-H_2O binary system. It is known from the literature that a weak hydrogen bond may occur between H_2S molecules compared with H_2O molecule. Although the weak influences

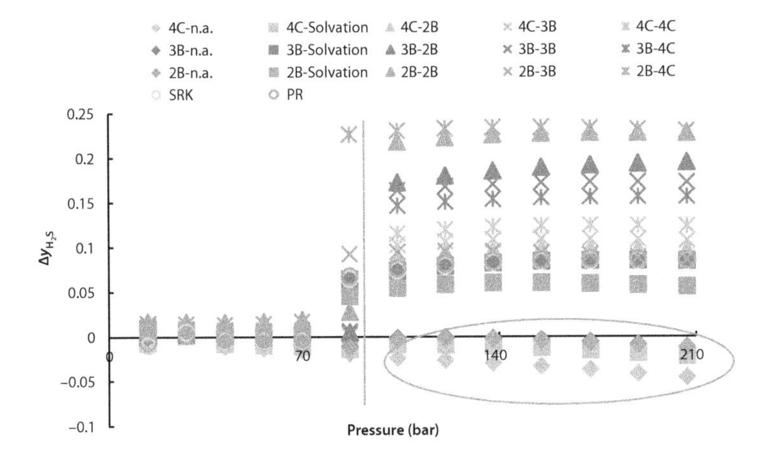

Figure 6.11 The Absolute Deviation between Simulated Solubility of H_2S and Experimental Data at 377 K.

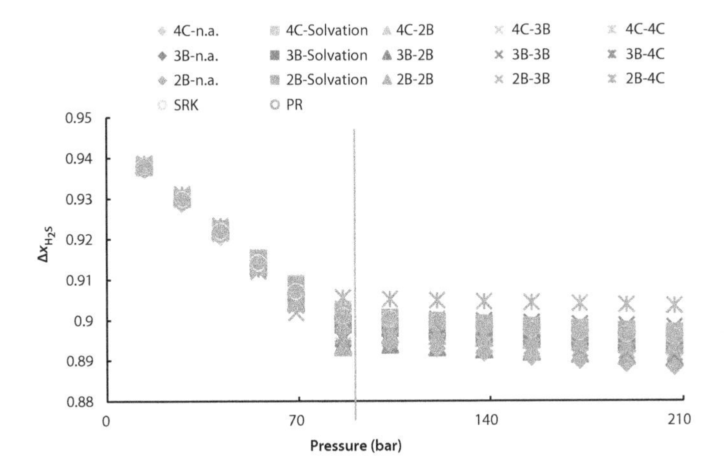

Figure 6.12 The Absolute Deviation between Simulated Solubility of H_2S and Experimental Data at 377 K.

of this hydrogen bonding interaction, the conventional equation of states still cannot simulate the thermodynamic properties accurately. The CPA gives more satisfactory results against experimental data for the solubilities of H_2S and H_2O than SRK and PR EoS.

One T- dependent BIP correlation is developed by nonlinear least square method in a wide range of temperature from 333 K to 453 K. Satisfactory prediction results are obtained with a typical deviation below 1.5% for the mole fraction of H_2S in the vapour phase.

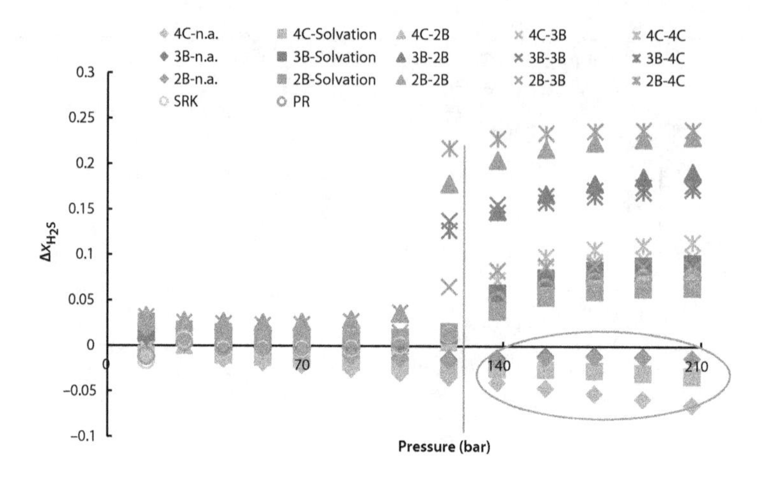

Figure 6.13 The Absolute Deviation between Simulation Solubility of H_2S and Experimental Data at 410 K.

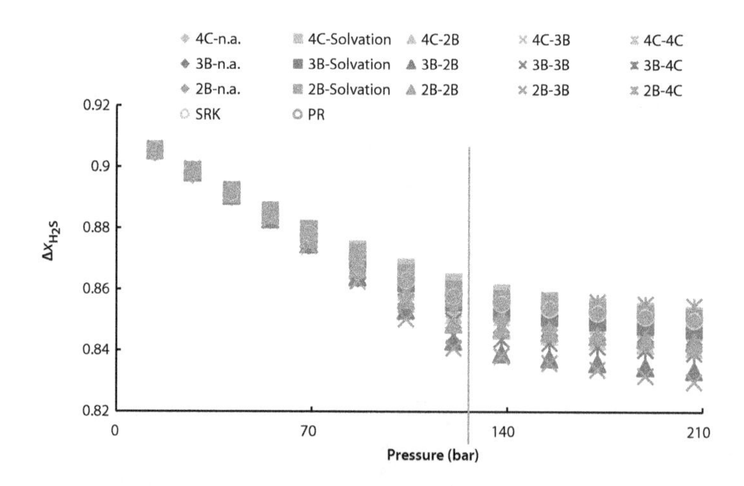

Figure 6.14 The Absolute Deviation between Simulation Solubility of H_2S and Experimental Data at 410 K.

In conclusion, this work provides a satisfactory simulation equilibria dataset that covers all types of association schemes between H_2S and H_2O that may occur. This fulfilled the research blank that H_2O is usually modelled as 4C site. The best CPA simulation results are achieved by assuming H_2S is a non-self-association compound and H_2O as 2B site. This best scenario was followed considering 3B site for H_2O and finally 4C site at

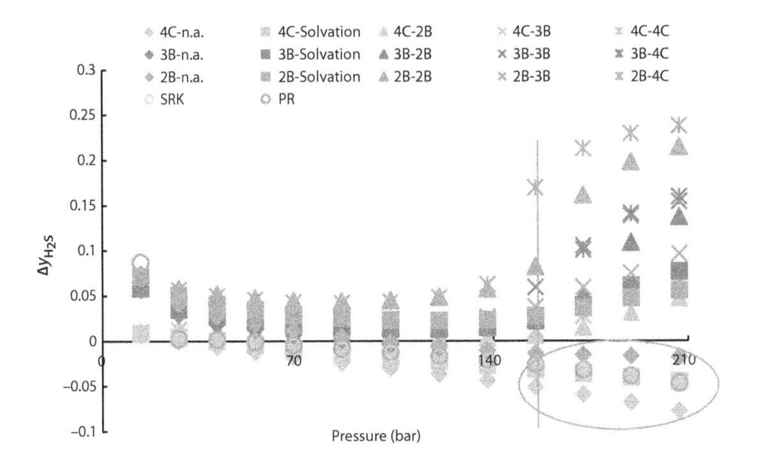

Figure 6.15 The Absolute Deviation between Simulation Solubility of H_2S and Experimental Data at 444 K.

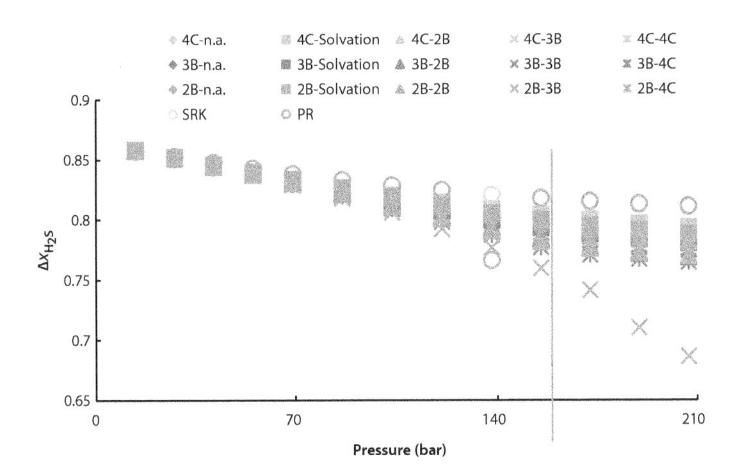

Figure 6.16 The Absolute Deviation between Simulation Solubility of H_2S and Experimental Data at 444 K.

temperature 377 and 410 K. At 444 K, the best association scheme is when 2B or 3B site for H_2S and 4C site for H_2O are considered.

To conclude, the current work proves that CPA EoS as a new tool for petroleum and chemical engineering that can accurately predict the complex phase behaviour and identify the best candidate association scheme for multicomponent mixtures which involve H_2S and H_2O.

Table 6.10 a Comparison for the solubilities for H_2O in this work and in literature (H_2O using 4C site).

Association sites in H_2S	K_{ij}		Liquid phase AAD (%) in X_{H2O}		Vapor phase AAD(%) in Y_{H2O}	
	[10] (310–444K)	In this work (377–444K)	[10] (310–444K)	In this work (377–444K)	[10] (310–444K)	In this work (377–444K)
0	−0.0098	−0.0413	1.7	1.7	36	41
2d-0	0.0985	0.2531	1.2	1.1	24	23
2B	0.2367	0.2888	1.3	1.5	>100	55
3B	0.2601	0.2939	1.3	1.6	>100	64
4C	0.2643	0.0414	1.5	1.6	>100	72

Acknowledgment

This work is supported by the China Scholarship Council (No. 201708510114)."

References

1. Santos, L.C.D., Abunahman, S.S., Tavares, F.W., Modelling water saturation points in natural gas streams containing CO_2 and H_2S comparisons with different equations of state. *Ind. Eng. Chem. Res.*, 54, 2015.
2. Nietodraghi, C., Mackie, A.D., Avalos, J.B., Transport coefficients and dynamic properties of hydrogen sulfide from molecular simulation. *J. Chem. Phys.*, 123, 2015.
3. Pecul, K., Study of the hydrogen bond and of the proton transfer between two H_2S molecules. *Theor. Chim. Acta*, 44(1), 77–83, 1977.
4. Cabaleiro-Lago, E.M., Rodríguez-Otero, J., Peña-Gallego, A., Computational study on the characteristics of the interaction in naphthalene...(H2X)n=1,2 (X = O,S) clusters. *J. Phys. Chem. A*, 112(28), 6344–6350, 2008.
5. Sennikov, P.G., Shkrunin, V.E., Tokhadze, K.G., Intermolecular interactions of hydrogen sulphide and hydrogen selenide with some proton donors and proton acceptors in liquid phase. *J. Mol. Liq.*, 46, 29–38, 1990.

6. Leś, A., A pseudopotential study of the hydrogen bond in $H_2O{\cdot}H_2S$, $H_2S{\cdot}H_2S$ and $H_2O{\cdot}H_2Se$ systems. *Theor. Chim. Acta*, 66(6), 375–393, 1985.
7. Novaro, O., Leś, A., Galván, M., Conde, G., Theoretical study of three-body nonadditive interactions for the H_2S-$(H_2O)2$ system. *Theor. Chim. Acta*, 64(2), 65–81, 1983.
8. Del Bene, J.E., Anab initio molecular orbital study of the structures and energies of neutral and charged bimolecular complexes of NH_3 with the hydrides AH_n (A = N, O, F, P, S, and Cl. *J. Comput. Chem.*, 10(5), 603–615, 1989.
9. Sennikov, P.G., V.E., S., Raldugin, D.A., Weak hydrogen bonding in ethanol and water solutions of liquid volatile inorganic hydrides of group IV-VI elements (SiH_4, GeH_4, PH_3, AsH_3, H_2S, and H_2Se). 1. IR spectroscopy of H bonding in ethanol solutions in hydrides. *J. Phys. Chem.*, 100, 6415–6420, 1996.
10. Tsivintzelis, I., Kontogeorgis, G.M., Michelsen, M.L., Stenby, E.H., Modeling phase equilibria for acid gas mixtures using the CPA equation of state. I. Mixtures with H2S. *AIChE J.*, 56(11), 2965–2982, 2010.
11. Soave, G., Equilibrium constants from a modified Redlich-Kwong equation of state. *Chem. Eng. Sci.*, 27(6), 1197–1203, 1972.
12. Kontogeorgis, G.M., Folas, G.K., Muro-Suñé, N., Roca Leon, F., Michelsen, M.L., Solvation phenomena in association theories with applications to oil & gas and chemical industries. *Oil &. Gas Science and Technology - Rev. IFP.*, 63(3), 305–319, 2008.
13. Kontogeorgis, G.M., Voutsas, E.C., Yakoumis, I.V., Tassios, D.P., An equation of state for associating fluids. *Ind. Eng. Chem. Res.*, 35(11), 4310–4318, 1996.
14. ZareNezhad, B., Ziaee, M., Accurate prediction of H_2S and CO_2, containing sour gas hydrates formation conditions considering hydrolytic and hydrogen bonding association effects. *Fluid Phase Equilib.*, 356, 321–328, 2013.
15. Huang, S.H., Radosz, M., Equation of state for small, large, polydisperse, and associating molecules: extension to fluid mixtures. *Ind. Eng. Chem. Res.*, 30(8), 1994–2005, 1990.
16. Ruffine, L., Mougin, P., Barreau, A., How to represent hydrogen sulfide within the CPA equation of state. *Ind. Eng. Chem. Res.*, 45(22), 7688–7699, 2006.
17. Ziaee, M., ZareNezhad, B., A wew approach for accurate prediction of the phase behaviour of the complex mixtures of acid gases, hydrocarbons, water, and methanol in the petroleum industry. *Pet. Sci. Technol.*, 33(17-18), 1633–1640, 2015.
18. Wertheim, M.S., Thermodynamic perturbation theory of polymerization. *J. Chem. Phys.*, 87(12), 7323–7331, 1987.
19. Kontogeorgis, G.M., Yakoumis, I.V., Meijer, H., Hendriks, E., Moorwood, T., Multicomponent phase equilibrium calculations for water–methanol–alkane mixtures. *Fluid Phase Equilib.*, 158-160, 201–209, 1999.
20. Voutsas, E.C., Yakoumis, I.V., Tassios, D.P., Prediction of phase equilibria in water/alcohol/alkane systems. *Fluid Phase Equilib.*, 158-160, 151–163, 1999.

21. Austegard, A., Solbraa, E., De Koeijer, G., Mølnvik, M.J., Thermodynamic models for calculating mutual solubilities in H_2O–CO_2–CH_4 mixtures. *Chemical Engineering Research and Design*, 84(9), 781–794, 2006.
22. Yakoumis, I.V., Kontogeorgis, G.M., Voutsas, E.C., Hendriks, E.M., Tassios, D.P., Prediction of phase equilibria in binary aqueous systems containing alkanes, cycloalkanes, and alkenes with the cubic-plus-association equation of state. *Ind. Eng. Chem. Res.*, 37(10), 4175–4182, 1998.
23. Zirrahi, M., Hassanzadeh, H., Abedi, J., Prediction of water solubility in petroleum fractions and heavy crudes using cubic-plus-association equation of state (CPA-EoS). *Fuel*, 159, 894–899, 2015.
24. Perfetti, E., Thiery, R., Dubessy, J., Equation of state taking into account dipolar interactions and association by hydrogen bonding: II-modelling liquid–vapour equilibria in the H_2O-H_2S, H_2O-CH_4, and H_2O-CO_2 systems. *Chem. Geol.*, 251(1-4), 50, --57, 2008. p.vol..
25. Duan, Z., Møller, N., Weare, J.H., Prediction of the solubility of H_2S in NaCl aqueous solution: an equation of state approach. *Chem. Geol.*, 130(1-2), 15–20, 1996.
26. Selleck, F.T., Carmichael, L.T., Sage, B.H., Phase behavior in the hydrogen sulfide-water system. *Ind. Eng. Chem.*, 44(9), 2219–2226, 1952.
27. Lee, J.I., Mather, A.E., Solubility of hydrogen sulfide in water. *Berichte der Bunsengesellschaft für physikalische Chemie*, 81(10), 1020–1023, 2010.
28. Dubessy, J., Tarantola, A., Sterpenich, J., Modellingof liquid-vapour equilibria in the H_2O-CO_2-NaCland H_2O-H_2S-NaClsystems to 270 °C. *Oil & Gas Science and Technology - Revue de l IFP*, 60, 339–355.
29. Chapoy, A., Mohammadi, A.H., Tohidi, B., Experimental measurement and phase behaviour modelling and literature review of the properties for the hydrogen sulfide-water binary system. *Ind. Eng. Chem. Res.*, 44, 7567–7574, 2005.
30. Bierlein, J.A., Kay, W.B., Phase-equilibrium properties of system carbon dioxide-hydrogen sulfide. *Ind. Eng. Chem.*, 45(3), 618–624, 1953.
31. Haghighi, H., Chapoy, A., Burgess, R., Mazloum, S., Tohidi, B., Phase equilibria for petroleum reservoir fluids containing water and aqueous methanol solutions: Experimental measurements and modelling using the CPA equation of state. *Fluid Phase Equilib.*, 278(1-2), 109–116, 2009.
32. Li, Z., Firoozabadi, A., Cubic-plus-association equation of state for water-containing mixtures: Is "cross association" necessary? *AIChE J.*, 55(7), 1803–1813, 2009.
33. Yan, W., Kontogeorgis, G.M., Stenby, E.H., Application of the CPA equation of state to reservoir fluids in presence of water and polar chemicals. *Fluid Phase Equilib.*, 276(1), 75–85, 2009.
34. Hajiw, M., Chapoy, A., Coquelet, C., Hydrocarbons - water phase equilibria using the CPA equation of state with a group contribution method. *Can. J. Chem. Eng.*, 93(2), 432–442, 2015.

7

High Pressure H_2S Oxidation in CO_2

S Lee and RA Marriott*

Department of Chemistry, University of Calgary, Calgary, AB, Canada

Abstract

If an acid gas is low-quality (< 1% H_2S in CO_2) the conventional Claus plant cannot be used and most available low-level recovery methods produce poor quality non-commercial sulfur. Another consideration is that conventional aqueous amine separation of any low-H_2S fluid will result in a low-pressure acid gas stream. Alternatively, newer cryogenic separation processes would result in CO_2 rich acid gas fluids in liquid form, which do not require re-compression but may still require H_2S removal. There are no commercial methods to oxidize low-level H_2S to elemental sulfur (S_8) in a high-pressure CO_2 fluid. Converting H_2S into S_8 while maintaining the dense-phase CO_2, would not only allow a producer to capture the economic benefits of sulfur recovery, but also to conserve a high-pressure CO_2 stream that would require minimal to no compression before sale. For such a technique to be developed, we have previously studied the solubility of S_8 within CO_2 and developed a model to predict the solubility over a range of temperatures and pressures. Thermodynamic information for H_2S, SO_2, COS, and water within CO_2 also have now been studied. From these studies, we are now able to calculate fugacity coefficients of most species that are involved in H_2S oxidation equilibrium, i.e., we now have the departure information required to calculate for sulfur recovery in real high-pressure fluids. This study describes Gibbs Energy Minimization calculations utilizing the calculated fugacity coefficients to predict the sulfur conversion % from initial H_2S concentrations. Results show an increased reaction of $H_2S + O_2$ to produce S_8 at high-pressures compared to atmospheric pressure, that is, sulfur conversion is more favorable at high-pressure and $T < 600$ °C.

Keywords: Oxidation, hydrogen sulfide, elemental sulfur

**Corresponding author:* rob.marriott@ucalgary.ca

Ying Wu, John J. Carroll and Yongle Hu (eds.) The Three Sisters, (91–96) © 2019 Scrivener Publishing LLC

7.1 Introduction

Many unconventional natural gas sources contain significant amounts of low-quality acid gas e.g., Horn River in N.E. BC (~10% CO_2 and ~500 ppm H_2S) [1]. Some of the commercially available methods to separate such acid gas and to remove H_2S are non-regenerable scavengers or liquid redox processes combined with alkonolamines [2]. The conventional acid gas removal techniques discharge CO_2 at low-pressure, which requires recompression to transfer at significant costs. In addition, many liquid redox processes produce low-quality elemental sulfur, which becomes an environmental liability rather than a marketable product.

An alternative methods to separate acid gas can include cryogenic separation of acid gas from the raw natural gas. [3–5] Cryogenic separation results in high-pressure acid gas fluids, which do not require recompression for geo-sequestration or Enhanced Oil Recovery (EOR). Further distillation can be performed to remove H_2S from such streams, but it is expensive [6] and there is currently no commercial method to produce elemental sulfur from H_2S in the dense CO_2. Oxidation of H_2S to S_8 within the dense-phase CO_2, would allow producers to recover sulfur, while conserving a high-pressure CO_2 stream that requires minimal to no compression or purification.

Using the experimental solubility of S_8 in CO_2, a Fluctuation Solution Theory (FST) correlation has been calibrated to calculate the solubility/fugacity coefficient of S_8 in dense CO_2 so that the best reaction conditions for catalysis can be estimated [7]. The previous S_8 solubility study suggested that it was possible to oxidize an acid gas stream that contained less than 0.1% H_2S in CO_2 while remaining as a single phase, meaning that produced elemental sulfur would stay dissolved within the CO_2 stream. Alternatively, larger concentrations of H_2S can be recovered in a sub dew-point catalyst system with periodic catalyst regeneration. Post catalyst bed, a temperature drop below sulfur dew-point will allow for the separation of S_8 and CO_2.

The fugacity coefficients of the Claus equilibrium reactants/products such as elemental sulfur [7], H_2S, carbonyl sulfide (COS), [8] water [9], carbon disulfide (CS_2), and sulfur dioxide (SO_2) [10] within CO_2 can be used to obtain theoretical equilibrium conversions within Gibbs Energy Minimization (GEM) routines when simulating equilibrium limits at high pressures.[8] This study describes those calculations and a high-pressure catalyst rig which has been commissioned to evaluate overall recovery and catalyst materials.

7.1.1 Experimental Section

Gibbs Energy Minimization (GEM) is commonly utilized in calculating the theoretical equilibrium conversions of the Claus reaction.[8, 11, 12] For our high-pressure system, the sum of all chemical Gibbs energies is first minimized in ideal gas conditions, where fugacity coefficients, $\phi_i = 1$. From here, the Gibbs energies are minimized again by replacing the fugacity coefficients with real fluid conditions and again changing the concentration of each species until the minimum total energy was reached.[7–10] Only two iterations are necessary for convergence, which is monitored by stable concentrations.

The experimental oxidation of H_2S to S_8 within dense-phase CO_2 is being studied by using a custom-built high-pressure catalytic reactor shown in Figure 7.1. High-pressure mixtures of H_2S/CO_2 and O_2/CO_2 are flown at desired flow rates using two syringe pumps into the catalytic reactor filled with alumina catalyst. The temperature is controlled in a modified gas chromatography oven (Thermostated Zone) and is measured with several

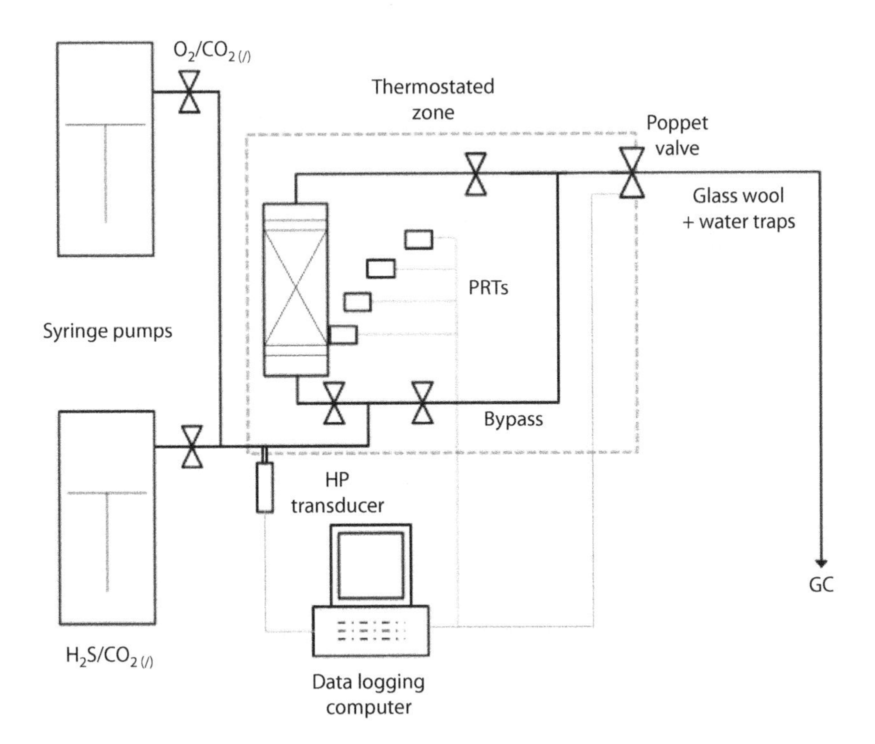

Figure 7.1 Schematic of the In-House Built High-Pressure Heterogeneous Catalytic Reactor.

platinum resistance thermometers. The pressure is controlled using the transducer measurements in conjunction with an automatic on/off poppet valve, which opens and closes quickly to obtain gas sample, but still maintain the desired pressure of the system. Released reaction effluents are quantified using gas chromatography equipped with a thermal conductivity detector. The results of the experimental oxidation of H_2S to S_8 in high-pressure CO_2 will be later compared against the equilibrium concentrations calculated by GEM.

7.1.2 Results and Discussion

The equilibrium limit of $H_2S + O_2$ has been successfully calculated by the GEM routine utilizing the fugacity coefficients of chemical species within CO_2 at high pressures as shown in Figure 7.2. The calculated results indicated an improved conversion from H_2S to elemental sulfur in comparison to calculations with ideal gas conditions where $\phi_i = 1$ [9]. The effect of pressure on the equilibrium at lower temperatures ($T < 600$ °C) can be explained by Le Chatelier's principle due to more moles of gas on the reactant side shifting the equilibrium towards the product side with less moles of gas.

$$3H_2S(g) + 3/2O_2(g) \rightleftarrows 3H_2O(g) + 3/8S_8(l).$$

Note that for $T > 600$ °C, the products are $S_2(g)$, which reverses the effect of pressure, i.e., a pressurized Claus thermal reactor would yield lower recovery with an ideal gas calculation.

The high-temperature region ($T > 600$ °C) shows the sulfur recovery of real fluid calculation fall below the ideal gas calculation. This is the effect of the high-pressure suppressing the S_2 formation from S_8 in the

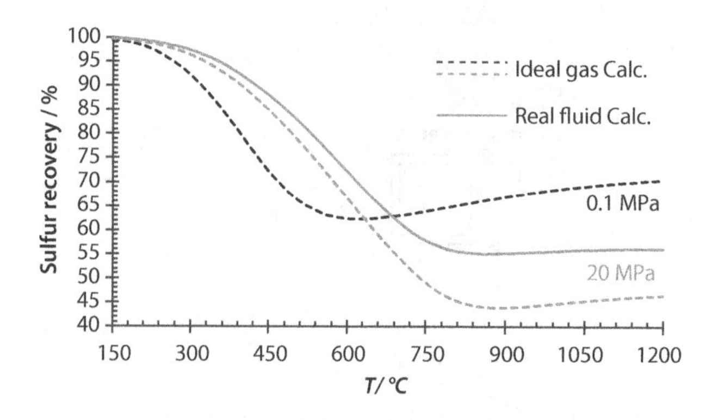

Figure 7.2 H_2S Oxidation Equilibrium Calculation with $H_2S/O_2/CO_2 = 0.25/0.125/0.625$.

high-temperature condition; therefore, limiting the equilibrium shift towards the sulfur formation. However, this process is aimed at lower temperature conditions to maximize the sulfur conversion, thus the lower conversion at high-temperature should not be an issue.

Due to the exothermic nature of the Claus reaction equilibrium, lower temperature conditions for the catalysis are favorable as well as high-pressure conditions. However, temperature conditions below sulfur dew-point can be problematic as the elemental sulfur can deposit on the catalyst surface.

7.1.3 Conclusion and Future Direction

There are now enough available fugacity coefficients to perform GEM calculations on partial H_2S oxidation to elemental sulfur at pressures relevant to CO_2 transportation and injection. The real fluid GEM calculations suggested that a higher H_2S conversion to elemental sulfur at $T < 600$ °C could be achieved in dense-phase CO_2 compared to the prediction using the ideal gas assumption. Pressure favours the more formation of S_8 (higher recovery), where S_8 and H_2O have a lower vapour pressure when compared to the reactants, H_2S and O_2. As both direct partial oxidation and modified Claus requires a catalyst at this temperature, a high-pressure catalyst rig has been described which will be used to evaluate (i) catalyst performance and (ii) the newly calculated equilibrium limits. Experimental oxidation of H_2S within high-pressure CO_2 is currently being carried out to compare the results to the GEM calculations.

References

1. BC oil and gas commission. Hydrocarbon and by-product reserves report January-December, 2012. Available from: http://www.bcogc.ca/node/11111/download [accessed April 2, 2018].
2. Campbell, J.M., Gas conditioning and processing. *Campbell Petroleum Series*, 4, 1982.
3. Kelley, B.T., Valencia, J.A., Northrop, P.S., Mart, C.J., Controlled Freeze ZoneTM for developing sour gas reserves. *Energy Procedia*, 4, 824–829, 2011.
4. Lallemand, F., Lecomte, F., Streicher, C., *In Highly Sour Gas Processing: H_2S Bulk Removal with the Sprex Process*. Doha, Qatar, International Petroleum Technology Conference, 2005.
5. Terrien, P., Dubettier, R., Leclerc, M., Meunier, V., In engineering of air separation and cryocapTM units for large size plants, oxyfuel combustion conference, ponferrada, Spain, 9-13 Sep, 2013.

6. Guvelioglu, G.H., Higginbotham, P., Palamara, J.E., Arora, G., Mamorsh, D.L., Fisher, K.S. 2015. In H_2S Removal from CO_2 by Distillation. Laurance Reid Gas Conditioning Conference, Norman: Oklahoma.
7. Lee, S., Marriott, R.A., Solubility of elemental sulfur in dense phase carbon dioxide from T = 324 to 424 K and p = 10 and 20 MPa. *J. Natural Gas Eng*, 2018. in press.
8. Deering, C.E., Saunders, M.J., Commodore, J.A., Marriott, R.A., The volumetric properties of carbonyl sulfide and carbon dioxide mixtures from T = 322 to 393 K and p = 2.5 to 35 MPa: application to COS hydrolysis in subsurface injectate streams. *J. Chem. Eng. Data*, 61(3), 1341–1347, 2016.
9. Deering, C.E., Cairns, E.C., McIsaac, J.D., Read, A.S., Marriott, R.A., The partial molar volumes for water dissolved in high-pressure carbon dioxide from T = (318.28 to 369.40) K and pressures to p = 35 MPa. *J. Chem. Thermodyn.*, 93, 337–346, 2016.
10. Commodore, J.A., Deering, C.E., Marriott, R.A., Phase behaviour and reaction thermodynamics involving dense-phase CO_2 impurities. In: Wiley, ed. *Carbon Dioxide Capture and Acid Gas Injection, Wu, Y. and Carroll, J*, 2017.
11. Gamson, B., Elkins, R., Sulfur from hydrogen sulfide. *Chemical Engineering Progress*, 49(4), 203–215, 1953.
12. Dowling, N. I., Marriott, R.A., Primak, A., Manley, S., The kinetics of H_2S oxidation by trace O_2 and prediction of sulfur deposition in acid gas compression systems. *Sour Gas and Related Technologies*. John Wiley & Sons. pp. 183–214, 2012.

8

Water Content of Carbon Dioxide – A Review

Eugene Grynia[1,*] and Bogdan Ambrożek[2]

[1]Gas Liquids Engineering Ltd., Calgary, Alberta, Canada
[2]West Pomeranian University of Technology, Szczecin, Poland

Abstract

The system carbon dioxide-water is important to many branches of science. The high pressure and high temperature region is especially important in geochemistry where these conditions are often encountered.

A thorough literature review was conducted to find all of the measured data for water content of carbon dioxide. Only experimental data was collected and no limit was imposed on pressure and temperature. The paper did not include the solubility of carbon dioxide in water. Experimental data for more complex mixtures, for example ternary mixtures of H_2O, CO_2 and nitrogen, CO_2 in brine (water + NaCl), mixtures of H_2O, CO_2 and inerts, were not reviewed at this time.

The system water + carbon dioxide belongs to type III of the phase behavior classification introduced by Scott and van Konynenburg. This type of phase behavior is briefly discussed in the paper.

The solubility data were converted to the same set of units: (i) mole fraction and (ii) g/Sm^3 (standard conditions of 60°F and 1 atm).

No predictions were attempted, but the data from the various sources were compared and notes were made which one deviated from the rest. The data were plotted (isotherms and isobars) and least-squares method was used to measure deviations from the norm.

The data were classified by the nature of the phases present: (1) water phase: aqueous liquid, hydrate, ice and (2) CO_2-rich phase: gas, liquid, supercritical.

The experimental methods used by the researchers to obtain phase equilibrium data were classified in accordance with the classification systems proposed by Dohrn *et al.*, [48]. This classification depends on how the composition of the

Corresponding author: egrynia@gasliquids.com

Ying Wu, John J. Carroll and Yongle Hu (eds.) The Three Sisters, (97–184)

two coexisting phases is determined; the methods are divided into analytical and synthetic.

Keywords: Water, carbon dioxide, solubility, experimental methods

8.1 Introduction

Only studies on the solubility of water in pure CO_2 have been reviewed; studies on more complex mixtures, e.g., ternary mixtures of H_2O, CO_2 and nitrogen, CO_2 in brine (water + NaCl), mixtures of H_2O, CO_2 and inerts, were not reviewed. Experimental investigations of the phase behavior of the binary systems carbon dioxide – water (three and four phase equilibria including fluid, hydrate and ice phases) were not reviewed. Studies on the densities of water saturated carbon dioxide were not reviewed. No modeling of the data collected was attempted.

Measuring water solubility in vapor carbon dioxide was probably first published by Pollitzer and Strebel [1]. The water content of compressed carbon dioxide was first measured by Wiebe and Gaddy [2]. Tödheide and Franck [3] were the first researchers to publish experimental data for the system carbon dioxide – water at very high pressures (3500 bar).

The solubility is of importance in the manufacture of dry ice (solid CO_2) as the moisture causes freeze-ups in the plant due to solid water formation. It is also important in fire extinguishers using carbon dioxide, where water presence may block CO_2 flow as a result of Joule-Thomson effect.

Water is frequently extracted alongside the natural products in near-critical extraction operations with CO_2 as solvent, which is most extensively used in fluid extraction processes. For that reason the solubility data for the CO_2-H_2O system is of relevance.

Knowledge of the water content of CO_2 is critical for carbon dioxide transport as part of Carbon Capture and Storage (CCS) or Enhanced Oil Recovery (EOR); it should be controlled because of risk of corrosion and hydrate formation in the CO_2 transporting pipelines. If carbon dioxide is not sufficiently dry, then water may condense when pressure and temperature change.

Knowledge of the water content of CO_2 is also essential in coupling Enhanced Gas Recovery (EGR) in depleted shale gas reservoirs with CO_2 storage, as dissolved water can affect the efficiency and economic favorability of CO_2 EGR in a target shale reservoir (Loring *et al.*, [68]).

To avoid corrosion of the pipeline, the presence of free water in CO_2 needs to be avoided and to avoid it, the water content in the CO_2 stream must be below the solubility of water under various operating conditions.

The worst case pipeline condition for hydrate formation in the CO_2 pipeline has been defined as 17 to 28 MPa and −46 °C (Jasperson *et al.*, [66]).

Scott and Konynenburg [4] classified various types of binary phase diagrams according to the nature of their P, T projections. In particular by the presence or absence of three phase lines and azeotrope lines and by the way critical lines connect with these. Nine major types were distinguished by them. Type III has two critical lines: C_1 to UCEP (G-L) and C_m to C_2 (L-L to G-L), where:

C_1 is the critical point of the component with lower critical temperature (CO_2)

C_2 is the critical point of the component with higher critical temperature (H_2O)

C_m is a liquid-liquid critical point at infinite pressure

UCEP is the upper critical end point

Type III behavior usually occurs for mixtures with large immiscibility. The (CO_2-H_2O) system exhibits type III behavior, with a discontinuous vapor-liquid critical curve, a wide region of liquid-liquid coexistence below the critical temperature of CO_2 and very limited mutual solubility in the regions of two- and three-phase equilibria [5]. It can also be stated as follows: the critical-temperature curve of the CO_2-rich phase starts at the critical temperature of pure CO_2 and comes to the upper critical end point. The critical temperature curve of the water-rich phase starts with the critical temperature of pure water and reaches the area of limited solubility of gases near 539 K (266 °C) and 200 MPa [6].

The water content of carbon dioxide was the subject of the papers reviewed as the main topic or as part of investigating more complex mixtures, where the solubility of water in pure CO_2 was examined to validate the experimental methods.

The results were presented as water content, water dew points or water enhancement factors. More information on enhancement factors can be found in Heck and Hiza [7] and Koglbauer and Wendland [8].

8.2 Literature Review

The water content of carbon dioxide was previously reviewed and data assembled by a number of authors: Carroll [62], Spycher *et al.*, [9], Chapoy *et al.*, [10], Mohammadi *et al.*, [63], Hu *et al.*, [11], Burgass *et al.*, [12], Aavatsmark and Kauffman [13], Aasen *et al.*, [14], Rowland *et al.*, [15].

The authors of this paper identified 54 research papers regarding water content of carbon dioxide, the first of which was published in 1924 and the last in 2017. In most of the papers water content of CO_2 was reported directly as concentration (1304 data points). In one paper water content was reported as water vapor concentration enhancement (63 data points) and two papers discussed synthetic fluid inclusions (126 data points).

A brief description, in the form of a timeline, of the papers dealing with water content of carbon dioxide, follows:

8.2.1 1924

Pollitzer and Strebel (Universität Karlsruhe) studied the relations between the saturated water vapor and various gases. They reported the concentrations of water vapor in gaseous carbon dioxide at 50 and 70 °C at pressures to 8.815 MPa (summarized by Takenouchi and Kennedy [16]).

8.2.2 1941

Wiebe and Gaddy (Fertilizer Research Division, Bureau of Plant Industry, U.S. Department of Agriculture) were the first to measure water in liquid CO_2.

8.2.3 1943

Stone (University of California, Los Angeles) reported the solubility of water in liquid CO_2 for several temperatures from −29 to 22.6 °C. The values reported indicated a fairly regular increase in the solubility from 0.02% water at the lower temperature to 0.1% at the higher.

8.2.4 1959

Malinin (Laboratory of Magmatogenic Processes, Institute of Geochemistry and Analytical Chemistry of the Academy of Sciences of the USSR) studied the equilibrium relations in the system water – carbon dioxide up to 300 °C and 59 MPa, to supplement the data only available at lower temperatures.

8.2.5 1963

Tödheide and Franck (Institute of Physical Chemistry, University of Göttingen) determined the boundary of the two-phase region in the system water - carbon dioxide between the critical temperatures of the two components and up to 350 MPa. They noted a density reversal of the phases at higher pressures, where the vapor phase is denser than the liquid phase.

8.2.6 1964

Takenouchi and Kennedy (Institute of Geophysics and Planetary Physics, University of California, Los Angeles) studied the system water-carbon dioxide to pressures of 160 MPa and temperatures between 110° and 350 °C, and determined the critical curve of the binary system. They, too, noticed an inversion in density of the CO_2-rich phase when the pressure of the system is increased.

8.2.7 1971

Coan and King (Department of Chemistry, University of Georgia) measured the solubility of water in compressed carbon dioxide between 0.101 and 6.08 MPa, and 25 and 100 °C. They provided an evidence for hydration of carbon dioxide in the gas phase.

8.2.8 1981

Zawisza and Malesińska (Institute of Physical Chemistry of the Polish Academy of Sciences) determined the solubility of water in gaseous carbon dioxide in the range of 0.466 to 3.35 MPa and 100 to 200 °C.

8.2.9 1982

Josef Chrastil (Physical Chemical Department, Technical Center, General Foods Corporation, Tarrytown, NY) determined the solubility of water in supercritical carbon dioxide at 50–80 °C and 10–25 MPa.

Gillespie and Wilson (Wiltec Research Co., Inc., Provo, UT) measured phase equilibrium data on the water-carbon dioxide system at 15–260 °C and 5–20 MPa. They determined that water is more soluble in liquid CO_2 than in vapor CO_2. They also determined that CO_2 is more soluble in water than water in CO_2.

8.2.10 1984

Song, Kobayashi and Marsh (Rice University, Houston, TX) presented initial data obtained on the CO_2-hydrate system in the GPA Research Report RR-80. Most of the data related to liquid CO_2-hydrate equilibria. Contrary to most hydrocarbon and related components, the reported data showed that the water content of liquid CO_2 changes substantially with pressure in the region adjacent to the three-phase bubble point boundary.

8.2.11 1986

Song and Kobayashi (Rice University, Houston, TX) prepared the GPA Research Report RR-99, which extended the work reported in RR-80 on the CO_2-water-hydrate system. The final data for this system ranged from −28 to 31°C and from 2.1 to 7.4 MPa.

8.2.12 1987

Song and Kobayashi (Rice University, Houston, TX) presented experimentally measured water content in CO_2-rich fluid in the gaseous or liquid state in equilibrium with liquid water or hydrate for pressures 0.69 to 13.79 MPa and temperatures from −28 to 25 °C. Most of the data were presented in 1984 and 1986. New data are for 8.28 MPa and −17 to +25 °C.

Briones *et al.*, (Department of Chemical Engineering, Clemson University, Clemson, SC) presented equilibrium compositions in the binary system water-carbon dioxide at 50 °C, between 6.82 and 17.68 MPa.

Nakayama *et al.*, (Research and Development Division, JGC Corporation, Yokohama, Japan, and Department of Chemical Engineering, Tohoku University, Sendai, Japan) designed and constructed a new static apparatus for the measurement of high pressure liquid-liquid equilibria up to 127 °C and 20 MPa. The experimental apparatus and procedure were checked by measuring the phase equilibria for the H_2O-CO_2 system at 25 °C which has been reported by other workers using different techniques.

Patel *et al.*, (Department of Chemical Engineering, Texas A&M University, College Station, TX) used a Burnett-isochoric apparatus and presented a comprehensive experimental study of the P-ρ-T properties for the CO_2-H_2O binary at five different compositions. They determined the dew points for each mixture by noting the change in slope of the isochores as they traversed from the single vapor phase to the two-phase region.

8.2.13 1988

Müller *et al.*, (Lehrstuhl für Technische Thermodynamik der Universität Kaiserlautern, Germany) reported experimental results for the vapor-liquid equilibrium for water-carbon dioxide system as part of the investigation of the ternary system ammonia-carbon dioxide-water.

D'Souza *et al.*, (School of Chemical Engineering, Georgia Institute of Technology, Atlanta, GA) measured phase equilibria in the carbon dioxide – n-hexadecane and carbon dioxide-water systems. The latter was just a test system.

Ohgaki *et al.*, (Osaka University, Faculty of Engineering, Department of Chemical Engineering, Japan) focused on the entrainer effect of water and ethanol on α-Tocopherol (vitamin E) extraction by compressed CO_2.

8.2.14 1991

Sako *et al.*, (National Chemical Laboratory for Industry, Ibaraki, Japan, and Department of Industrial Chemistry, Nihon University, Chiba, Japan) measured the isothermal high pressure vapor-liquid equilibria for two binary systems (CO_2-water and CO_2-furfural) and a ternary system (CO_2-water-furfural) as part of the phase equilibrium study of extraction and concentration of furfural produced in reactor using supercritical carbon dioxide.

Sterner and Bodnar (Department of Geological Sciences, Virginia Polytechnic Institute & State University, Blacksburg, VA) determined experimentally the P-V-T-x relations in the CO_2-H_2O system from 200 to 600 MPa and 400 to 700 °C for fluid compositions between 12.5% to 87.5% CO_2 using the synthetic fluid inclusion technique. This technique was used to determine the temperature of the phase transition.

8.2.15 1992

Mather and Franck (Institut für Physikalische Chemie und Elektrochemie, Universität Karlsruhe, Germany) measured dew points for three CO_2-H_2O mixtures (53.0, 62.5% and 78.9% CO_2) at 225–275 °C and 114–311 MPa. These data appeared to have resolved the discrepancy that had existed for more than 25 years in the results of carbon dioxide-water phase behavior at high temperatures and pressures.

King *et al.*, (School of Chemical Engineering, University of Birmingham, UK) presented new data for phase compositions at 15, 20, 25, 35 and 40 °C, and the pressures to a little over 20 MPa. This extended the coverage of the classic work of Wiebe and Gaddy.

8.2.16 1993

Dohrn *et al.*, (Technische Universität Hamburg-Harburg) measured vapor-liquid equilibrium data for the ternary and quaternary systems of glucose, water, CO_2 and ethanol with a novel apparatus. To ensure the reliability of the experimental data, they studied the binary system water-CO_2 at 50 °C and compared their data to that of Wiebe and Gaddy [2] and noted that their data showed a higher solubility of water in the vapor phase.

8.2.17 1995

Jackson *et al.*, (Chemical Sciences Department, Pacific Northwest Laboratory, Richland, WA) used a relatively fast and inexpensive technique to measure water solubilities using a simple long path length optical cell in an FT-IR spectrometer. They used the method to obtain solubility information for water in supercritical CO_2 and refrigerants.

8.2.18 1996

Fenghour *et al.*, (Department of Chemical Engineering and Chemical Technology, Imperial College, London, UK, and National Engineering Laboratory Executive Agency, Glasgow, Scotland, UK) determined the dew points of 11 mixtures of water + carbon dioxide using a fully automated isochoric instrument. At the dew point, when the fluid becomes a single phase, there is a discontinuity in the line relating the pressure, in the isochoric system, to its temperature.

8.2.19 1997

Frost and Wood (CETSEI, Department of Geology, University of Bristol, UK) determined experimentally the volumes of H_2O-CO_2 mixtures between 950 and 1940 MPa and 1100–1400 °C. They used a synthetic fluid inclusion technique to capture H_2O-CO_2 fluids of known compositions in pre-cracked corundum.

8.2.20 2000

Bamberger and Maurer (Lehrstuhl für Technische Thermodynamik, Universität Kaiserlautern) reported experimental results for high pressure – up to 14 MPa – (vapor + liquid) of binary systems of carbon dioxide +either water or acetic acid. Such data were needed for the design of supercritical extraction processes using carbon dioxide.

8.2.21 2002

Sabirzyanov *et al.*, (Kazan State University of Technology, Kazan, Tatarstan, Russian Federation) developed a new experimental circulation facility to investigate the solubility of liquids in supercritical fluids - they investigated the solubility of water in supercritical carbon dioxide.

8.2.22 2004

Valtz *et al.*, (Centre d'Energétique, Ecole National Supérieur des Mines de Paris, France, and Department of Chemical Engineering, Vanderbilt University, Nashville, TN, USA) reported new experimental VLE data of CO_2-water binary system over a wide temperature range, using a static-analytic apparatus, taking advantage of two pneumatic capillary samplers. The experimental solubility data were in very good agreement with previous data reported in literature.

Jarne *et al.*, (Departamento de Química Orgánica y Química Física. Facultad de Ciencias, Universidad de Zaragoza, Spain; Laboratoire de Chimie Physique de Marseille. Faculté des Sciences de Luminy, Université de la Méditerranée, Marseille, France) determined dew points for four CO_2-H_2O mixtures and eight CO_2-H_2O-methanol mixtures as part of the research aiming at investigating the influence of these components on the VLE of natural gas within the usual pressure and temperature conditions of natural gas transport by pipeline. The presence of solid hydrates and of liquid CO_2 was avoided.

Iwai *et al.*, (Department of Chemical Engineering, Faculty of Engineering, Kyushu University, Japan), measured the solubilities of palmitic acid in supercritical CO_2 and supercritical CO_2 + water using a static recirculation method combined with on-line Fourier transform infrared (FTIR) spectroscopy.

8.2.23 2008

Koglbauer and Wendland (Institut für Verfahrens- und Energietechnik, Universität für Bodenkultur Wien, Austria) developed a new method to measure the dew point of compressed humid gases, expressed as vapor concentration enhancement factor, by Fourier transform infrared spectroscopy.

8.2.24 2009

Chapoy *et al.*, (Hydrafact Ltd., Heriot-Watt University, Progressive Energy Ltd.) studied the phase behavior of CO_2 in presence of common impurities like N_2, H_2, CO, H_2S and water, and evaluated the risk of hydrate formation in a rich carbon dioxide stream. These impurities are found in CO_2 produced by carbon capture processes.

Youssef *et al.*, (Institut Francais du Pétrol, Rueil-Malmaison, France, and Université Claude Bernard Lyon, Villeurbanne, France) developed a new experimental procedure combining an equilibrium cell with a water measurement by a Karl Fischer coulometer. They used the procedure to measure the hydrate dissociation temperatures of methane, ethane, and CO_2 in the absence of any aqueous phase.

8.2.25 2011

Tabasinejad *et al.*, (Department of Chemical and Petroleum Engineering, University of Calgary, Canada, and Apache Corporation, Midland and Houston, TX, USA) conducted a series of experiments to measure the water solubility in supercritical nitrogen and carbon dioxide at experimental conditions up to 210 °C and 134 MPa.

Seo *et al.*, measured the water solubility in CO_2-rich liquid phase in equilibrium with gas hydrates using an indirect method by determining the temperature at which hydrates dissolved completely with predetermined compositions. Their results were compared with other data sets. The measured solubility was smaller at 10.1 MPa but larger at 6.1 MPa than previous experimental results, and showed weak pressure dependence.

8.2.26 2012

Chapoy *et al.*, (Hydrates, Flow Assurance & Phase Equilibria, Institute of Petroleum Engineering, Heriot-Watt University, Edinburgh, Scotland, UK, and Hydrafact Ltd., Edinburgh, Scotland, UK) reported new experimental data for the water content of liquid carbon dioxide in equilibrium with hydrates at 13.9 MPa and –20 to +4 °C.

Kim *et al.*, (Department of Chemical and Biological Engineering, Korea University, Seoul, and Department of Chemical and Biological Engineering, Korea National University of Transportation, Chungju) determined visually the equilibrium solubility of water in carbon dioxide at 8, 10 and 20 MPa and 10 to 39 °C. They observed small droplets of water which completely disappeared when the temperature was slowly raised.

8.2.27 2013

Hou *et al.*, (Qatar Carbonates and Carbon Storage Research Centre, Department of Chemical Engineering, Imperial College London, UK) made the measurements on the binary system (CO_2 + H_2O) at temperatures 15–175 °C and pressures 1.5–18 MPa. They used an analytical apparatus that they designed to study the phase behavior of fluid mixtures of relevance to CO_2-enhanced oil recovery and carbon dioxide storage in deep aquifers or depleted oil fields.

Chapoy *et al.*, (Hydrates, Flow Assurance & Phase Equilibria Research Group, Institute of Petroleum Engineering, Heriot-Watt University, Edinburgh, Scotland, and MINES ParisTech, CTP-Centre Thermodynamique des Procédés, Fontainebleau, France) presented experimental results on the phase behaviour and thermophysical properties of carbon dioxide in the presence of O_2, Ar, N_2 and water (water content of CO_2 was measured at 15 MPa and −40 to +15 °C). They also determined experimentally saturation pressures and hydrate stability of the CCS stream.

Cox *et al.*, (Department of Chemical and Biomolecular Engineering, Rice University, Houston, TX, USA) repeated the measurements of Song and Kobayashi using an updated apparatus.

Wang *et al.*, (Pacific Northwest National Laboratory, Richland, WA, USA, and Department of Chemistry, University of Alabama, Tuscaloosa, AL, USA) used near-infrared (NIR) spectroscopy to investigate the dissolution and chemical interaction of water dissolved into supercritical carbon dioxide and the influence of $CaCl_2$ in the co-existing aqueous phase at four temperatures: 40, 50, 75 and 100 °C and at 9.119 MPa.

8.2.28 2014

Burgas *et al.*, (Centre for Gas Hydrate Research, Institute of Petroleum Engineering, Heriot-Watt University, Edinburgh, Scotland) presented experimental equipment, methods and results for the water content of carbon dioxide in equilibrium with hydrates at −50 to −10 °C and 1.0 MPa to 10.0 MPa.

8.2.29 2015

Jasperson *et al.*, (Wiltec Research Company, Provo, UT, USA; Department of Chemical and Biological Engineering, Korea University, Seoul, South Korea; Alaska Gasline Development Corporation, Anchorage, AK, USA;

Fluor/Worley Parsons Arctic Solutions, Aliso Viejo, CA, USA) focused on pipelines with CO_2-rich fluids in cold environments and in CO_2 capture, and specifically on the equilibrium water concentration for hydrate formation in CO_2-rich mixtures and pure CO_2. Measuring and confirming water content of carbon dioxide at low temperatures (down to –46 °C) and high pressures (up to 28 MPa) were required for the ASAP project (The Alaska Stand Alone Pipeline). Two laboratories (Wiltec Research Company and Korea University) made the measurements independently, which was considered to be a Round Robin Testing.

Foltran *et al.*, (School of Chemistry, The University of Nottingham, UK; Department of Chemical and Environmental Engineering, The University of Nottingham, UK; Department of Chemical and Environmental Engineering, University of Nottingham, Ningbo, China) measured the solubility of water in both pure CO_2 and CO_2 and N_2 mixtures using Fourier Transform Infra Red (FTIR) spectroscopy and demonstrated that this method is a suitable technique to determine the concentration of H_2O in CO_2.

Meyer and Harvey (Sensor Science Div., National Institute of Standards and Technology, Gaithersburg, MD; Applied Chemicals and Materials Div., National Institute of Standards and Technology, Boulder, CO) constructed a flow apparatus to measure the water content at saturation in compressed carbon dioxide (water dew point). The knowledge of the water dew point is important to avoid condensation when transporting CO_2 for sequestration, which can lead to corrosion.

8.2.30 2016

Caumon *et al.*, (Université de Lorraine, CNRS, CREGU, GeoRessources Laboratory, Vandoeuvre-lès-Nancy, France) built a new experimental device to measure mutual solubility in the CO_2-H_2O system without sampling by coupling a batch reactor with Raman immersion probes.

Comak *et al.*, (Department of Inorganic Chemistry, School of Chemistry, The University of Nottingham, UK; Department of Chemical and Environmental Engineering, University of Nottingham Ningbo China) developed a new synthetic-dynamic method for studying phase behavior using Attenuated Total Reflection (ATR) spectroscopy to provide relevant information on the solubility of water in CO_2. They determined the dew point of water with mole fraction of H_2O between 0.01 and 0.04. The data obtained filled the gap in literature and could be of high importance in the context of CCS technology.

8.2.31 2017

Loring *et al.*, (Pacific Northwest National Laboratory, Richland, Washington, USA; OLI Systems Inc., Morris Plains, New Jersey, USA; Department of Chemical Engineering, Imperial College London, UK) used in situ high-pressure infrared spectroscopic titrations to quantify the solubility of water in six CO_2-CH_4 mixtures ranging from pure CO_2 to pure CH_4, at shallow shale reservoir conditions of 50 °C and 9 MPa.

8.3 Data Analysis

Table 8.1 summarizes the experimental investigations of the saturated water content of carbon dioxide.

The first reviewed work on water content of carbon dioxide, often cited by other researchers, is that by Wiebe and Gaddy [2]. The authors presented their work on vapor phase composition of carbon dioxide – water mixtures at various temperatures and pressures to 700 atmospheres.

Figure 8.1 presents their results in metric units; water content in g/Sm^3 and pressure in MPa.

Wiebe and Gaddy noted a minimum in water content of carbon dioxide, as opposed to water-hydrogen and water-nitrogen composition.

Stone [57] reported the solubility of water in liquid carbon dioxide for four temperatures between – 29 and + 22.6 °C and corresponding vapor pressure of liquid CO_2 between 1.52 and 6.08 MPa. The values of water content are shown in Figure 8.2. Stone calculated the following average deviation from the mean:

For −29 °C data: 5.1%

For 5.08 °C data: 3.2%

For 15.0 °C data: 3.0%

For 22.6 °C data: 7.5%

Stone concluded that the carbon dioxide liquid and gaseous phases in equilibrium with water at the same temperature and pressure contain different percentages of water.

Stone also noted that considerable uncertainty existed with regard to the solubility of water in liquid carbon dioxide. He mentioned the work of Lowry and Erickson [39], who attempted to determine the solubility from the density of the gaseous phase coexisting with the liquid carbon dioxide in the presence and absence of water and then computing the solubility by Raoult's law. Since the differences they observed were less than the experimental error, they concluded that water could not be soluble to more than

Table 8.1 Experimental investigations of the water content of carbon dioxide.

Source	Temp. [°C]	Pressure [MPa]	NP	Phases present
Pollitzer and Strebel [1]	50–70	3.09–8.82	20	CO_2 gas [a)]
Wiebe and Gaddy [2]	25–75	0.101–70.93	39	CO_2-rich gas in equilibrium with phase rich in water
Stone [57]	–29–22.6	1.52–6.08	4	Liquid CO_2 [b)]
Sidorov *et al.*, [17]	25	3.6–6.4	?	[c)]
	0–75	2.53312–30.3975	?	[d)]
Malinin [40]	200–330	19.61–58.84	9	Supercritical CO_2
Tödheide and Franck [3]	50–350	20–350	208	Water-rich liquid and CO_2-rich vapor
Takenouchi and Kennedy [16]	110–350	10–150	116	Water-rich liquid and CO_2-rich vapor
Coan and King [18]	25–100	1.733–5.147	22	CO_2-rich vapor
Kobayashi *et al.*, [58]	7.2–25	2.53–12.41	9	Internal ARCO Report [e)]
Zawisza and Malesińska [6]	100–200	0.466–3.350	14	CO_2-rich vapor

(Continued)

Table 8.1 Cont.

Source	Temp. [°C]	Pressure [MPa]	NP	Phases present
Chrastil [19]	50–80	10.132–25.331	7	Supercritical CO_2
Gillespie and Wilson [20]	15.6–260	0.689–20.265	40	CO_2 liquid or vapor phase with aqueous liquid phase
Gillespie *et al.*, [54]	16–260	0.7–13.8		AIChE paper 34-b unavailable
Song *et al.*, [58]	−17.8–25	3.45–13.79	30	Gas or liquid CO_2 in equilibrium with liquid water or hydrates
Song and Kobayashi [21]	−17.8–25	3.45–13.79	30 Table 8.1	Same as Song *et al.*, [58]
	−28–31.06	0.689–7.384	26 Table 8.4	Gas or liquid CO_2 in equilibrium with liquid water or hydrates, gas CO_2 with ice
Song and Kobayashi [21]	−28–31.06	0.689–13.790	56 (five new)	Liquid CO_2 in equilibrium with liquid water or hydrate
Briones *et al.*, [42]	50	6.82–17.68	8	Vapor or supercritical CO_2 in equilibrium with liquid water
Nakayama *et al.*, [22]	25	3.63–10.99	8	Vapor CO_2 in equilibrium with liquid water and liquid CO_2 phases

(Continued)

Table 8.1 Cont.

Source	Temp. [°C]	Pressure [MPa]	NP	Phases present
Patel *et al.*, [43]	39.36–208.87	0.0852–9.6114	47	Vapor CO_2
Müller *et al.*, [44]	100–200	0.325–8.11	49	Vapor CO_2 in equilibrium with liquid water
d'Souza *et al.*, [45]	50–75	10.133–15.2	4	Vapor CO_2 in equilibrium with liquid water
Ohgaki *et al.*, [23]	25–40	7.85–14.7	16	Liquid and supercritical CO_2
Sako *et al.*, [46]	75.05–148.25	10.18–20.94	8	Vapor CO_2 in equilibrium with liquid water
Sterner and Bodnar [24]	400–700	200–600	107	Synthetic fluid inclusions
Mather and Franck [59]	225.3–273.3	114.4–311.1	11	Vapor CO_2
King *et al.*, [47]	15–40	5.17–20.27	41	Liquid and supercritical CO_2
Dohrn *et al.*, [48]	50	10.1–30.1	3	Vapor CO_2 in equilibrium with water

(Continued)

Table 8.1 Cont.

Source	Temp. [°C]	Pressure [MPa]	NP	Phases present
Jackson *et al.*, [25]	50–75	34.48	2	Supercritical CO_2
Fenghour *et al.*, [26]	131.85–339.85	5.7–24	11	Vapor CO_2 (at dew point)
Frost and Wood [60]	1100–1400	950–1940	19	Synthetic fluid inclusion
Bamberger *et al.*, [27]	50–80	4.05–14.11	29	Vapor or supercritical CO_2 in equilibrium with water
Sabirzyanov *et al.*, [28]	40–50	8.31–20.54	14	Supercritical CO_2
Valtz *et al.*, [29]	5.07–45.07	0.464–7.963	30	Vapor CO_2
Jarne *et al.*, [30]	−21.25–15.05	0.12–4.11	99	Vapor CO_2
Iwai *et al.*, [31]	40	15.0	1	Supercritical CO_2
Koglbauer and Wendland [8]	25–100	0.1–5.5	63	Vapor CO_2

(Continued)

Table 8.1 Cont.

Source	Temp. [°C]	Pressure [MPa]	NP	Phases present
Chapoy *et al.*, [32]	14.7–16.0	2.048–8.584	5	Vapor or liquid CO_2 (Detector calibr.)
	−2–4	4.944–21.698	6	Liquid CO_2 in equilibrium with hydrates
Youssef *et al.*, (2009)	−2.05–3.15	2.0	3	CO_2 hydrate dissociation temperatures in absence of aqueous phase
Tabasinejad *et al.*, [33]	149.83–205.20	3.85–129.19	40	Supercritical CO_2
Seo *et al.*, [64]	1.2–21.4	6.1–10.1	28	Liquid CO_2 in equilibrium with hydrates or liquid water
Chapoy *et al.*, [34]	−20.0–4.0	13.79	13	Liquid CO_2 in equilibrium with hydrates
Kim *et al.*, [49]	17.7–38.9	8.1–20.1	26	Liquid CO_2

(Continued)

Table 8.1 Cont.

Source	Temp. [°C]	Pressure [MPa]	NP	Phases present
Hou *et al.*, [5]	25.0–75.0	1.089–17.551	43	Vapor, liquid and supercritical CO_2
Chapoy *et al.*, [35]	−40–15	15	7	Liquid CO_2 in equilibrium with hydrates
Cox *et al.*, [38]	−20	0.689–20.684 (read from graph)	11	Liquid CO_2 in equilibrium with liquid water or hydrate
Wang *et al.*,.[65]	40–100	9.119	4	Supercritical CO_2
Burgass *et al.*, [12]	−50–−10	1–10	23	Vapor and liquid CO_2 in equilibrium with hydrates
Jasperson *et al.*, [66] - Wiltec	−45–5	4.15–27.66	26	Liquid CO_2 in equilibrium with hydrates
Jasperson *et al.*, [66] Korea University	−16.02–5.48	6.12–23.07	29	Liquid CO_2 in equilibrium with hydrates
Foltran *et al.*, [50]	40	8.27–12.41	5	Supercritical CO_2
Meyer and Harvey [67]	10–80	0.5–5	58	Vapour CO_2
Caumon *et al.*, [36]	100	0.49–20.1	21	Vapour CO_2
Comak *et al.*, [51]	74–117.8	4.05–6.03	15	Vapour CO_2

(Continued)

Table 8.1 Cont.

Source	Temp. [°C]	Pressure [MPa]	NP	Phases present
Loring *et al.*, [68]	50	9.0	1	Supercritical CO_2

NP – number of data points

[a] data unavailable to the authors; the numbers were reported by Bamberger and Maurer [27] and by Mohammadi *et al.*, (2004).

[b] only the pressure range was provided, exact pressures for the solubility data unknown.

[c] data unavailable to the authors; the numbers were reported by Burgass *et al.*, [12].

[d] data unavailable to the authors; the numbers were reported by Mohammadi *et al.*, (2004), who obtained the data from the Dortmund Data Base.

[e] data from Song *et al.*, [37].

Table 8.2 Experimental data for water content in equilibrium with hydrates [12].

Source	**T [°C]**	**P [MPa]**	**Equilibrium type**
Song and Kobayashi [58]	−28–31.06	0.69–13.79	$V - H/L - H/L - L_w$
Youssef *et al.*, (2009)	−12.45–3.15	2.0–2.1	$V - H$
Seo *et al.*, [64]	1.35–21.35	6.1–10.1	$L - H/L - L_w$
Chapoy *et al.*, [34]	−20–4	13.79	$L - H$
Cox *et al.*, [38]	−20	0.69–20.6	$V - H/L - H$

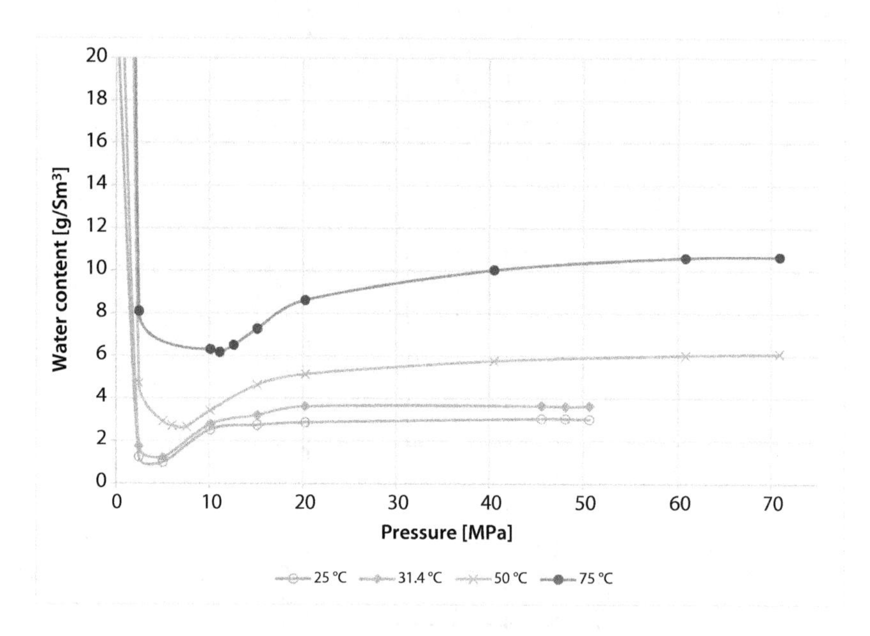

Figure 8.1 Water content of carbon dioxide – Wiebe and Gaddy [2].

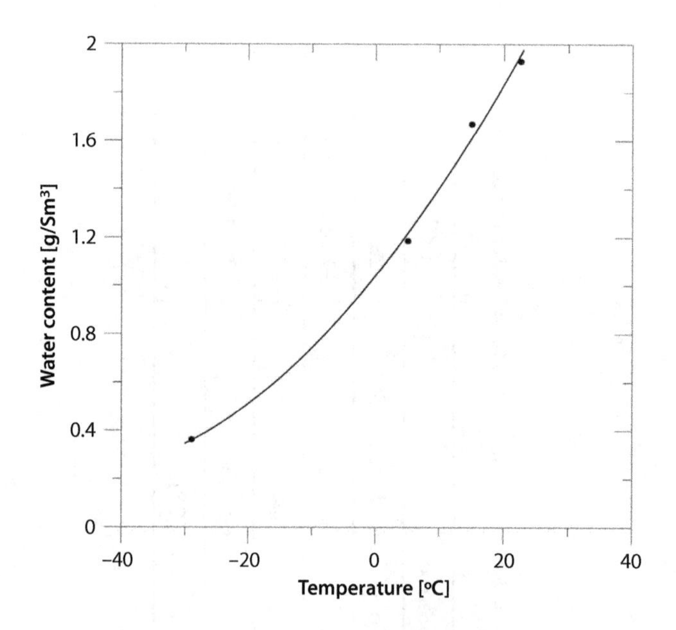

Figure 8.2 Water Content of Carbon Dioxide – Stone (1943). Pressures 1.52 to 6.08 MPa.

0.05% (0.38 g/Sm³) in liquid carbon dioxide over a temperature range of – 5.8 to + 22.9 °C.

The authors of this paper did their own calculations of the average relative deviation from the mean, which was 4.48%. The following equation was used for the correlation of the experimental data:

$$y = e^{\left(-46.743 - \frac{1.10498}{(T+273)} + 8.33714 \cdot ln(T+273)\right)} \tag{8.1}$$

Where: y - water content in g/Sm³

T - temperature in °C

Malinin [40] studied the system water – carbon dioxide at high temperatures (200–330 °C) and pressures (19.61 MPa to 49.03 MPa). He commented on the experimental method by saying that "we had to forsake the dynamic method of saturation (as in the work of Wiebe and Gaddy) and to use the static method". The solubility data measured by Malinin are presented in Figure 8.3.

Tödheide and Franck [3] determined the boundary of the two-phase region in the CO_2 – H_2O system between the critical temperatures of the two components (+31 and 374 °C) and at pressures up to 350 MPa. They

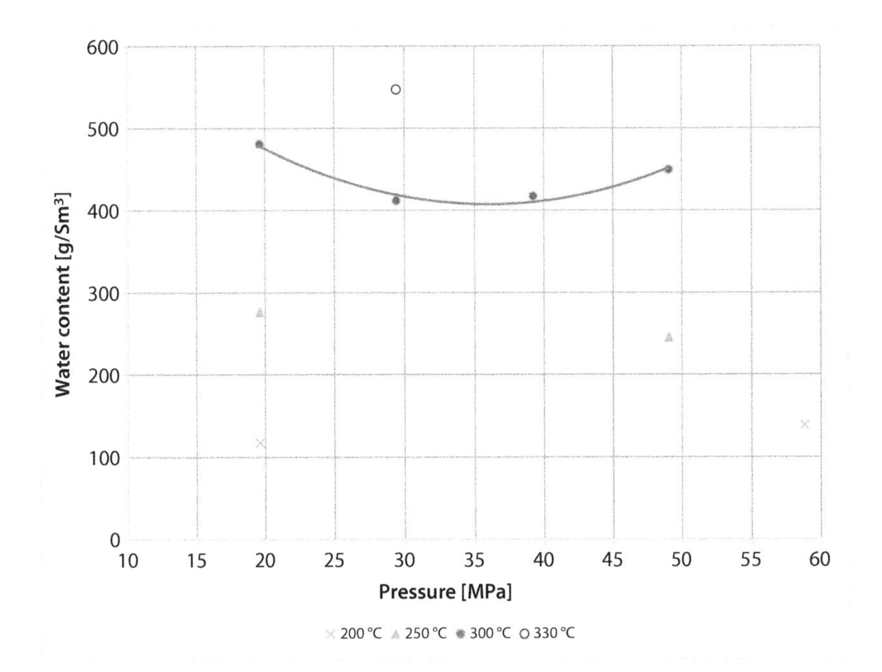

Figure 8.3 Water content of carbon dioxide – Malinin (1953).

were the first to publish experimental data for this system at very high pressure. The solubility data from their work is shown in Figure 8.4.

The authors also made an observation that the system CO_2 – H_2O exhibits the so-called density reversal of the equilibrium phases, when the vapour phase becomes denser than the liquid phase with isothermal pressure increase. It takes place at 50 °C and around 80 MPa, at 100 °C and around 125 MPa, and at 250 °C and around 200 MPa.

Takenouchi and Kennedy [16] studied the water-carbon dioxide system to pressures of 160 MPa and over a temperature range of 110 to 350 °C. Their experimental data are shown in Figure 8.5. Their data are in good agreement with the results of Malinin [40].

They also did a preliminary work up to 300 MPa in the system H_2O-CO_2, providing coexisting compositions of the CO_2-rich phase and the H_2O-rich phase in the critical region of this system.

Similar to Tödheide and Franck [3], Takenouchi and Kennedy offered the following comment: "It is interesting to note that inversion takes place in the system. At high pressures the water-rich phase floats on top of the carbon dioxide-rich phase, whereas at lower pressures the carbon dioxide-rich phase floats on top of the water-rich phase", and: "One of the more unusual aspects of the system H_2O-CO_2 is the inversion in density where the CO_2-rich phase at low pressures is the lightest and at high pressures

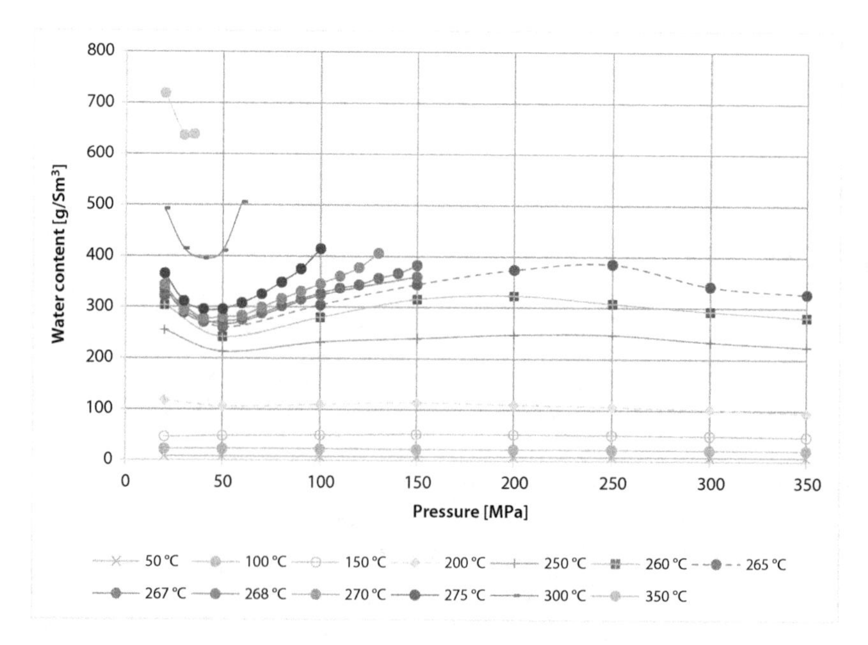

Figure 8.4 Water content of carbon dioxide – Tödheide and Franck [3].

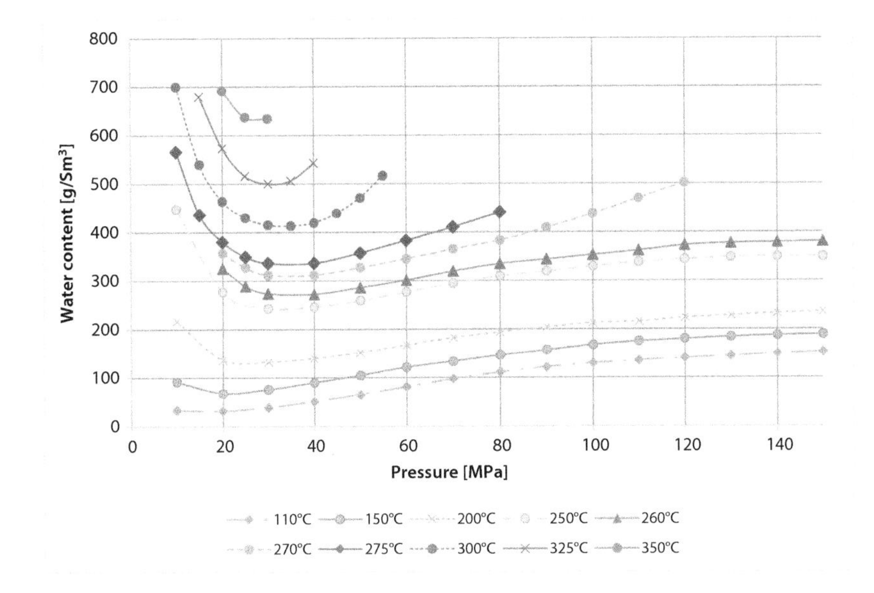

Figure 8.5 Water Content of Carbon Dioxide – Ohgaki *et al.* [Takenouchi and Kennedy [16]].

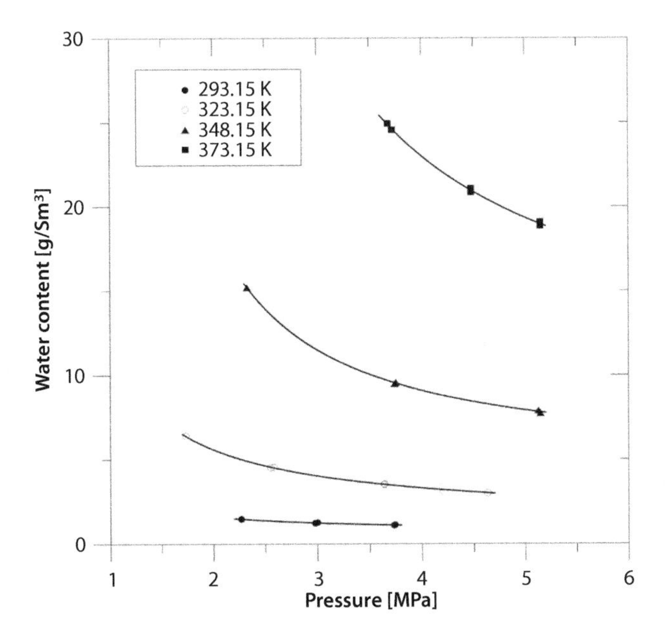

Figure 8.6 Water Content of CO_2 – Coan and King [18].

the densest of the two coexisting phases. The pressure of density inversion depends upon the temperature of the system and increases with temperature".

An interesting discussion was presented by the authors regarding the time required for achieving equilibrium between water and carbon dioxide. Their investigation showed that at least 3 days were required for equilibrium at a temperature of 200 °C and substantially more than 1 week was required at 110 °C.

Coan and King [18] measured the solubility of water in compressed carbon dioxide (as well as in nitrous oxide and ethane) over a pressure range of 1.733 to 5.147 MPa and temperatures ranging from 25 to 100 °C. The data were plotted (isotherms) and shown in Figure 8.6. They stated that water and carbon dioxide were allowed to come to equilibrium for at least 12 hr at the desired temperature.

A least-squares method was used to measure deviations from the norm. The following equation was used for the correlation of the experimental data:

$$y = e^{A+\frac{B}{P}+C \cdot ln(P)} \tag{8.2}$$

where: y is the water content

A, B and C are constants:

T[°C]	A	B	C
25	0.09146	1.05802	-0.19975
50	1.51666	0.97328	-0.40583
75	1.27231	3.17613	0.10245
100	1.78562	4.33102	0.19291

P is the pressure

The average relative error was 1.132% for 25 °C solubility values, 0.391% for the 50 °C values, 0.548% for the 75 °C values and 0.459% for 100 °C solubility values.

Zawisza and Malesińska [6] determined the solubility of water in gaseous carbon dioxide in the range 0.466–3.350 MPa and 100–200 °C. The experimental points are shown in Figure 8.7.

The authors commented that this work had been undertaken because of lack of data in the range of 100–200 °C below 10 MPa.

Chrastil [19] determined the solubility of solids and liquids in supercritical CO_2, including the solubility of water. The results of his investigation are shown in Figure 8.8.

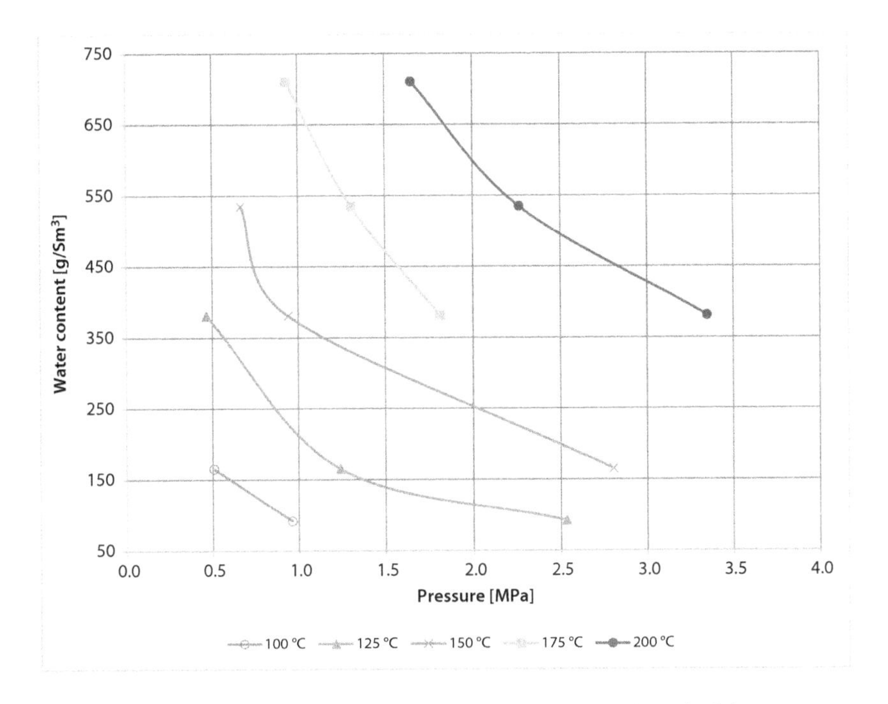

Figure 8.7 Water Content of Carbon Dioxide – Zawisza and Malesińska [6].

The following equation was used for the correlation of experimental data, with the constants listed below the equation:

$$y = e^{A+\frac{B}{P}+C \cdot ln(P)} \tag{8.3}$$

T[°C]	A	B	C
50	5.812157	−23.3294	−1.00245
80	3.806193	−15.7930	−0.274594

The average relative error for 50 °C data is 0.00005%, and for 80 °C:1.28%

Gillespie and Wilson [20] prepared the GPA Research Report RR-48 on vapor-liquid and liquid-liquid equilibria of water-methane, water-carbon dioxide, water-hydrogen sulfide, water- n-pentane and water-methane-n-pentane. The results of their work for the water-carbon dioxide portion is presented in Figure 8.9.

As a result of their investigations, they came up with the following conclusions:

1. Water is more soluble in liquid carbon dioxide than in vapor carbon dioxide.

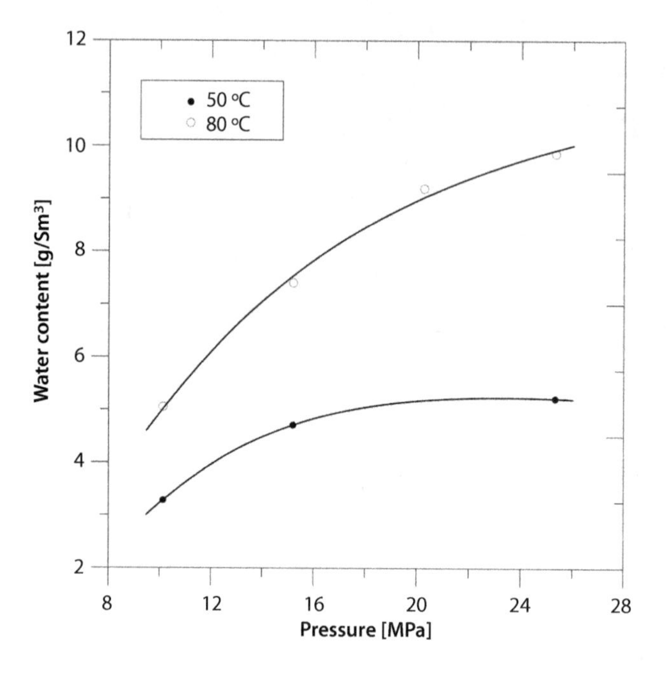

Figure 8.8 Water Content of Carbon Dioxide – Chrastil [19].

2. Carbon dioxide is more soluble in water than water in carbon dioxide.
3. The carbon dioxide-water system exists as two liquid phases at temperatures below the critical point of carbon dioxide at pressures above its vapor pressure.
4. At temperatures up to the critical point of carbon dioxide, the solubility of water in carbon dioxide vapor decreases with increasing pressure.
5. As the pressure increases to the vapor pressure of carbon dioxide, the carbon dioxide condenses to a liquid phase and a discontinuity occurs. Water is more soluble in carbon dioxide liquid than in gas, and the solubility of water in liquid carbon dioxide increases with increasing pressure.

Song *et al.*, [37] studied the water content of CO_2-rich fluids in equilibrium with liquid water or hydrate, which was the subject of the GPA Research Report RR-80. For each of the data point they determined the equilibrium phases: CO_2 liquid or gas in equilibrium with hydrate or liquid water, listed in Table 8.3.

Their data in graphical form are presented in Figure 8.10.

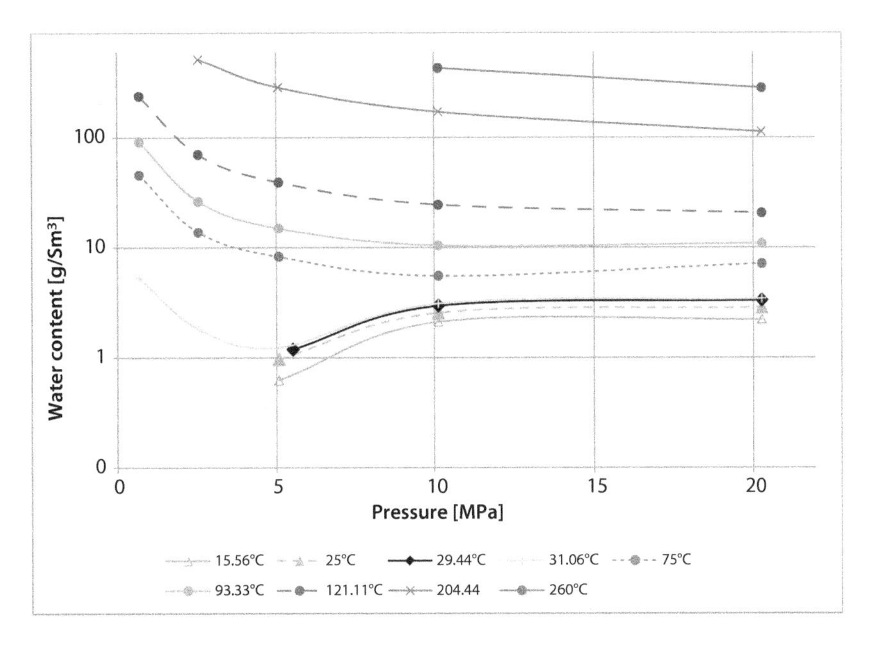

Figure 8.9 Water Content of Carbon Dioxide – Gillespie and Wilson [20].

Song and Kobayashi [21] prepared the GPA Research Report RR-99 on the water content of CO_2-rich fluids in equilibrium with liquid water and/or hydrates, which extended the work reported in RR-80.

Song and Kobayashi also determined the equilibrium phases for each data point. The list of the data points with associated equilibrium phases are shown in Table 8.4.

Figure 8.11 shows the data points obtained experimentally by the authors.

Song and Kobayashi [21] repeated most of their previous work [37] [21], but added a few new experimental points at 8.274 MPa shown in Figure 8.12. The following equation was used for curve fitting:

$$y = a_0 + a_1 \cdot P + a_2 \cdot P^2 + a_3 \cdot P^3 \tag{8.4}$$

$\mathbf{a_0}$	**1.580773**
a_1	0.051653
a_2	−0.00056
a_3	−4.06E-05

with the average relative error of 2.81E-12%.

Table 8.3 Original experimental water content data in CO_2-Rich Phase – From Song *et al.*, [37].

P	T		H_2O	
MPa	**°C**	**Equilibrium phases**	**Mol fraction**	**G/Sm³**
3.45	−17.78	CO_2 (l) - H_2O (H)	0.000261	0.199
	−13.0	CO_2 (l) - H_2O (H)	0.000322	0.245
	−3.5	CO_2 (l) - H_2O (H)	0.000458	0.348
	1.0	CO_2 (g) - H_2O (H)	0.000241	0.183
	5.5	CO_2 (g) - H_2O (H)	0.000379	0.288
	12.0	CO_2 (g) - H_2O (l)	0.000602	0.458
	20.0	CO_2 (g) - H_2O (l)	0.000999	0.760
4.83	−17.76	CO_2 (l) - H_2O (H)	0.000331	0.252
	−10.0	CO_2 (l) - H_2O (H)	0.000470	0.357
	−3.5	CO_2 (l) - H_2O (H)	0.000539	0.410
	3.0	CO_2 (l) - H_2O (H)	0.000717	0.546
	16.9	CO_2 (g) - H_2O (l)	0.000822	0.625
	25.0	CO_2 (g) - H_2O (l)	0.001277	0.971

(Continued)

Table 8.3 Cont.

P	T		H_2O	
MPa	°C	Equilibrium phases	Mol fraction	G/Sm³
6.21	−16.0	CO_2 (l) - H_2O (H)	0.000516	0.393
	−9.44	CO_2 (l) - H_2O (H)	0.000664	0.505
	7.0	CO_2 (l) - H_2O (H)	0.001094	0.833
8.27	−17.0	CO_2 (l) - H_2O (H)	0.001087	0.827
	−7.22	CO_2 (l) - H_2O (H)	0.001571	1.196
	−3.0	CO_2 (l) - H_2O (H)	0.001866	1.420
	13.78	CO_2 (l) - H_2O (l)	0.002781	2.115
	25.0	CO_2 (l) - H_2O (l)	0.003010	2.290
10.34	−17.0	CO_2 (l) - H_2O (H)	0.001272	0.967
	−9.0	CO_2 (l) - H_2O (H)	0.001648	1.254
	3.0	CO_2 (l) - H_2O (H)	0.002465	1.875
	25.0	CO_2 (l) - H_2O (l)	0.003368	2.563

(Continued)

Table 8.3 Cont.

P	T		H_2O	
MPa	**°C**	**Equilibrium phases**	**Mol fraction**	**G/Sm³**
13.79	−17.8	CO_2 (l) - H_2O (H)	0.001507	1.146
	−13.06	CO_2 (l) - H_2O (H)	0.001803	1.371
	−5.44	CO_2 (l) - H_2O (H)	0.002201	1.674
	2.72	CO_2 (l) - H_2O (H)	0.002740	2.084
	13.17	CO_2 (l) - H_2O (l)	0.003357	2.554

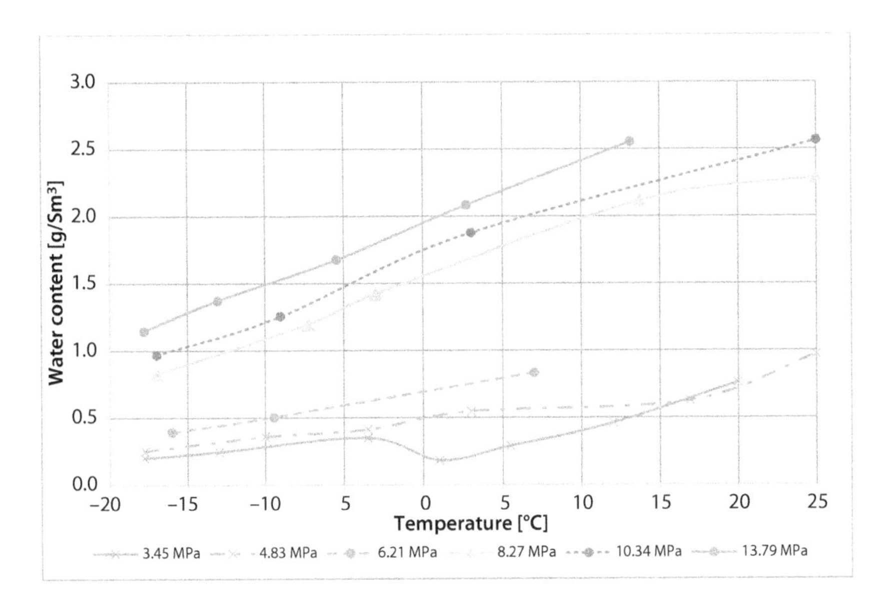

Figure 8.10 Water Content of Carbon Dioxide – Song *et al.*, [37].

The equilibrium phases for the experimental data points are listed in the Table 8.5:

The equilibrium water content data of natural gas and CO_2-rich liquids measured by Song and Kobayashi in this extensive research program for GPA (Song *et al.*, [21]) and in Song and Kobayashi [21] have been questioned by other researchers because the measured water solubility exhibited strong pressure dependence. Seo *et al.*, [64] measured data at high pressure (6.1 to 10.1 MPa) and at 1 to 21 °C and concluded that the pressure effect on the solubility is weak. Also Yang *et al.*, [61] and Li *et al.*, [41] questioned the strong pressure dependence by using theory and analysis (Jasperson *et al.*, [66]).

Briones *et al.*, [42] presented ternary phase equilibrium data for acetic acid-water mixtures with supercritical carbon dioxide at 40 and 50 °C and at 7.3, 10.4 and 13.9 MPa. The equilibrium compositions in the binary system water-carbon dioxide was also investigated at 50 °C and pressures from 5.6 to 17.7 MPa. The binary system data are shown in Figure 8.13.

The following equation was used for the correlation of the experimental data:

$$y = 96.2168 - 122.863 \cdot ln(P) + 52.4912 \cdot ln(P)^2 - 7.23724 \cdot ln(P)^3 \quad (8.5)$$

The mean relative error of predicting the experimental values is 4.11%.

Table 8.4 Recently measured data of water content in CO_2-Rich Phase – From Song and Kobayashi [21].

P	T	Equilibrium	H_2O	
MPa	°C	Phases	Mol fraction	g/Sm^3
2.068	−4.31	CO_2 (g) - H_2O (H)	0.000232	0.176
	−12.5	CO_2 (g) - H_2O (H)	0.000119	0.091
	−16.0	CO_2 (g) - H_2O (H)	0.000089	0.068
	−20.5	CO_2 (l) - H_2O (H)	0.000201	0.153
	−28.0	CO_2 (l) - H_2O (H)	0.000136	0.103
2.068	0.0	CO_2 (g) - H_2O (H)	0.000277	0.211
	2.5	CO_2 (g) - H_2O (H)	0.000436	0.332
	15.56	CO_2 (g) - H_2O (l)	0.001064	0.809
1.379	−18.0	CO_2 (g) - H_2O (H)	0.000114	0.087
	−15.2	CO_2 (g) - H_2O (H)	0.000147	0.112
	−11.0	CO_2 (g) - H_2O (H)	0.000220	0.167
	−2.2	CO_2 (g) - H_2O (H)	0.000488	0.371
	1.7	CO_2 (g) - H_2O (l)	0.000683	0.519

(Continued)

Table 8.4 Cont.

P	T	Equilibrium	H_2O	
MPa	°C	Phases	Mol fraction	g/Sm^3
0.689	−21.4	CO_2 (g) - H_2O (H)	0.000180	0.137
	−19.0	CO_2 (g) - H_2O (H)	0.000219	0.166
	−8.0	CO_2 (g) - ice	0.000556	0.423
	21.1	CO_2 (g) - H_2O (l)	0.004320	3.287
5.240	15.56	CO_2 (g) - H_2O (l)	0.000639	0.486
	15.56	CO_2 (l) - H_2O (l)	0.001118	0.851
5.792	20.20	CO_2 (g) - H_2O (l)	0.000898	0.683
	20.20	CO_2 (l) - H_2O (l)	0.001498	1.139
6.688	26.67	CO_2 (g) - H_2O (l)	0.001268	0.965
	26.67	CO_2 (l) - H_2O (l)	0.001951	1.484
7.171	29.50	CO_2 (g) - H_2O (l)	0.001496	1.138
	29.50	CO_2 (l) - H_2O (l)	0.002190	1.666
7.384	31.06	3-phase crit. end point	0.002104	1.601

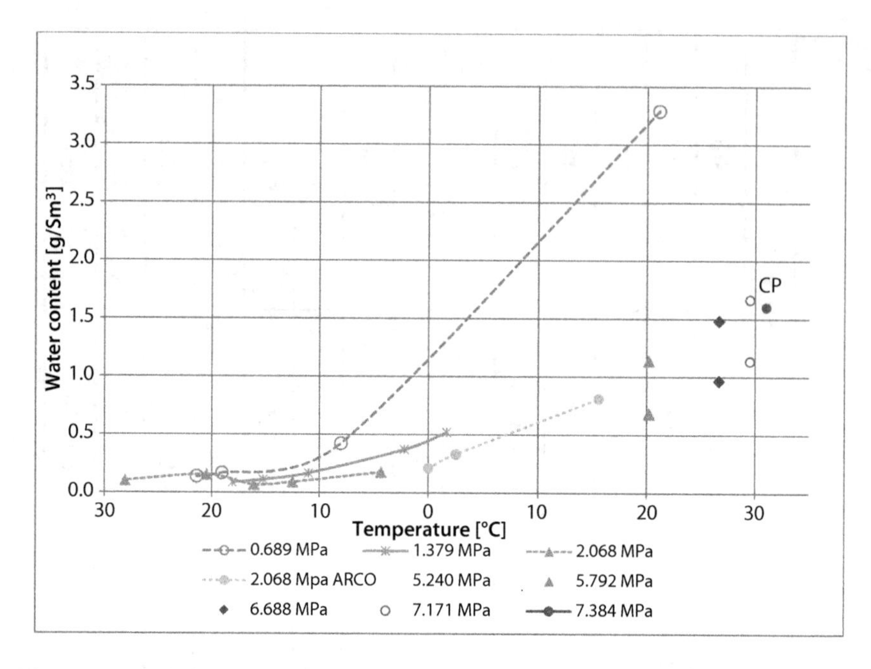

Figure 8.11 Water Content of Carbon Dioxide – Song and Kobayashi [21].

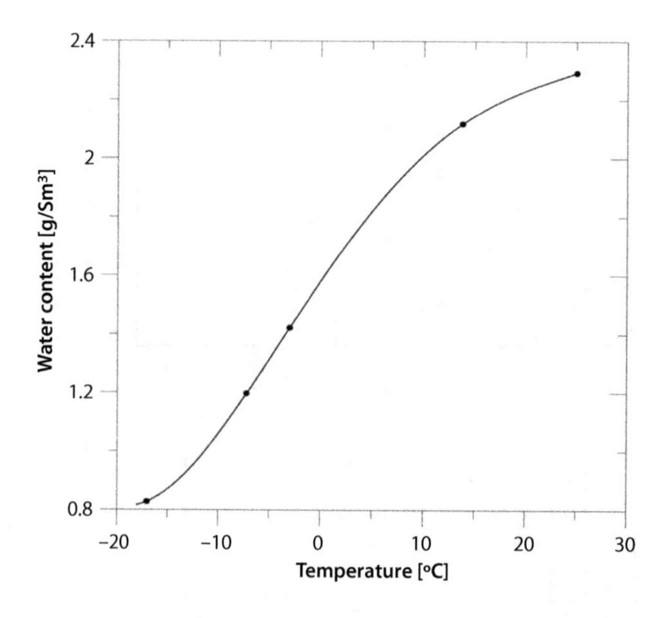

Figure 8.12 Water Content of Carbon Dioxide at 8.274 MPa – Song and Kobayashi [21].

Table 8.5 Experimental water content in CO_2-Rich Phase – From Song and Kobayashi [21].

T	P	Equilibrium Phase	Water
°C	MPa		g/Sm³
−17.0	8.274	CO_2(l)/H_2O(H)	0.828
−7.22		CO_2(l)/H_2O(H)	1.197
−3.0		CO_2(l)/H_2O(H)	1.422
13.78		CO_2(l)/H_2O(l)	2.118
25.0		CO_2(l)/H_2O(l)	2.293

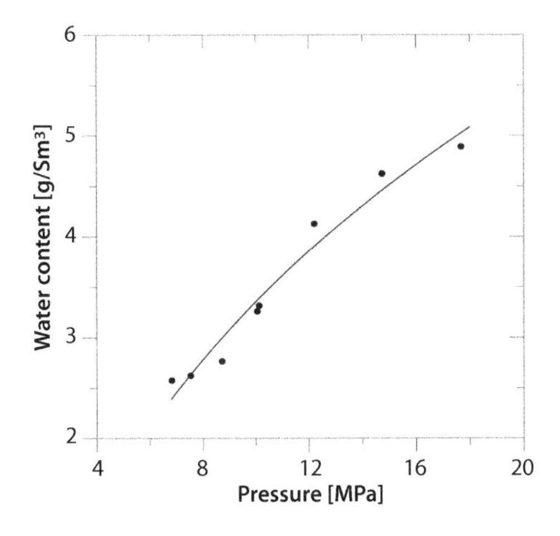

Figure 8.13 Water Content of Carbon Dioxide – Briones *et al.*, [42].

Nakayama *et al.*, [22] measured high pressure liquid-liquid equilibria for the system of water, ethanol and 1,1-difluoroethane, to study the latter's ability to extract ethanol from low concentration fermentation broths. They checked their experimental apparatus and procedure by measuring the phase equilibria for the H_2O-CO_2 system at 25 °C.

Figure 8.14 depicts the results of their findings.

They compared their results with those of Wiebe and Gaddy [2] and Coan and King [18] and concluded that their measurements were in good agreement with literature values.

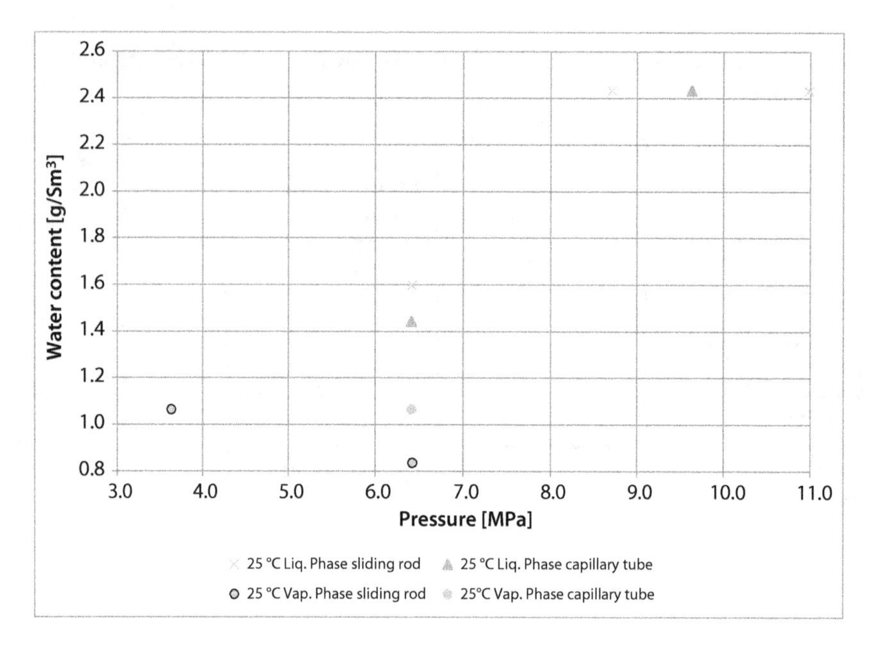

Figure 8.14 Water Content of Carbon Dioxide – Nakayama *et al.*, [22].

Patel *et al.*, [43] obtained vapor-phase P-ρ-T data for mixtures of 2, 5, 10, 25 and 50% water in carbon dioxide for temperatures from approximately 39 to 209 °C and pressures from 0.08 MPa to 9.6 MPa. The data were used to determine the compressibility factors for the above mixtures.

Figure 8.15 presents the experimental results (mixture dew points) obtained by Patel *et al.*, [43].

Müller *et al.*, [44] reported experimental results for the vapor-liquid equilibrium in the ternary system ammonia-carbon dioxide-water as well as for the binary gaseous electrolyte-water mixtures. The results for the water-carbon dioxide mixtures are shown in Figure 8.16.

D'Souza *et al.*, [45] measured high pressure phase equilibria in the carbon dioxide – n-hexadecane and carbon dioxide – water systems as part of the interest in supercritical fluid extraction as a potential separation and purification technique.

The carbon dioxide-water data is presented in Figure 8.17 and compared with the data of Wiebe and Gaddy [2]. Both sets of data are in good agreement with one another.

Ohgaki *et al.*, [23] measured the solubilities of α-tocopherol (vitamin E) in the CO_2-rich phase for α-tocopherol-H_2O-CO_2 and α-tocopherol-ethanol-CO_2 systems to investigate the entrainer effect in supercritical fluid extraction, water and ethanol being the entrainers.

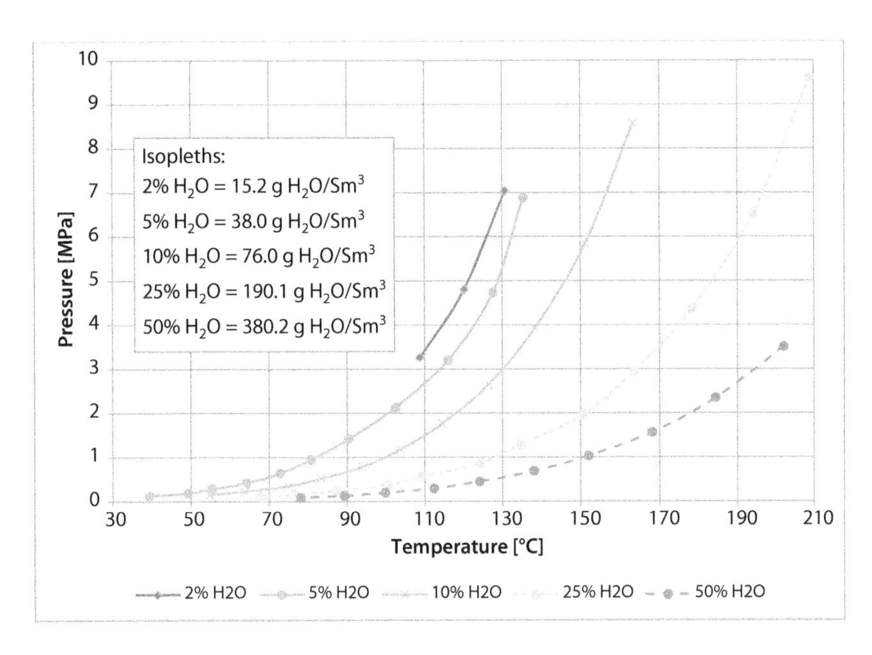

Figure 8.15 CO_2-H_2O Mixture Dew Points – Patel *et al.*, [43].

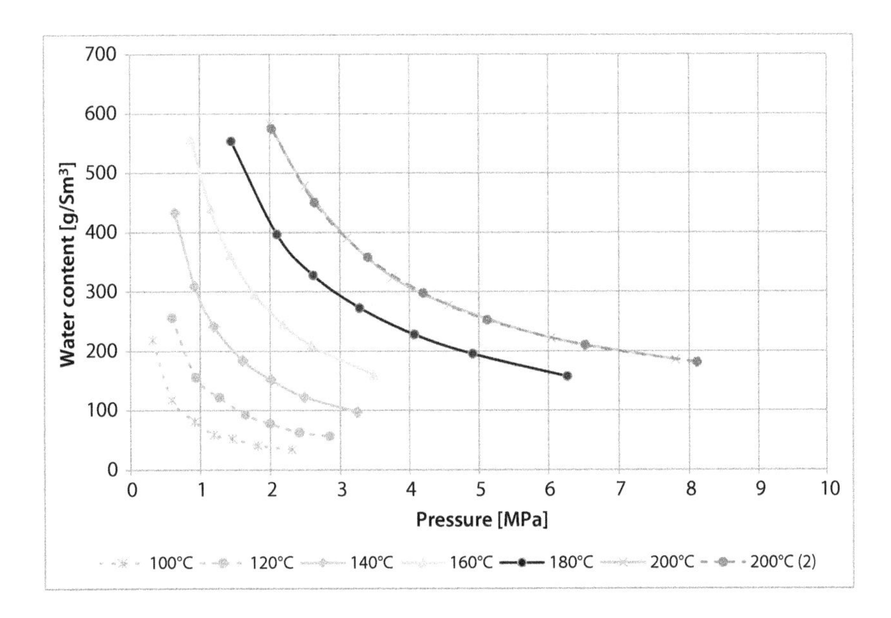

Figure 8.16 Water Content of Carbon Dioxide – Müller *et al.*, [44].

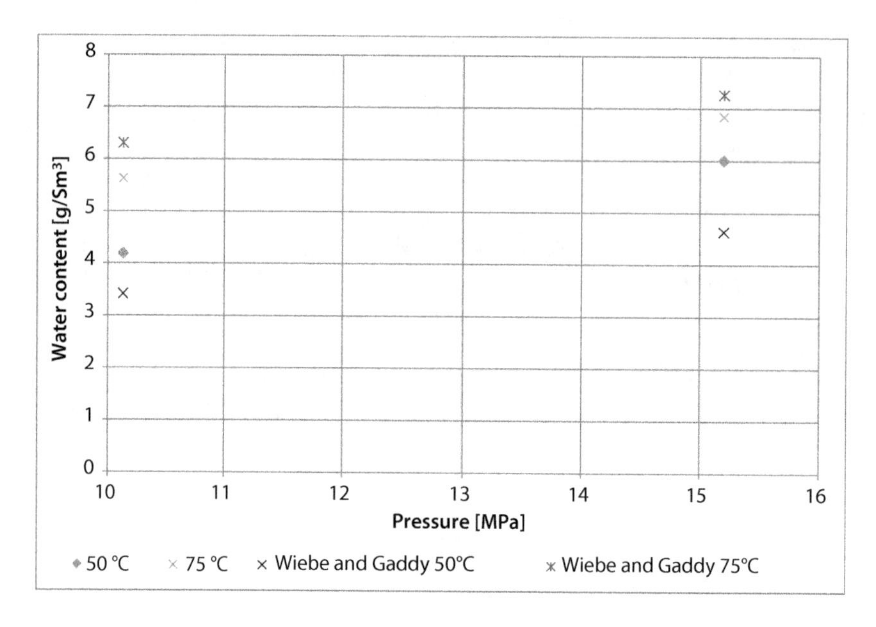

Figure 8.17 Water Content of Carbon Dioxide – d'Souza *et al.*, (1988).

The data obtained by the authors for the water-CO_2 system is shown in Figure 8.18, in which the 25 °C data from Wiebe and Gaddy [2] are shown for comparison. Again, the two sets of data for 25 °C are in good agreement.

$$\text{For } 25^\circ\text{C}: \; y = e^{\left(0.482482+\frac{0.798455}{P}+0.165606\cdot ln(P)\right)} \text{(mean relative error = 1.71\%)} \tag{8.6}$$

$$\text{For } 40^\circ\text{C}: \; y = e^{\left(11.6132-\frac{40.2139}{P}-2.79942\cdot ln(P)\right)} \text{(mean relative error = 2.56\%)} \tag{8.7}$$

Sako *et al.*, [46] measured isothermal high pressure vapor-liquid equilibria for two binary systems (CO_2-H_2O and CO_2-furfural) and a ternary system (CO_2-H_2O-furfural) to obtain data for the extraction and concentration of furfural produced in a reactor by using critical CO_2.

The VLE data for the CO_2-H_2O system is shown in Figure 8.19. The equation used for curve fitting of the 75 °C data is of the following form:

$$y = e^{\left(1.50548-\frac{3.4437}{P}+0.168769\cdot ln(P)\right)} \tag{8.8}$$

The mean relative error of forecasting of the experimental values is 7.49E-5 %. The equation used for curve fitting of the 148 °C data is of the following form:

$$y = e^{\left(0.701199+\frac{15.2071}{P}+0.70513\cdot ln(P)\right)} \tag{8.9}$$

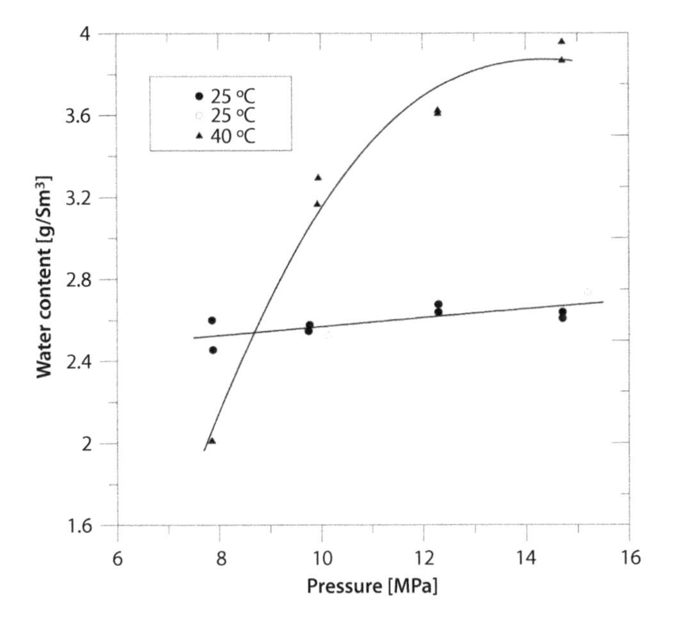

Figure 8.18 Water Content of Carbon Dioxide – Ohgaki *et al.*, [23]. The Two Empty Circles Represent Data from Wiebe and Gaddy [45].

The mean relative error of forecasting of the experimental values is 1.96%.

Sterner and Bodnar [24] used a synthetic fluid inclusion technique to determine the temperature of the phase transition. The procedure involved the healing of fractures in natural quartz in the presence of H_2O-CO_2 solutions of known compositions at predetermined temperatures and pressures within the one-fluid phase field. The CO_2 liquid-vapor homogenization temperatures and total homogenization temperatures of individual inclusions were measured.

Mather and Franck [59] noticed that for more than 25 years there had been disagreement about the phase behaviour in the system carbon dioxide-water at high temperatures and pressures, by comparing the work of Tödheide and Franck [3] and Takenouchi and Kennedy [16] and noting serious quantitative disagreement between the two. They presented new data for that system. They measured dew points for three mixtures 53.0, 62.5 and 78.9 mol% CO_2 at temperatures from 225 to 275 °C and pressures from 114 to 311 MPa. They stated that the data appeared to had resolved the discrepancy – the new data substantiated the data of Tödheide and Franck. Also, indirect evidence in support of Mather and Franck's work had been presented by Sterner and Bodnar [24].

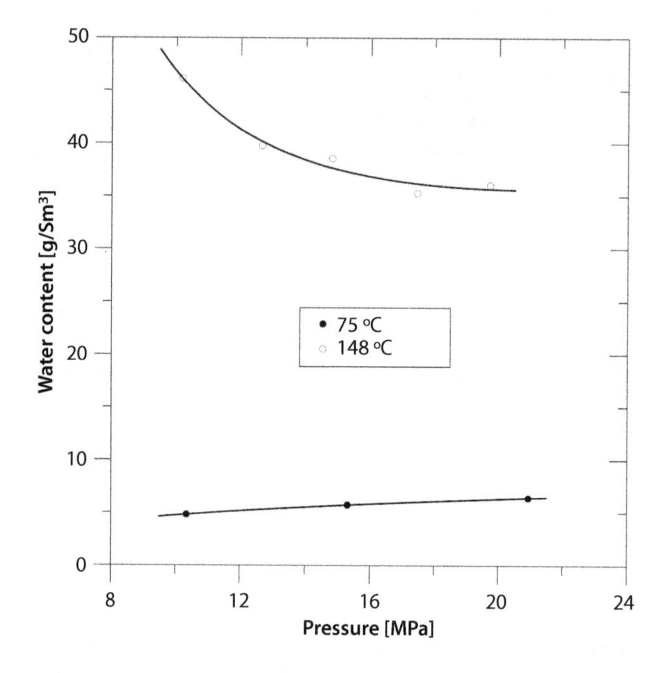

Figure 8.19 Water Content of Carbon Dioxide – Sako *et al.*, [46].

Figure 8.20 shows the experimental data obtained by Mather and Franck.

King *et al.*, [47] presented new data for phase compositions and phase densities at 15, 20, 25, 35 and 40 °C, and pressures a little over 20 MPa. The data extended the coverage of the work of Wiebe and Gaddy [2]. The overlap was at 25 °C, where the agreement in water solubility was within 2%.

Figure 8.21 shows the experimental data points obtained by the authors. The equations used for measuring deviations from the norm and mean relative errors are shown below:

$$15^\circ\text{C data}: \ y = e^{\left(-0.0970065 + \frac{0.951578}{P} + 0.267445 \cdot ln(P)\right)} \text{(mean relative error} = 0.49\%) \tag{8.10}$$

$$20^\circ\text{C data}: \ y = e^{\left(0.0696417 + \frac{0.855121}{P} + 0.271693 \cdot ln(P)\right)} \text{(mean relative error} = 1.26\%) \tag{8.11}$$

$$25^\circ\text{C data}: \ y = e^{\left(0.214365 + \frac{0.720028}{P} + 0.268082 \cdot ln(P)\right)} \text{(mean relative error} = 0.65\%) \tag{8.12}$$

$$35^\circ\text{C data}: \ y = e^{\left(0.761691 - \frac{1.35287}{P} + 0.211116 \cdot ln(P)\right)} \text{(mean relative error} = 0.56\%) \tag{8.13}$$

$$40^\circ\text{C data}: \ y = e^{\left(-466362 + \frac{2.3333}{P} + 0.609959 \cdot ln(P)\right)} \text{(mean relative error} = 0.23\%) \tag{8.14}$$

Dohrn *et al.*, [48] measured vapor-liquid equilibrium data for the ternary glucose-water-CO_2 system and for the quaternary glucose-water-ethanol-CO_2 system at temperatures 50–70 °C and pressures up to 30 MPa. The study was undertaken for engineering design purposes using a novel apparatus: the use of supercritical CO_2 as a solvent for the extraction

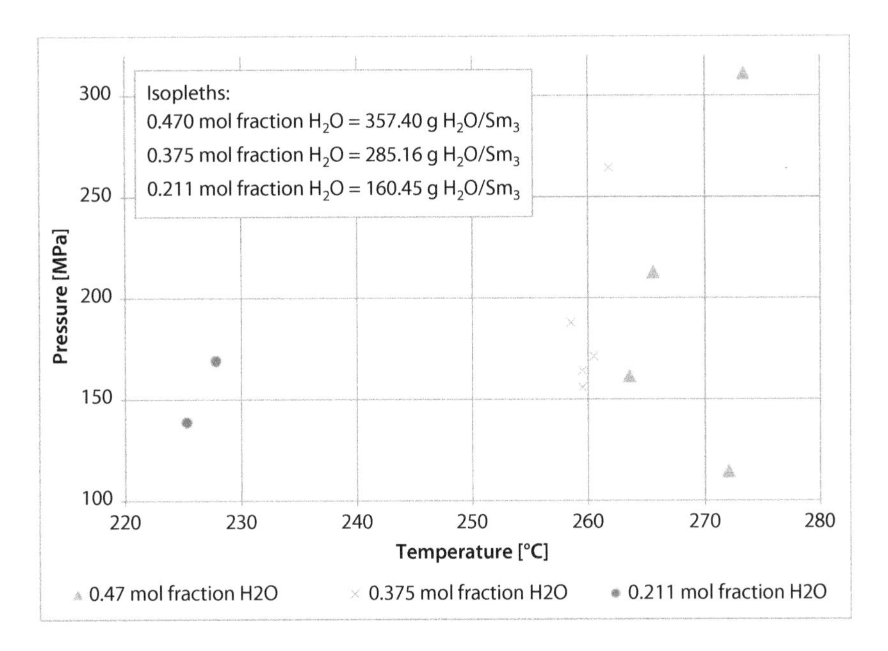

Figure 8.20 CO_2-H_2O Mixture Dew Points – Mather and Franck [59].

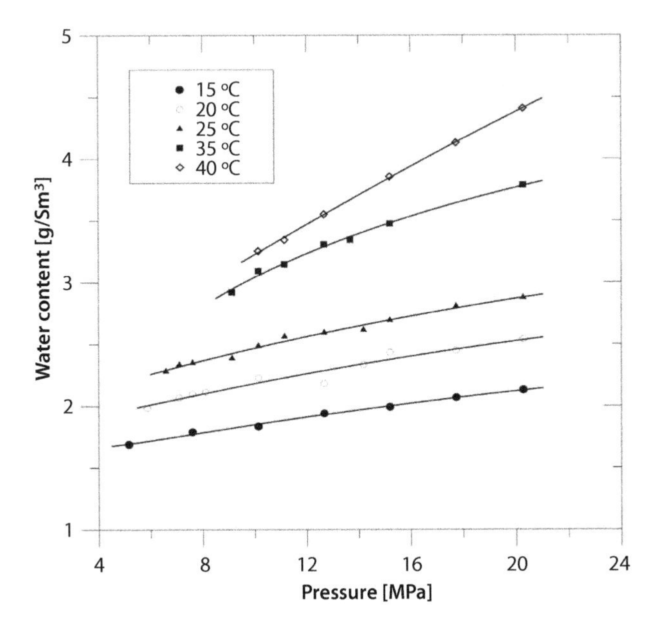

Figure 8.21 Water Content of Carbon Dioxide – King *et al.*, [47].

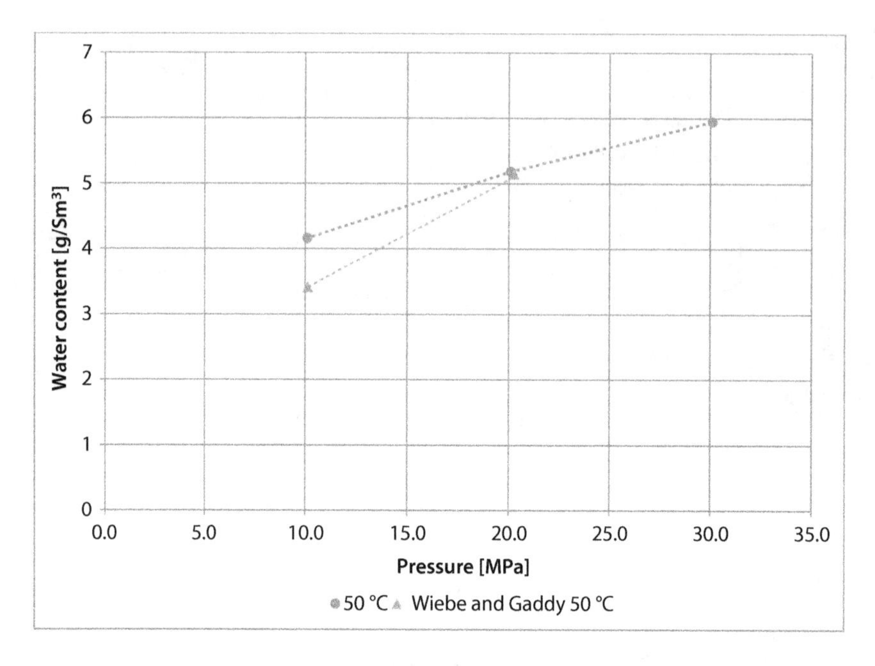

Figure 8.22 Water Content of Carbon Dioxide – Dohrn *et al.*, [48].

Table 8.6 Jackson *et al.*, [25] water solubility data in supercritical CO_2.

T	P	**Water**
°C	MPa	g/Sm^3
50.0	34.48	5.703
75.0	34.48	10.114

of natural substances and addition of polar substances to enhance the process of extraction.

To ensure the reliability of the experimental data obtained in this work, the authors studied the binary system water-CO_2 at 50 °C and compared the results with those of Wiebe and Gaddy [2]. Figure 8.22 shows the results for the binary system and comparison with Wiebe and Gaddy results.

Jackson *et al.*, [25] presented water solubility data in supercritical fluids and high pressure liquids (carbon dioxide and two types of refrigerants, R22 and R134a) using near infrared spectroscopy. Their water solubility data in supercritical in Table 8.6.

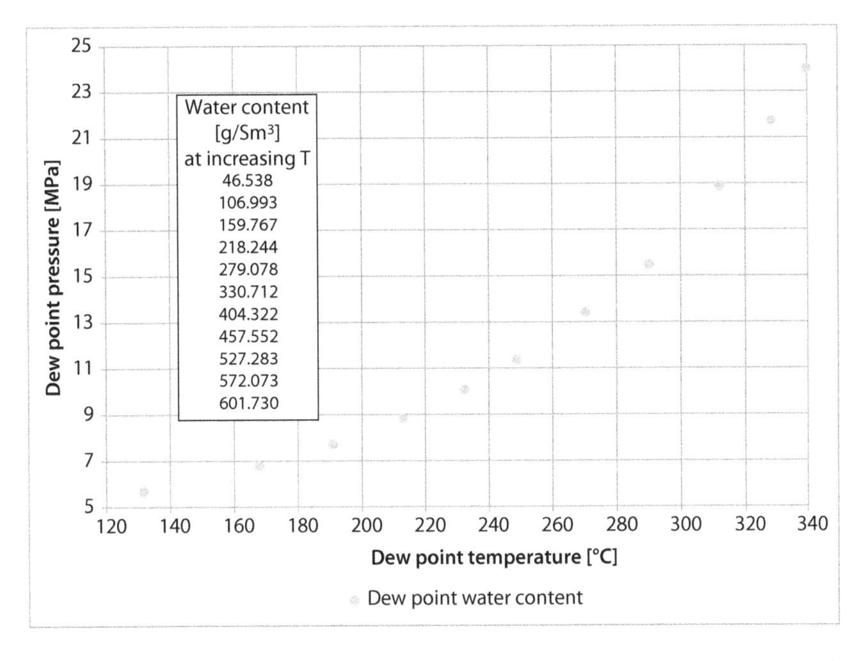

Figure 8.23 Water Dew Point Pressures and Temperatures for Known Compositions of H_2O-CO_2 Mixtures – Fenghour *et al.*, (1996).

A small amount of water added to a supercritical fluid can greatly increase the solubility of polar species in non-polar fluids. These modified supercritical fluids can greatly increase the use of the fluids in separations and reactions. Often the solubility data for water in a supercritical fluid are not available under a given set of pressure and temperature conditions, therefore the data obtained by the authors were useful for the researchers studying the subject.

The authors concluded that the water solubility was quite low in CO_2, as one would expect for a highly polar solute in a non-polar solvent.

Fenghour *et al.*, [26] reported new measurements of the density of eleven mixtures of (water + carbon dioxide) using a fully automated isochoric instrument. The dew points of the mixtures studied were also determined.

Figure 8.23 shows the water content of the eleven mixtures of H_2O and CO_2 and the corresponding dew point temperatures and pressures.

Figure 8.23a shows the same data in a different manner – as water content vs. dew point pressures. The equation used for correlation of experimental data is of the following form:

$$y = e^{\left(9.67409 - \frac{24.3945}{P} - 0.714549 \cdot ln(P)\right)} \tag{8.15}$$

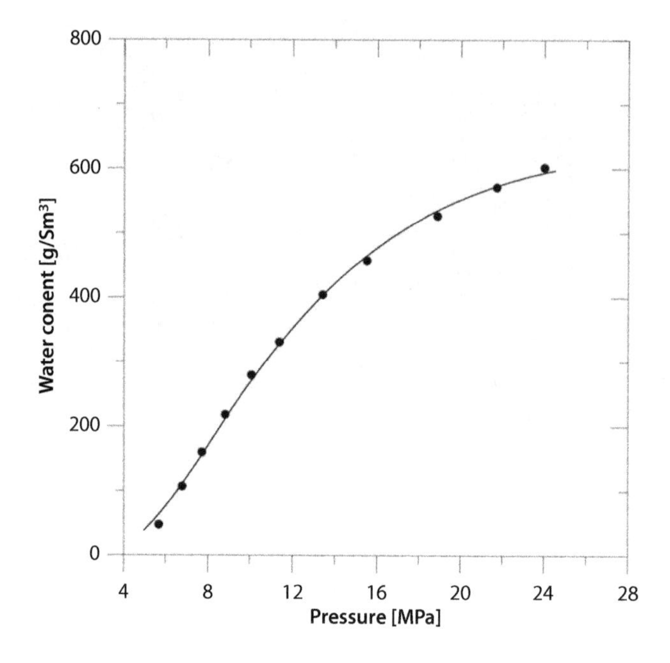

Figure 8.23a Water Dew Point Pressures for Known Compositions of H_2O-CO_2 mixtures - Fenghour *et al.*, [26].

and the mean relative error for the data obtained by Fenghour *et al.*, is 4.92%.

Frost and Wood [60] determined experimentally volumes of H_2O-CO_2 mixtures between 950 and 1940 MPa at 1100–1400 °C and reported PVTX properties of the mixtures. They used a synthetic fluid inclusion technique to capture H_2O-CO_2 fluids of known composition in precracked corundum, and determined the densities of the fluids from measurements of the homogenization temperatures along the CO_2 liquid-vapor equilibrium curve.

Bamberger *et al.*, [27] reported experimental results for the high pressure (vapor + liquid) equilibrium of binary systems of carbon dioxide + (either water or acetic acid) for temperatures from 40 to 80 °C and pressures from 1 to 14 MPa – useful for the design of supercritical extraction processes. The experimental results for the H_2O-CO_2 system are given in Figure 8.24.

Sabirzyanov *et al.*, [28] developed a new experimental circulation facility to investigate the solubility of liquids in supercritical carbon dioxide, which is most extensively used in fluid extraction processes. The solubility of water was experimentally investigated on the 40 and 50 °C isotherms

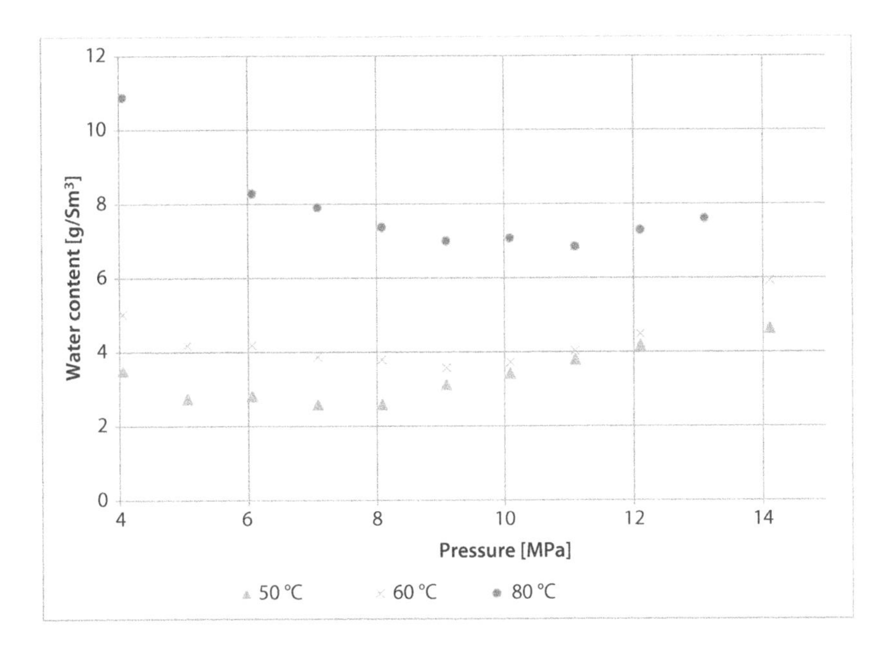

Figure 8.24 Water Content of Carbon Dioxide – Bamberger *et al.*, [27].

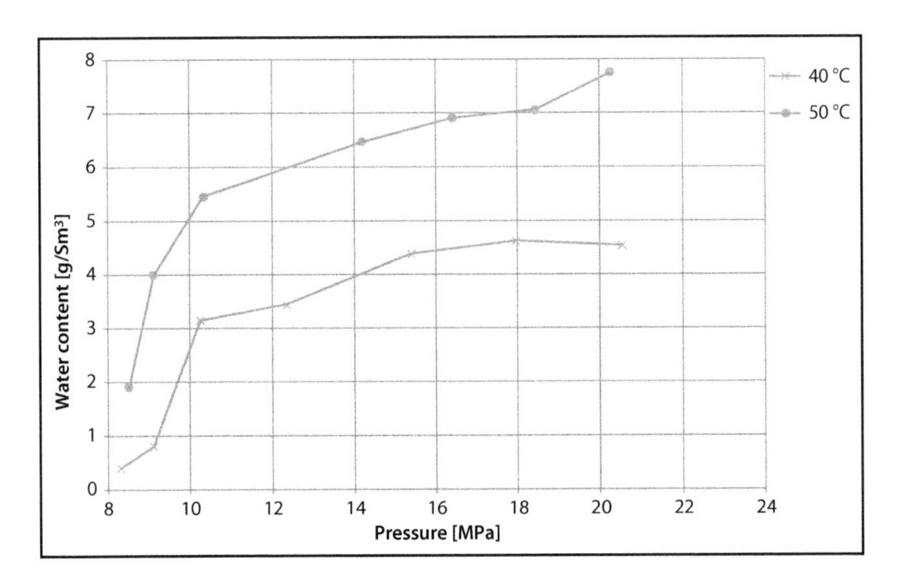

Figure 8.25 Water Content of Carbon Dioxide – Sabirzyanov *et al.*, [28].

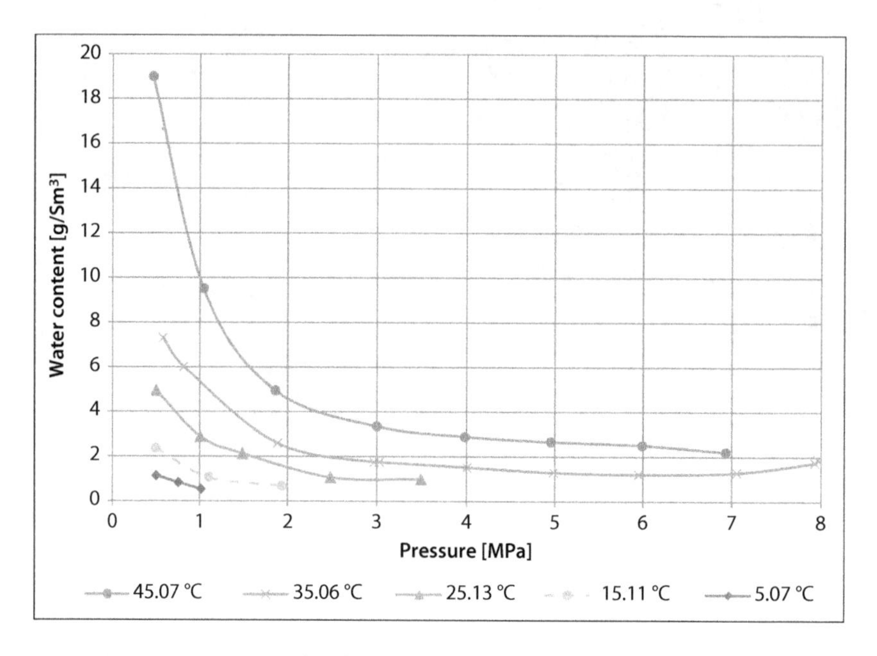

Figure 8.26 Water Content of Carbon Dioxide – Valtz *et al.*, [29].

in the pressure range of 8 – 20 MPa and the results are now presented in Figure 8.25.

Valtz *et al.*, [29] reported new experimental VLE data for the carbon dioxide-water system at temperatures between 5 and 45 °C and pressures up to 8 MPa, shown in Figure 8.26. They compared the experimental results with three models: model 1 – Peng-Robinson EoS + Henry's law, model 2 - PR-EoS/WS-NRTL, and model 3 - SAFT-VR EoS. The absolute average deviation (AAD) for model 1 was 11.7%, for model 2–6.4% and for model 3–12.0%.

Jarne *et al.*, [30] experimentally determined dew points for four carbon dioxide-water mixtures and eight carbon dioxide-water-methanol mixtures. Their work was part of a research aiming at investigating the effect of CO_2, H_2O, CH_3OH and heavy hydrocarbons of natural gases on the VLE of natural gas within the usual pressure and temperature conditions of natural gas transport by pipeline. This work was done because the experimental dew point data for these systems were not found in the literature in the temperature and pressure ranges of interest.

The authors avoided the presence of solid hydrates and liquid carbon dioxide. They concluded that for a given pressure value, the dew point temperature increases when the water amount also increases. The increase

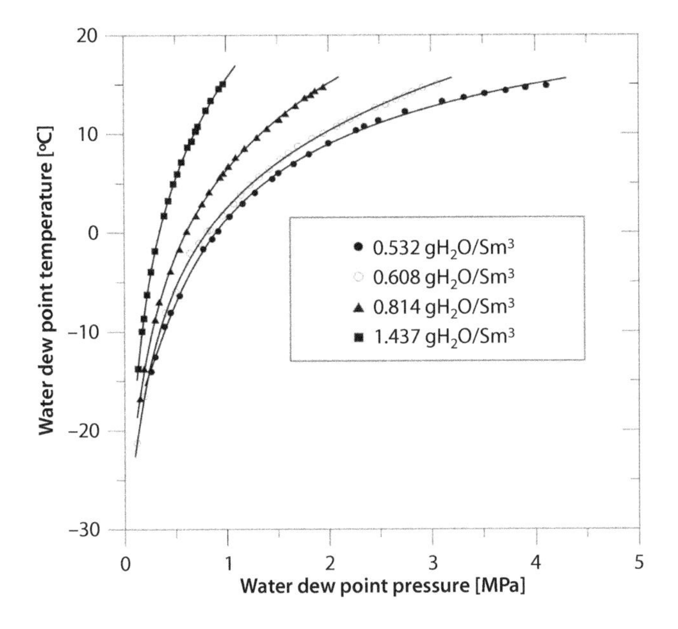

Figure 8.27 Water Dew Point Temperatures – Jarne *et al.*, [30].

is higher for high values of pressure than for low values. The conclusion is depicted in Figure 8.27.

Jarne *et al.*, analyzed the experimental results in terms of a predictive excess function-equation of state (EF-EOS) method, which reproduced the experimental data with absolute average deviation between 0.8 and 1.8 °C for the carbon dioxide-water system.

The authors of this paper used the following equations for the correlation of the experimental data of Jarne *et al.*,

$$\text{For } 0.532\ \text{g/Sm}^3: \quad y = e^{\left(3.45715 - \frac{0.53202}{P} + 0.109153 \cdot \ln(P) - 17.263\right)} \quad (\text{mean relative error} = 9.20\%) \tag{8.16}$$

$$\text{For } 0.608\ \text{g/Sm}^3: \quad y = e^{\left(3.53152 - \frac{0.17035}{P} + 0.223182 \cdot \ln(P) - 26.327\right)} \quad (\textit{mean relative error} = 7.64\%) \tag{8.17}$$

$$\text{For } 0.814\ \text{g/Sm}^3: \quad y = e^{\left(3.57583 - \frac{0.16953}{P} + 0.24597 \cdot \ln(P) - 23.788\right)} \quad (\textit{mean relative error} = 2.86\%) \tag{8.18}$$

$$\text{For } 1.437\ \text{g/Sm}^3: \quad y = e^{\left(3.88734 - \frac{0.08092}{P} + 0.253121 \cdot \ln(P) - 29.376\right)} \quad (\textit{mean relative error} = 1.37\%) \tag{8.19}$$

Iwai *et al.*, [31] measured the solubilities of palmitic acid in supercritical carbon dioxide and entrainer effect of water using a static recirculation method combined with online Fourier transform infrared (FTIR) spectroscopy. Since the solvent power of supercritical carbon dioxide is relatively poor, a small amount of entrainer, like water, is sometimes used to improve the solubility and selectivity of CO_2. The authors obtained

Table 8.7 Experimental data point for the system CO_2-H_2O – Iwai *et al.*, [31].

T	P	Water	
[°C]	[MPa]	g/Sm³	
		Iwai *et al.*,	King *et al.*, [47]
40	15.0	3.741	3.855

one data point for the CO_2-H_2O system and compared the result with the respective data point from the work of King *et al.*, [47]. Both experimental values are shown in Table 8.7.

Koglbauer and Wendland [8] developed a new method to measure the dew points, expressed as vapor concentration enhancement factors, by Fourier transform infrared (FTIR) spectroscopy, in compressed humid nitrogen, argon, and carbon dioxide. The measurements for CO_2 were made for pressures up to 5.5 MPa and temperatures between 25 and 100 °C.

The authors described the solubility of water in compressed gas as follows:

a) By the vapor pressure enhancement factor f_w

$$f_w\left(T,P\right) = \frac{p_w}{p_w^0} = x_w \cdot \frac{P}{p_w^0} \tag{8.20}$$

where: p_w is the partial pressure of water
p_w^0 is the vapor pressure of pure water
x_w is the mole fraction of water in compressed gas

b) By the vapor concentration enhancement factor g_w

$$g_w\left(T,P\right) = \frac{c_w}{c_w^0} \tag{8.21}$$

where: c_w is the water concentration (in moles or mass per unit volume) in saturated gas
c_w^0 is the saturated vapor density of pure water

Koglbauer and Wendland commented that the measurements of the mole fraction of water with increasing pressure are tedious and sensitive to errors and that measuring the water concentration (per unit volume), or g_w, would avoid these problems.

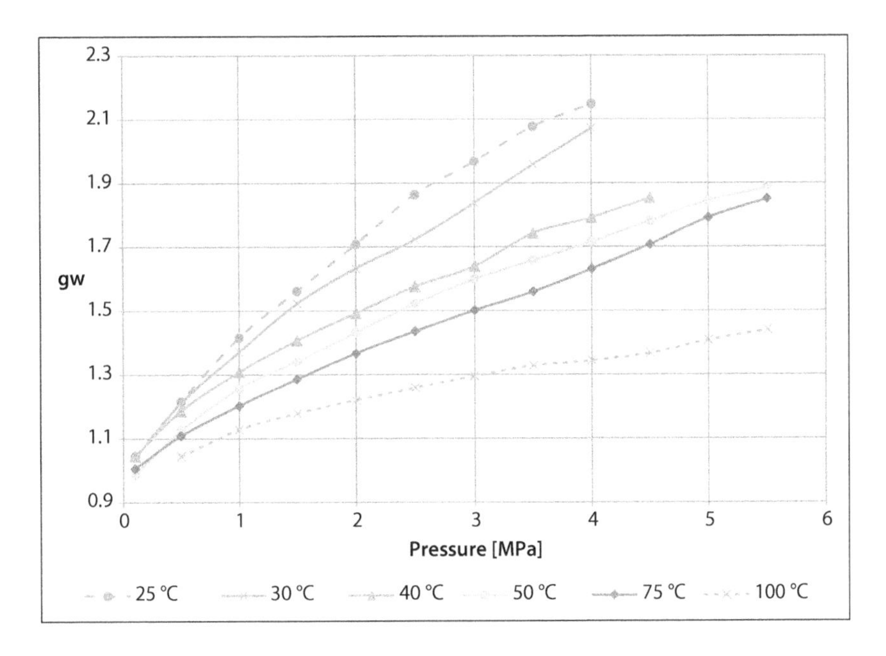

Figure 8.28 Vapor Concentration Enhancement Factors g_W for Compressed Humid CO_2 – Koglbauer and Wendland [8].

They developed an apparatus with a high-pressure view cell placed in the sample compartment of an FTIR spectrometer. The IR light absorbance could be measured in situ in the saturated gas phase inside the view cell. The vapor concentration enhancement factor can be determined rapidly and with good accuracy as a relationship of the absorbance in compressed humid gas to that in saturated pure vapor. The experimental data for g_W of compressed humid CO_2 which were collected by Koglbauer and Wendland, are plotted in Figure 8.28.

The authors stated that no experimental values for the vapor concentration enhancement factor g_W were found in the literature for compressed carbon dioxide; only experimental data for the mole fraction x_W or for quantities which can easily be converted to mole fraction are available.

Chapoy *et al.,* [32] studied the effect of common impurities, like nitrogen, hydrogen, carbon monoxide, hydrogen sulfide and water, on the phase behaviour of carbon dioxide rich systems.

The amount of water contained in carbon dioxide produced by carbon capture needs to be reduced to avoid hydrate formation and corrosion when CO_2 is compressed and sent by pipeline for sequestration. The authors developed an experimental methodology to measure water content in CO_2 rich phase in equilibrium with hydrates. The aim of the study was to analyze the

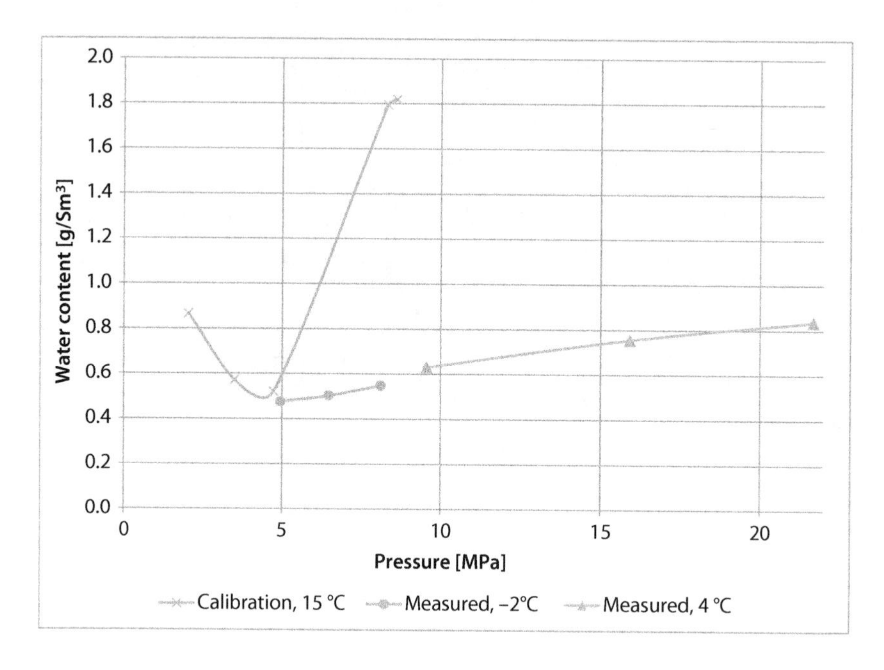

Figure 8.29 Water Content of Carbon Dioxide – Chapoy *et al.*, [32].

risk of hydrate formation in a carbon dioxide rich stream in a pipeline where the stream could be cooled down to −2 °C after a choke.

First, they measured water content of CO_2 at 15 °C in the vapor-rich CO_2 and in the liquid region to validate the experimental setup and calibrations. Then they measured the water content of pure CO_2 in equilibrium with hydrates at simulated pipeline conditions of 4 °C and pressures between 9.5 and 21.7 MPa, and after simulated choke conditions of −2 °C and pressures between 5 and 8 MPa. The results of the experiments are shown in Figure 8.29.

Youssef *et al.*, (2009) measured hydrate dissociation temperature of methane, ethane, and carbon dioxide in the absence of any aqueous phase. They developed a new experimental procedure combining an equilibrium cell with a water measurement by a Karl Fischer coulometer.

The new experimental method to detect the hydrate dissociation temperature consisted of measuring the water content in the vapor phase as a function of the equilibrium cell temperature. Before the hydrate formation, the water content in the vapor phase is constant with temperature. When the quantity of water drops, it is assumed that the water was consumed to form hydrate. The hydrate dissociation temperature is the intersection point due to the slope change of the curve representing the amount

Table 8.8 Hydrate dissociation temperature of CO_2 – Youssef *et al.*, (2009).

Water	P	T
g/Sm^3	MPa	°C
0.271	2.0	3.2
0.213	2.0	−1.0
0.192	2.0	−2.0

of water in the vapor phase versus temperature. Table 8.8 shows the summary of experimental data.

Based on the experimental results, the authors observed that the classical Platteeuw and van der Waals model associated with the CPA equation of state correctly predicts the hydrate equilibrium with and without aqueous phase.

Tabasinejad *et al.*, [33] conducted a series of experiments to measure the water solubility in supercritical nitrogen and carbon dioxide at the experimental conditions of 150 to 205 °C and 3.9 to 129.2 MPa.

The authors commented that many experimental studies had been done in the past on the phase behavior of the water-CO_2 system. Most of these studies were devoted to low pressure and temperature conditions. Only a few studies had been performed at higher pressures and temperatures, but the reported experimental data were not consistent. The data collected by Tabasinejad *et al.*, are presented in Figure 8.30.

The authors compared the new measured data and those from Tödheide and Franck [3] and Takenouchi and Kennedy [16] to show that the reported data do not agree with each other. The measured data of Tödheide and Franck show higher water content values at the same temperatures compared with the results obtained by Tabasinejad *et al.*,[33]. But the data trend is almost similar at the same pressure range, which is in contrast with the experimental data obtained by Takenouchi and Kennedy. The results of the experiments done by Mather and Franck at elevated temperatures and pressures are closer to the data of Tödheide and Franck.

Seo *et al.*, [64] measured the solubility of water in the CO_2-rich phase in the presence of hydrate at temperature between 1.2 and 21.4 °C and pressures of 6.1 and 10.1 MPa. The temperatures at which two-phase L_{CO2}-H mixtures become a single liquid phase were visually determined at constant pressures for predetermined amounts of CO_2 and water. Constant

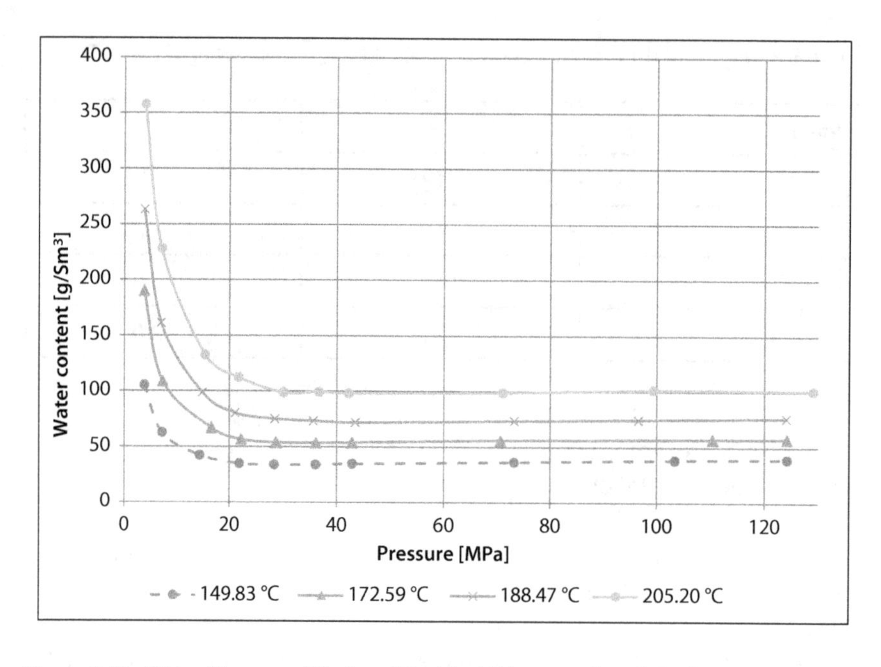

Figure 8.30 Water Content of Carbon Dioxide – Tabasinejad *et al.,* [33].

temperature and pressure were maintained for more than 6 hr to equilibrate the system. The measured solubility data are shown in Figure 8.31.

Song and Kobayashi [58] reported experimental data for this system and a large pressure dependence was observed at 6.21 and 10.34 MPa. Seo *et al.,*[64] compared their solubility data with published experimental data at the hydrate-free region and observed a weak pressure dependence compared with the data of Song and Kobayashi.

Chapoy *et al.,* [34] reported new experimental data for the water content of liquid carbon dioxide in equilibrium with hydrates at 13.9 MPa and – 20 to + 4 °C. The knowledge of maximum allowable water content in CO_2-rich fluids is important for safe transport of CO_2 to storage sites, as the presence of water may cause corrosion or form hydrates which could block the flow in pipelines. There is a disagreement between experts – some recommend the maximum water content of CO_2 of 500 ppm, others recommend 50 ppm with a good safety margin.

Figure 8.32 shows the new experimental data.

Kim *et al.,* [49] measured the solubility of water in carbon dioxide-rich phase over a temperature range of 10.1–38.9 °C and pressure of 8.1, 10.1 and 20.1 MPa.

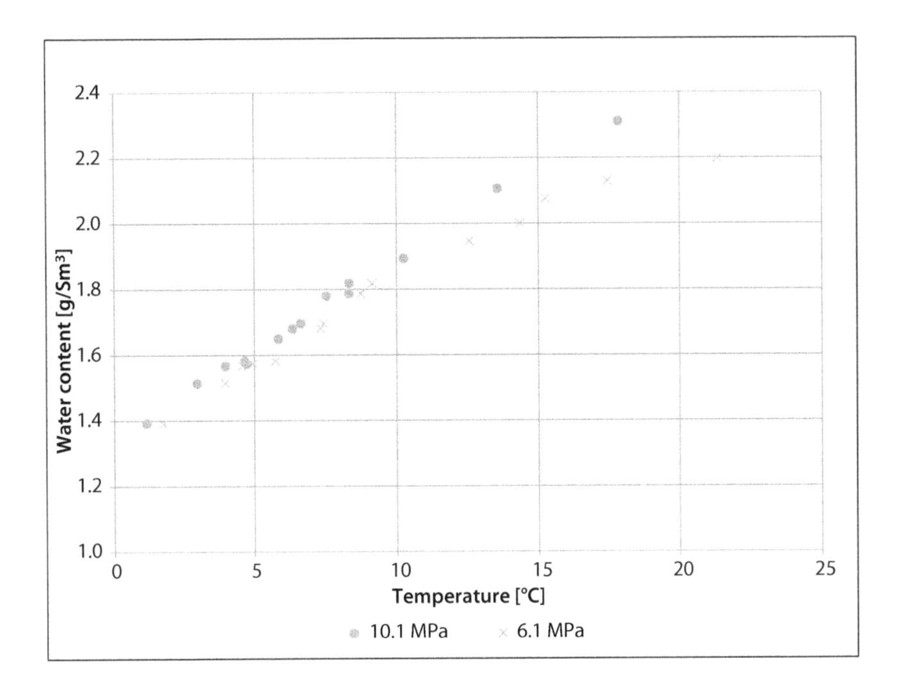

Figure 8.31 Water Content of Liquid CO_2 in Equilibrium with Hydrates – Seo *et al.*, (2011).

The authors noted that experimental data for the solubility of water in carbon dioxide-rich phase do not agree among researchers due to experimental difficulties since it is difficult to accurately determine water mole fraction in the order of 0.001. To avoid such difficulty, Wendland *et al.*, (1999) and Seo *et al.*, [64] used an indirect method for the determination of hydrate containing equilibrium of water-carbon dioxide system. This method does not require the conventional analytic composition analysis. For the mixture of known composition, the equilibrium condition is determined by visually observing the transition point at which the two-phase mixture forms a single phase.

In their work, Kim *et al.*, determined equilibrium solubility of H_2O in CO_2-rich phase visually, by observing small droplets of water which completely disappear when the temperature is slowly raised.

Kim *et al.*, compared their results with published data and commented that at 20.1 MPa the measured data were in good agreement with the data reported by King *et al.*, [47], Wiebe and Gaddy [2], and Gillespie and Wilson [20]. At 10.1 MPa, the measured solubility also agrees with the value reported by Gillespie and Wilson, however, becomes slightly larger

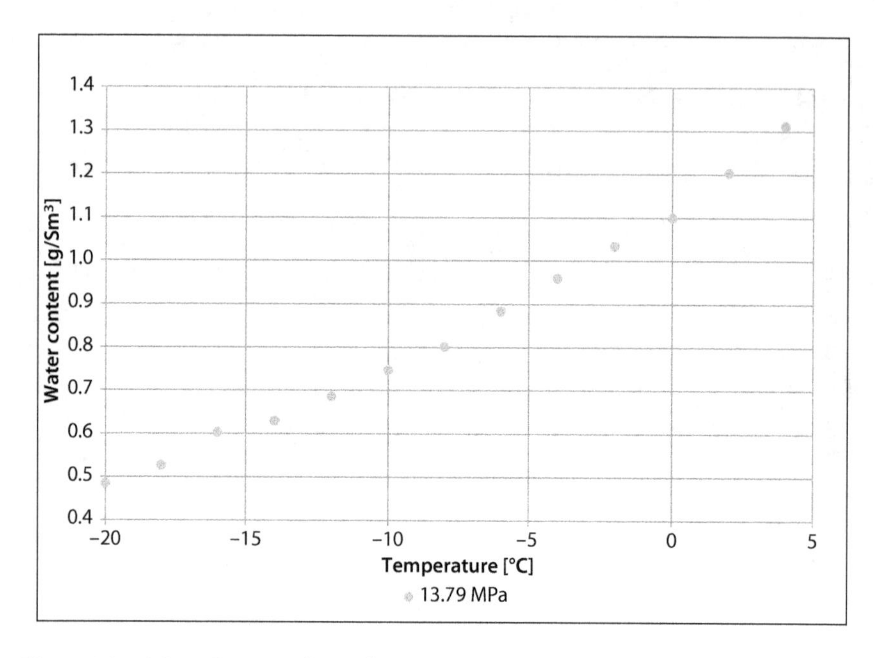

Figure 8.32 Water Content of Liquid CO_2 in Equilibrium with Hydrates – Chapoy *et al.*, [34].

than those by King *et al.*, and Wiebe and Gaddy. The deviation with the literature data become largest at 8.11 Mpa; however, it should be noted that these experimental conditions are near the critical point of pure CO_2, in which the analytic sampling is both difficult and often the principal source of experimental uncertainties.

The experimental data of Kim *et al.*, are depicted in Figure 8.33.

Hou *et al.*, [5] measured the phase behavior of the carbon dioxide-water mixture at temperatures from 25 to 175 °C and pressures from 1.1 to 17.5 MPa, as well as vapor-liquid-liquid and liquid-liquid equilibrium points at 25 °C, and compared their results with available literature data.

The authors commented that even though many experimental studies of this system had been conducted, significant uncertainty remains, especially at elevated temperatures and pressures owing to quite large discrepancies between literature sources.

Hou *et al.*, compared their results with literature data by first stating that for the vapor phase results, there are fewer data available with which to compare and no comprehensive evaluation has been published. Since the liquid phase data of Bamberger *et al.*, [27] and Zawisza and Malesińska [6] were recommended, Hou's results were compared with their vapor phase data, where available, and showed good consistency with these literature sources.

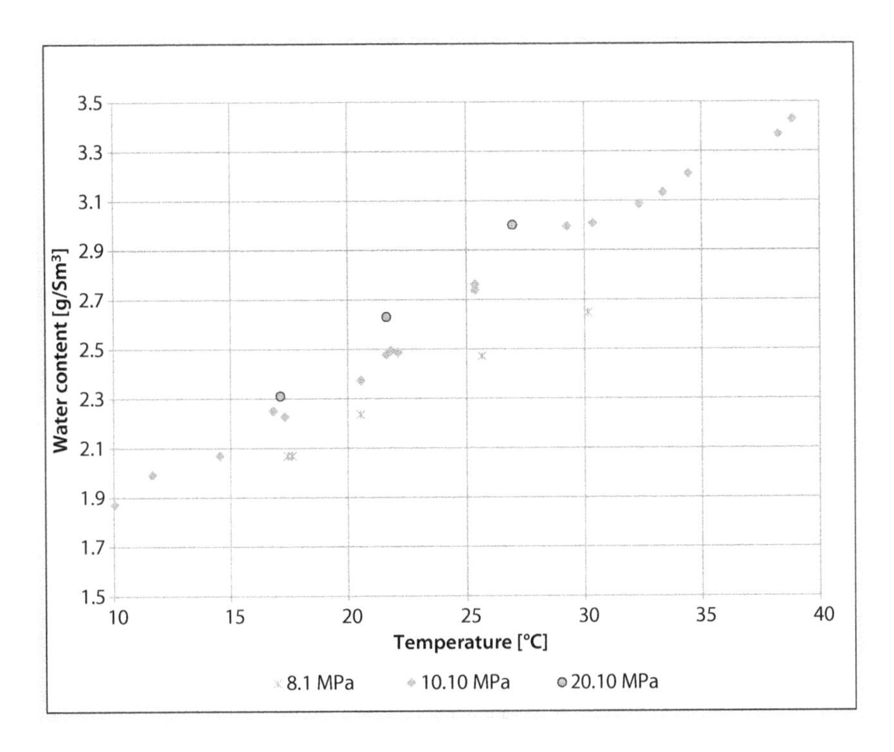

Figure 8.33 Water Content of Carbon Dioxide – Kim *et al.*, [49].

Various researches provided various times for the CO_2-H_2O system to reach equilibrium. Hou *et al.*, wrote that for all the isotherms studied, an equilibrium condition was reached within approximately 2 hr.

Figure 8.34 summarizes the results obtained by Hou *et al.*,

Chapoy *et al.*, [35] presented experimental results on the phase behavior and thermophysical properties of carbon dioxide in the presence of O_2, Ar, N_2 and water. They also determined experimentally saturation pressures and hydrate stability (in water saturated and undersaturated conditions) of the CCS stream.

New experimental data were reported for water content in the CO_2 liquid region in equilibrium with hydrates at 15 MPa and temperature range from −40 to + 15 °C. The data were compared with data predicted with the CPA EOS model, which were in good agreement, with AAD (Absolute Average Deviation) of around 5%.

The new water content data using TDLAS (Tunable Diode Laser Adsorption Spectroscopy) were in good agreement with the data generated by Chapoy *et al.*, [34] using a chilled mirror, hence validating the analytical methods used in the two studies.

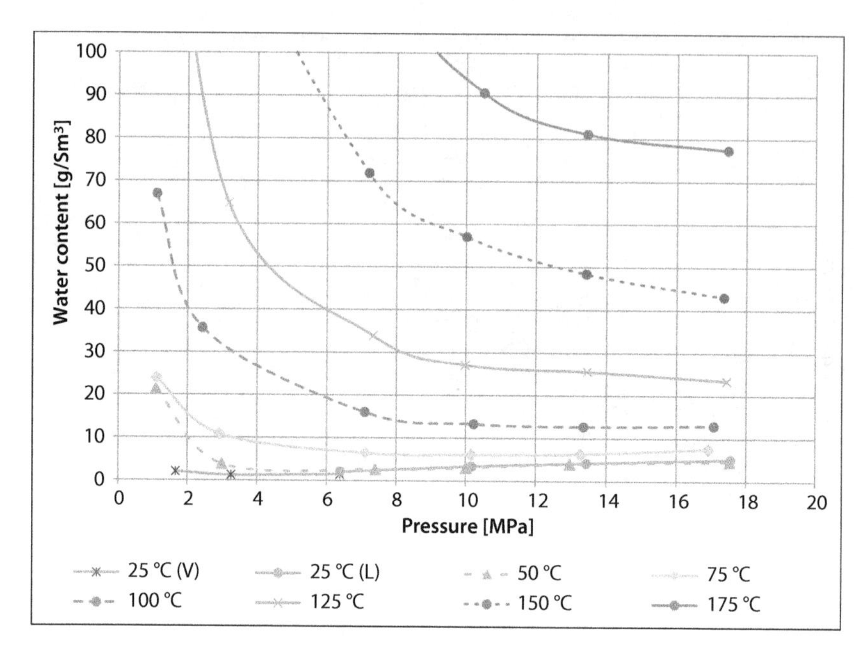

Figure 8.34 Water Content of Carbon Dioxide – Hou *et al.*, [5].

The water contents for pure CO_2 in equilibrium with hydrates at 15 MPa are shown in Figure 8.35.

The following equation was used by the authors of this paper for the correlation of the experimental data:

$$y = e^{\left(-55.7387 - \frac{154.492}{(T+273)} + 10.0451 \cdot ln(T+273)\right)} \quad \text{(mean relative error = 2.39\%)} \tag{8.22}$$

Cox *et al.*, [38] presented a paper titled "Water content of liquid CO_2 in equilibrium with liquid water or hydrate", in which they reviewed some previous work (Song and Kobayashi, [58]) and compared it with repeated experiments. The results of the experiments were presented in graphical form and no digital water content data were provided in the presentation.

Wang *et al.*, [65] used near-infrared spectroscopy to investigate the dissolution and chemical interaction of water dissolved into supercritical carbon dioxide and the influence of $CaCl_2$ in the co-existing aqueous phase at 40, 50, 75 and 100 °C and at 9.1 MPa.

At the temperatures and pressures at depths relevant to geological carbon sequestration (GCS), the injected CO_2 exists in the supercritical state. The interaction between supercritical CO_2 and the host rock, caprock and saline aquifer are critical to predicting the long-term stability and end state of the injected CO_2.

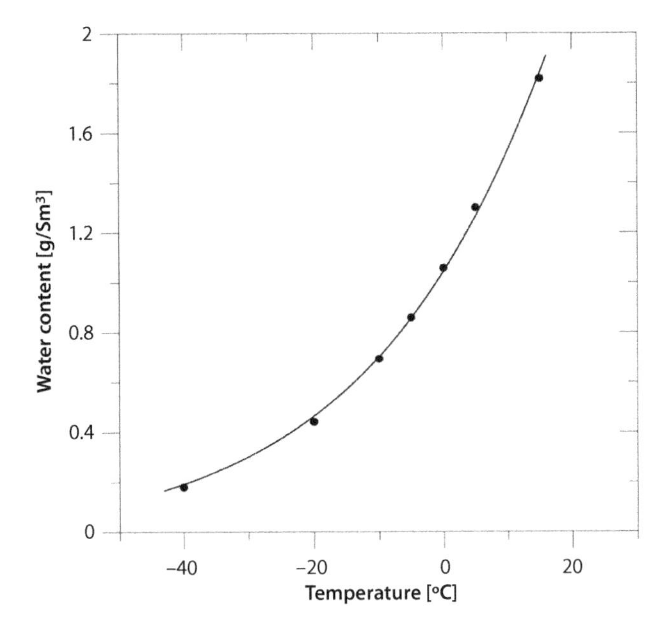

Figure 8.35 Water Content of Carbon Dioxide in Equilibrium with Hydrates – Chapoy *et al.*, [35].

Under the conditions of most GCS operations, two liquid phases exist, an aqueous phase saturated with dissolved CO_2 and a supercritical CO_2 phase saturated with dissolved water. So far, there have been extensive studies of the solubility and reactivity of CO_2 dissolved in aqueous solutions. However, little has been reported on the solubility of water in supercritical carbon dioxide. There is evidence suggesting that supercritical CO_2 saturated with water is highly reactive with some natural rocks and minerals such as basalt and forsterite.

The water solubility values obtained by Wang *et al.*, (see Figure 8.36) are close to available data from either experimental measurements or model simulations under similar temperature and pressure conditions, although there is a spread of measured water solubility values around the values obtained by Wang *et al.*, However, a comprehensive set of experimental water solubility data currently does not exist, particularly for the conditions of the work done by Wang *et al.*, and/or conditions relevant to GCS.

The spread of the measured water solubility values mentioned above may be due to the use of different measurement methods. Most previous methods were based on condensation of the dissolved water after reaching thermal equilibrium and quantifying the supercritical liquid out-flow. The present method of Wang *et al.*, is based on direct measurement of the optical

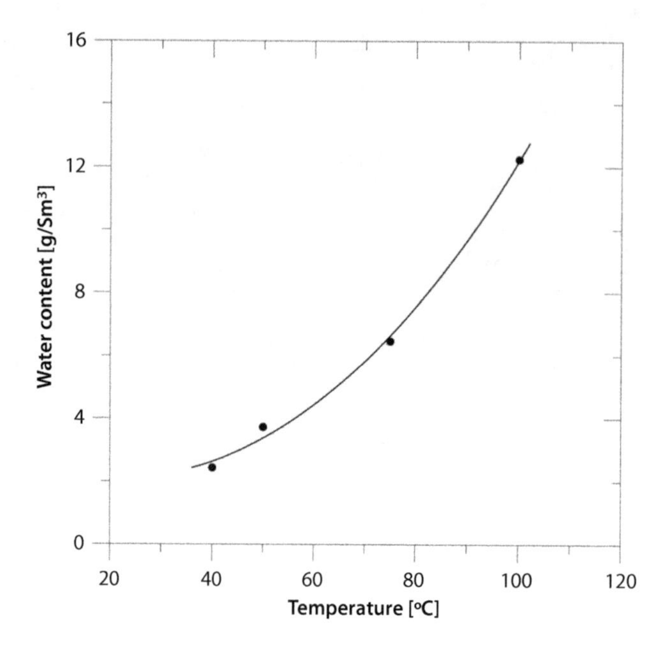

Figure 8.36 Water Content of Supercritical Carbon Dioxide – Wang *et al.*, (2013).

absorption in the NIR (near infrared) region. The present method exerts the least disturbance to the dissolution equilibrium and offers a more reliable approach to the measurement of the water concentration in supercritical CO_2.

The authors concluded that the dissolution of water, and subsequent dissolution of mineral phases (e.g., metal silicates) in supercritical CO_2, has profound consequences on mineral transformation processes involved in geological carbon sequestration. Such dissolution and precipitation may thus alter the pore structures, mineral volumes, and mineral compositions of both the caprock and bedrock, and thus affect the stability and integrity of the repository.

The following equation was used for the correlation of the experimental data:

$$y = e^{\left(-13.1168 + \frac{92.6802}{T} + 3.19023 \cdot ln(T)\right)} \text{(mean relative error } = \text{ 5.16\%)} \quad (8.23)$$

Burgass *et al.*, [12] presented experimental equipment, methods and results for the water content of carbon dioxide in equilibrium with hydrates at temperatures between −50 and −10 °C and pressures between 1.0 and 10.0 MPa, hence in both vapor and liquid phases.

The experimental values for the water content of pure CO_2 in equilibrium with hydrates are presented in Figure 8.37.

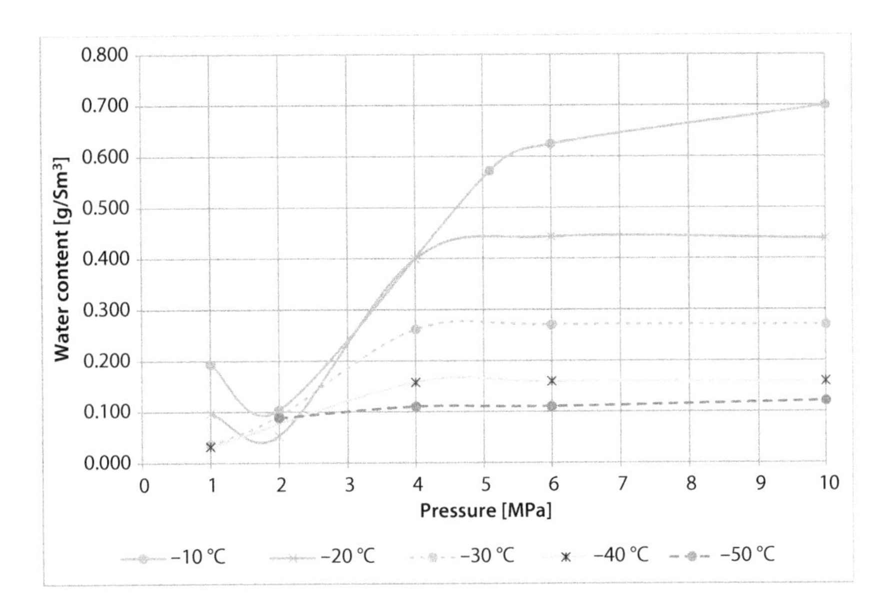

Figure 8.37 Water Content of CO_2 in Equilibrium with Hydrates – Burgass *et al.*, [12].

The data is of significant importance in relation to establishing the correct drying requirements for CO_2 being transported in pipelines, where hydrate formation is a serious flow assurance issue. The new data, being for pure CO_2, can act as a baseline for comparison with data generated for CO_2 with impurities coming from capture processes.

In this paper the experimental data have been compared with literature data and predictions made using a previously developed thermodynamic model. The good match between experimental data and predicted values further validates the model, which has previously been validated at conditions without hydrates.

Jasperson *et al.*, [66] determined experimentally the equilibrium water content of CO_2 at high pressure and low temperature. Their paper focused on pipelines with CO_2-rich fluids in cold environments (Alaska) and in CO_2 capture, and more specifically, on the equilibrium water concentration for hydrate formation in CO_2-rich mixtures.

Since there is a considerable uncertainty in the available literature data for the equilibrium water content of CO_2, a round-robin testing program was structured by Fluor/Worley Parsons Arctic Solutions to quantitatively establish the water solubility in CO_2-rich mixtures. For the CO_2 pipeline that is part of the ASAP project (The Alaska Stand Alone Pipeline) the worst case pipeline condition for hydrate formation had been defined as 17–28 MPa and –46 °C. If the water concentration of the CO_2 product from the gas purification plant exceeds the solubility limit, a dehydration unit must be added to process design.

Two laboratories were chosen to independently make the measurements, Wiltec Research Company and Korea University. The experimental procedures used by Wiltec and Korea University were quite different, The Wiltec one was analytical and the Korea University one was synthetic.

The Wiltec data suggest that the pressure dependence is very weak and the slope of the solubility vs. pressure curves decreases as the temperature is lowered. It is also evident from Korea University data that the variation of water solubility with temperature is far greater than the variation with pressure.

Jasperson *et al.*, compared the pressure dependence of the data of [21] to other data and concluded that the strong pressure variation of the Song and Kobayashi measurements agrees with the newer data from Rice University [38]; however, the new Wiltec data and the measurements from Heriot-Watt University [34], and [12] indicate that the water solubility varies weakly with pressure. The authors stated that the new data from Wiltec are clearly in broad agreement with the Heriot-Watt data, and this is further evidence that both sets of Rice University data have some kind of experimental bias.

Based on the round robin program data, the data from Seo *et al.*, [64], Chapoy *et al.*, [34] and Burgass *et al.*, [12], Jasperson *et al.*, concluded that the pressure dependence of the water content of CO_2-rich mixtures is weak and the water solubility at –46 °C and 28 MPa is 0.00017 mole fraction with an expanded combined uncertainty (three standard deviations) of 20%.

Figure 8.38 shows the Wiltec data for water content of CO_2 and Table 8.9 summarizes the data collected by the Korea University.

The following equations were used for the correlations of the experimental data:

$$\text{For } -5^\circ\text{C}: \; y = e^{\left(-0.765588+\frac{1.51346}{P}+0.205655\cdot ln(P)\right)} \text{(mean relative error = 3.76\%)} \tag{8.24}$$

$$\text{For } -15^\circ\text{C}: \; y = e^{\left(0.27076-\frac{3.74048}{P}-0.18807\cdot ln(P)\right)} \text{(mean relative error = 2.44\%)} \tag{8.25}$$

$$\text{For } -25^\circ\text{C}: \; y = e^{\left(-8.52258+\frac{35.5865}{P}+1.9041\cdot ln(P)\right)} \text{(mean relative error = } 1.49\cdot 10^{-5}\text{ \%)} \tag{8.26}$$

$$\text{For } -35^\circ\text{C}: \; y = e^{\left(-5.57728+\frac{15.1939}{P}+1.10075\cdot ln(P)\right)} \text{(mean relative error = 2.05\%)} \tag{8.27}$$

$$\text{For } -45^\circ\text{C}: \; y = e^{\left(-2.84361+\frac{3.2934}{P}+0.260576\cdot ln(P)\right)} \text{(mean relative error = 7.38\%)} \tag{8.28}$$

Foltran *et al.*, [50] investigated the solubility of water in carbon dioxide and nitrogen mixtures (at 40 °C and 8.3–12.4 MPa for H_2O in CO_2). The motivation for this work was to aid the understanding of water solubility in complex CO_2-based mixtures, which is required for the safety of anthropogenic CO_2 transport via pipeline for CCS technology.

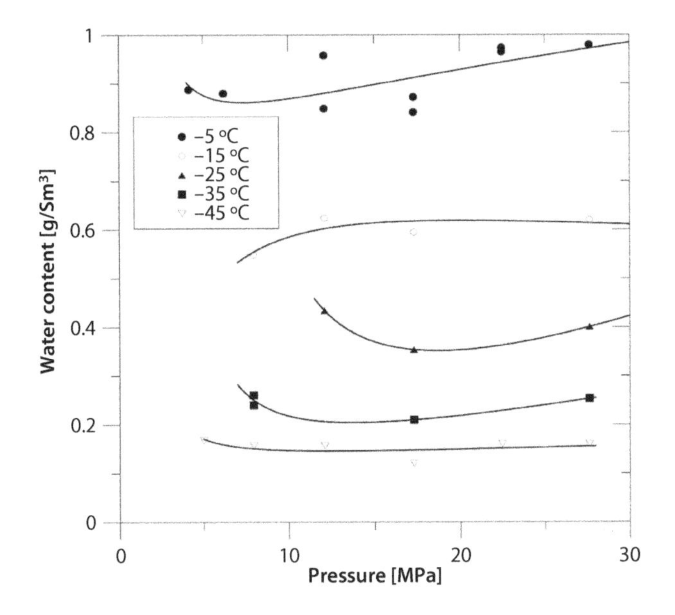

Figure 8.38 Water Content of Carbon Dioxide – Jasperson *et al.*, [66], Wiltec data.

The authors made an observation that there had been many studies on the solubility of water in pure CO_2, but there were very few reports on the solubility in more complex mixtures and none of these studies addressed the solubility of water in anthropogenic CO_2.

All of their measurements have been carried out in situ using custom-built high pressure equipment coupled to an FTIR spectrometer; infrared spectroscopy is highly sensitive to water, making the technology highly suitable to determine trace amounts of water in compressed gases. They validated their method by measuring the solubility of water in pure CO_2, where the results can be benchmarked against existing literature. Figure 8.39 shows the results of the validation experiments.

Foltran *et al.*, found that water content at 40 °C increased with the pressure; their measurements were compared to the data reported by King *et al.*, [47] and found to be in good agreement with errors 3% or less.

The authors of this paper used the following equation to correlate the experimental data:

$$y = e^{\left(-0.670424 + \frac{2.11882}{P} + 0.714444 \cdot ln(P)\right)} \text{(mean relative error } = 1.70\%) \qquad (8.29)$$

Meyer and Harvey [67] constructed a flow apparatus to measure the water content at saturation in a compressed gas. A saturator humidifies the flowing gas by equilibrating it with liquid water. Then, a gravimetric hygrometer measures the water mole fraction of the humid gas.

Table 8.9 Water content of carbon dioxide – Jasperson *et al.*, [66], Korea University Data.

T	P	Water
°C	MPa	g/Sm³
5.48	6.12	1.525
4.68	10.13	1.525
2.02	17.05	1.364
1.54	19.06	1.364
0.74	17.05	1.287
0.32	19.06	1.287
−0.31	21.06	1.287
−0.57	23.07	1.287
−0.42	17.05	1.179
−0.92	19.06	1.179
−1.81	21.06	1.179
−2.34	23.07	1.179
−4.88	17.05	1.049
−5.44	19.06	1.049
−5.7	21.06	1.049
−5.91	23.07	1.049
−5.24	17.05	1.020
−5.78	19.06	1.020
−6.07	21.06	1.020
−6.28	23.07	1.020
−6.06	19.06	0.992
−6.51	21.06	0.992
−6.65	23.07	0.992

(Continued)

Table 8.9 Cont.

T	P	Water
-8	19.06	0.923
−10.21	19.06	0.807
−15.51	17.05	0.640
−15.75	19.06	0.640
−15.91	21.06	0.640
−16.02	23.07	0.640

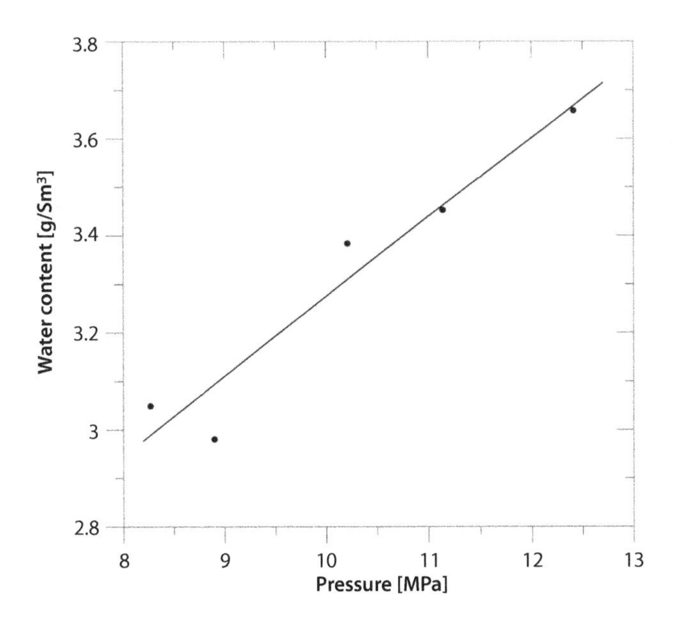

Figure 8.39 Water Content of Carbon Dioxide at 40 °C – Foltran *et al.*, [50].

Meyer and Harvey reported dew point data for H_2O in CO_2 on six isotherms between 10 and 80 °C at pressures from 0.5 to 5.0 MPa. The relation between the dew-point temperature, water content, and pressure is important for determining the degree to which CO_2 must be dried before transportation in pipelines for carbon capture and sequestration.

The authors acknowledged that while this work had greatly improved knowledge of the dew points of the mixtures up to 5 MPa, that upper limit is still below pressures that might be encountered in sequestration

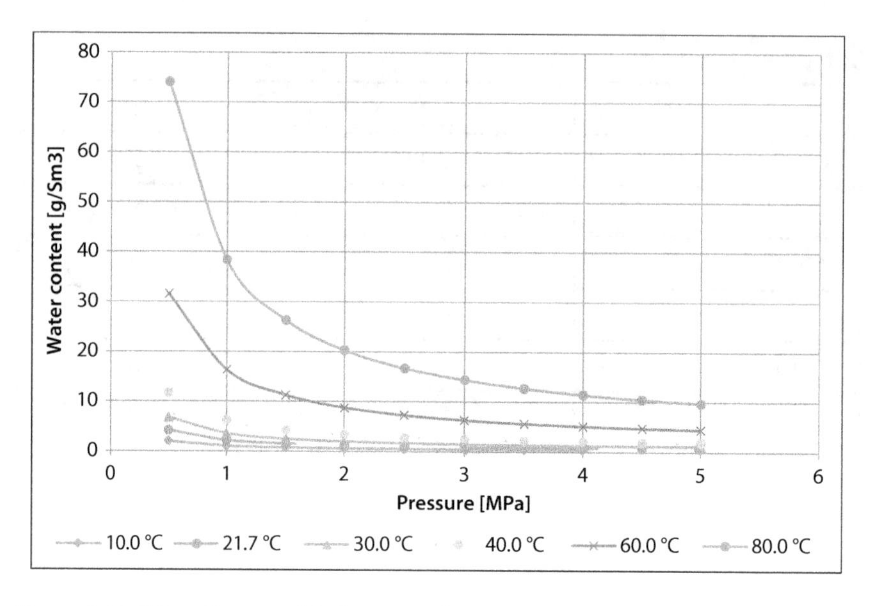

Figure 8.40 Water Content of Carbon Dioxide – Meyer and Harvey (2015).

operations. Such operations might be as high as 20 MPa in the pipeline and 50 MPa in compression for geologic sequestration.

The six isotherms obtained experimentally by Meyer and Harvey are shown in Figure 8.40.

Caumon *et al.*, [36] built a new experimental device to measure mutual solubility in the H_2O-CO_2 system up to 20 MPa and at 100 °C without sampling, by coupling a batch reactor with Raman spectrometer and immersion probes.

As the authors stated, the geological storage of CO_2 would be efficient if it is in a critical state, that is at temperature above 31.1 °C and pressure above 7.38 MPa. Considering a hydrostatic pressure, these conditions are reached below 800 m. At this depth, assuming thermal gradient in the crust of 25 °C/km, the temperature is 35 °C. These pressure and temperature conditions have to be considered as minima for an efficient CO_2 storage, so most of the storage reservoirs should be deeper, implying higher temperatures, 35–150 °C, and pressures, 8–25 MPa. This was the reason for selecting the experimental conditions of up to 20 MPa and 100 °C.

Figure 8.41 summarizes the experimental results of Caumon *et al.*, [36].

Comak *et al.*, [51] developed a new synthetic method for measuring water solubility in high pressure CO_2 using attenuated total reflection Fourier transform infrared (ATR-FTIR) spectroscopy. They determined dew points of water at three different pressures, 4.05, 5.05 and 6.03 MPa, with mole

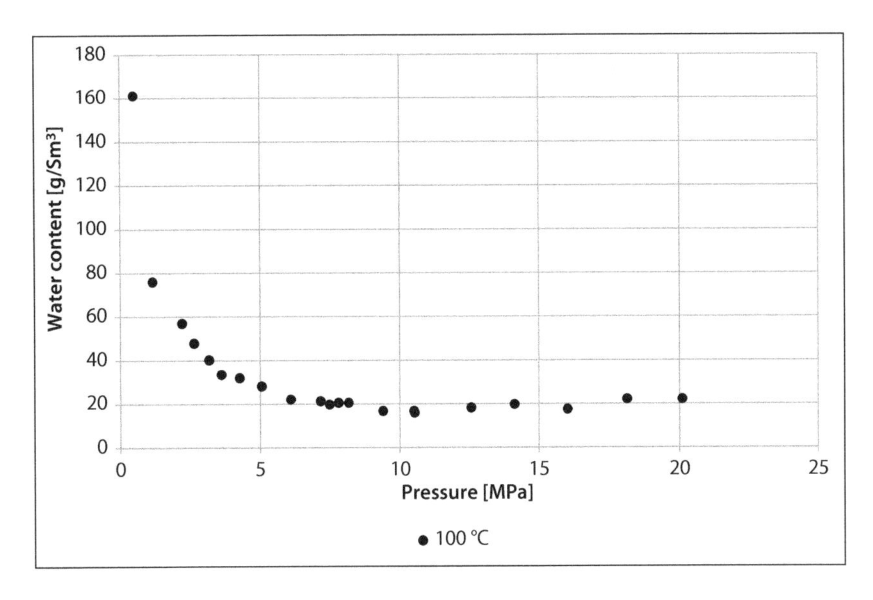

Figure 8.41 Water Content of Carbon Dioxide - Caumon *et al.*, [36].

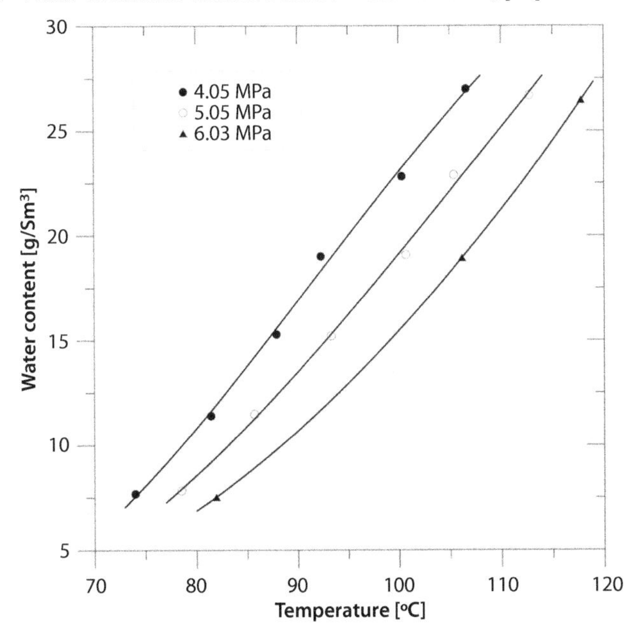

Figure 8.42 Dew Points of Water in Carbon Dioxide - Comak *et al.*, [51].

fractions of water between 0.01 and 0.04. The authors stated that the data obtained filled the gap in the literature in these regions of pressures and temperatures and could be of high importance in the context of CCS.

Table 8.10 Comparison of water content of CO_2 between Comak *et al.,* and other literature sources.

Source	P [MPa]	T [°C]	H_2O [g/Sm^3]	A%D
Comak *et al.,* [51]	5.05	100.7	19.087	0.6
Coan and King [18]	5.15	100	18.859	
Coan and King [18]	5.15	100	19.087	
Comak *et al.,* [51]	5.05	93.3	15.209	1.5
Gillespie and Wilson [20]	5.07	93.3	14.981	

The dew points of water in CO_2 collected at various pressures by Comak *et al.,* are shown in Figure 8.42. The authors compared their data with the data available in the literature at the same conditions of temperature and pressure; the comparison is presented in Table 8.10.

The following equations were used for the correlation of experimental data:

$$\text{For 4.05 MPa}: \ y = e^{\left(28.518 - \frac{666.04}{T} - 4.0649 \cdot ln(T)\right)} \text{(mean relative error} = 1.90\%) \tag{8.30}$$

$$\text{For 5.05 MPa}: \ y = e^{\left(13.2192 - \frac{435.95}{T} - 1.283 \cdot ln(T)\right)} \text{(mean relative error} = 1.43\%) \tag{8.31}$$

$$\text{For 6.03 MPa}: \ y = e^{\left(-3.0472 - \frac{177.47}{T} + 1.64175 \cdot ln(T)\right)} \text{(mean relative error} = 0.001\%) \tag{8.32}$$

Loring *et al.,* [68] used in situ high pressure infrared spectroscopic titrations to measure the solubility of water in six CO_2-CH_4 mixtures, ranging from pure CO_2 to pure CH_4 at shallow shale reservoir conditions of 50 °C and 9 MPa. The concentration of water dissolved in CO_2-CH_4 supercritical fluids is an important parameter that can control shale permeability, affect methane transmissivity and impact the enhanced gas recovery (EGR) operations.

The benefit of using CO_2 for EGR in depleted shale gas reservoirs is that at the same time CO_2 can be stored underground. Carbon dioxide has a higher affinity for the organic and inorganic components of shales and will therefore displace adsorbed methane.

Table 8.11 shows the only experimental data obtained by Loring *et al.,* for water content in pure carbon dioxide. The authors compared that result with data in literature at similar conditions:

8.4 Experimental Methods

The methodologies used to obtain phase equilibrium data are based on the classification systems proposed by Dohrn *et al.,* [52]. Depending on how

Table 8.11 Water content of pure CO_2 – Loring *et al.*,

Source	T	P	Water
	°C	MPa	g/Sm³
Loring *et al.*, [68]	50	9.000	2.844
Briones *et al.*, [42]	50	8.72	2.768
Bamberger *et al.*, [27]	50	9.09	3.118

the composition of the two coexisting phases is determined, the methods are divided into analytical and synthetic.

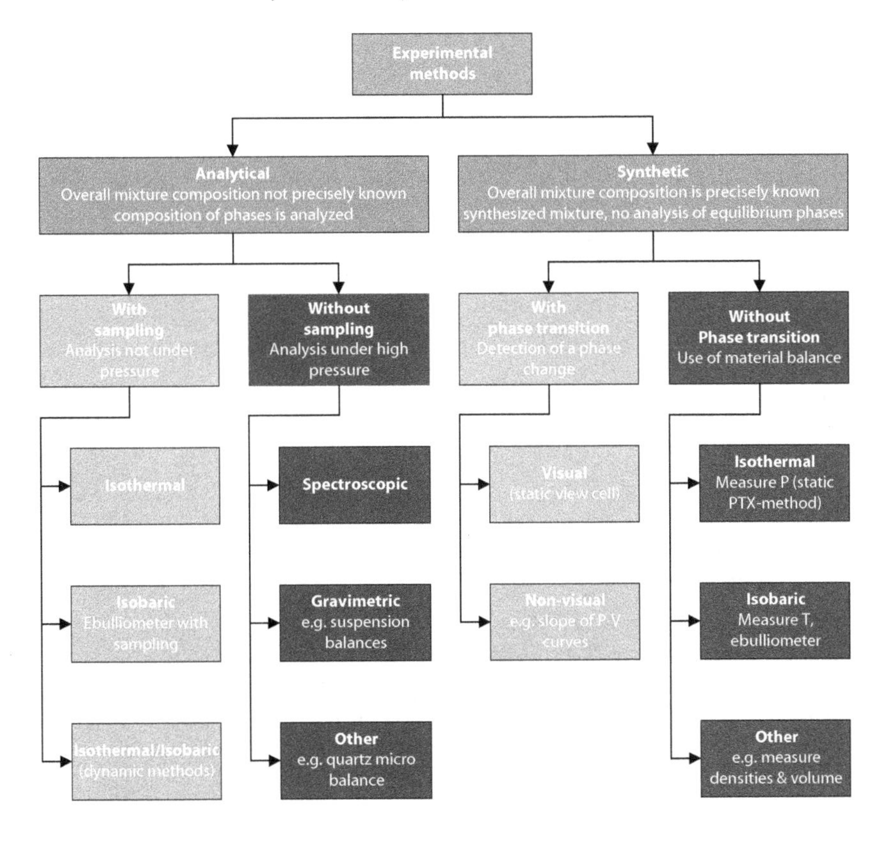

Mohammadi *et al.*, [63] summarized experimental methods used for measuring water content and water dew point of gaseous systems. Those can be divided into two categories:

Table 8.12 Measurement principles of direct methods:.

Method	Measuring principle
Dew point mirror (chilled mirror)	Temperature at which water or ice appears on a cooled surface
Karl Fischer titration	Titration of absorbed water vapor with iodine
Gravimetric hygrometer	Increase in weight by absorption of water

a) Direct (absolute) methods: there is a direct relationship between the measured quantity and water content. Direct methods include:

- Dew Point Mirror
- Karl Fischer Titration
- Gravimetric Hygrometer

b) Indirect methods: physical properties are measured, which are a function of water content, which are then used to calculate water content or water dew point. Indirect methods include:

- Spectroscopic
- Chromatographic
- Hygroscopic
- Electrolytic
- Capacitance
- Change in Mass
- Conductivity

They tabulated measurement principles of the two categories:

In addition, experimental methods can be divided into dynamic and static, depending whether or not the sample is moving through the measuring device.

Most experiments described in literature on the solubility of water in carbon dioxide are based on analytical methods, where water is absorbed from the CO_2-rich phase using a compound such as $Mg(ClO_4)_2$ (magnesium perchlorate) or is condensed from the saturated CO_2 mixture. Few of these experiments are based on synthetic methods, where, for example, Burnett-isochoric apparatus was used for mixtures of different

Table 8.13 Measurement principles of indirect methods.

Method	Measuring principle
Spectroscopic	Water content is measured by detecting the energy absorption due to the presence of water vapor
Gas chromatography	Size of the peak is proportional to the amount of analyzed sample
Hygroscopic	Sensor responds to water vapor pressure in the gas phase
Electrolytic	The current due to electrolysis of the absorbed water at known constant gas flowrate is directly proportional to water vapor concentration
Capacitance	Dielectric constant of Al_2O_3 is a function of water vapor concentration
Change in mass (quartz crystal)	Hygroscopic coating adsorbs water; crystal resonant frequency is a function of mass and thus related to water vapor concentration
Conductivity	Salt/glycerol solution absorbs water vapor; conductivity is a function of water vapor concentration

compositions, or a view cell was used for the detection of H_2O solubility in CO_2, or Near-IR spectroscopy was used to measure the water saturation levels of CO_2. A new method, using Fourier Transform Infrared Spectroscopy (FTIR), measures the solubility of H_2O in mixtures at relatively low temperatures and high pressures. A new FTIR method, using Attenuated Total Reflection (ATR), can be used at much higher temperatures. This method could be used to measure water solubility in CO_2-rich mixtures for all conditions likely to be found in CCS processes [51].

In liquefied gases, at low temperatures and ambient pressures, water is normally quantified using Karl-Fischer titration, gravimetric methods (phosphorus pentoxide, P2O5), or electric conductometric methods. At higher temperatures and pressures these methods become difficult to perform and more elaborate systems were proposed like direct fluid injection mass spectrometry (DFI MS), Raman and internal reflectance IR spectroscopy [25].

A new method was developed to measure the dew point, expressed as vapor concentration enhancement factor, by FTIR spectroscopy. The water enhancement factor, g_w, is the ratio of the water concentration c_w (in moles or mass per unit volume) in the saturated compressed humid gas to the saturated vapor density c_w^0 of pure water.

One important condition of obtaining reliable data for water content of carbon dioxide is to reach equilibrium in the water-carbon dioxide system. However, there is a disagreement between researchers regarding the time required for CO_2 and H_2O to attain equilibrium. The time reported in various papers ranges from 1 hr to 24 hr. At higher pressures and lower temperatures even longer period was required for equilibrium, wherethe system was heterogeneous, due to increasing viscosity and decreasing diffusion speed. Takenouchi and Kennedy [16] noted that 3 days wererequired for equilibrium at a temperature of 200 °C and substantially more than 1 week at 110 °C. Mohammadi *et al.*, [63] observed that experimental data on water content for hydrocarbons and non-hydrocarbon gases (including carbon dioxide) at low temperature conditions are scarce. It seems, they concluded, that achieving equilibrium at low temperature conditions, especially near and inside hydrate forming conditions, is a very slow process and requires a long time. Another problem is the determination of water traces in gases – accurate measurements require very specialized techniques.

Table 8.14 summarizes the experimental methods used by the authors of the reviewed papers to obtain the information on water content of carbon dioxide.

8.5 Conclusions

Most of the experimental methods were the analytical ones. Ten experimental methods were synthetic.

The water solubility results were presented as isotherms (31 graphs), isobars (10 graphs), isopleths (same water content – three graphs); 10 graphs were of a mixed nature (one-point data etc.)

The lowest solubility of water in carbon dioxide was determined to be 0.032 g/Sm^3 (at 1 MPa and −40 °C).

The highest solubility of water in carbon dioxide was determined to be 718.6 g/Sm^3 (at 20 MPa and 350 °C).

The authors used thermodynamic models or other authors' results to validate their experimental data.

Table 8.14 Experimental methods used for studying water content of carbon dioxide.

Source	Experimental method	Water content determination
Pollitzer and Strebel [1]	?	?
Wiebe and Gaddy [2]	Analytical, static method. Description in Wiebe and Gaddy [2].	Gas analysis apparatus, see Shepherd [53]. Weight gained by a train of adsorbents.
Stone [57]	Analytical. Mixing H_2O and CO_2 for 8 hr to achieve equilibrium.	Adsorption of H_2O on magnesium perchlorate, adsorption of CO_2 on soda lime.
Sidorov *et al.*, [17]	?	?
Malinin [40]	Analytical. Static method of saturation. Shaking stove.	Weight method. Anhydride used to catch water, $CaCl_2$ for drying of CO_2, 40% KOH for collecting of CO_2
Tödheide and Franck [3]	Analytical - equilibrium cell	Adsorbent weight increase.
Takenouchi and Kennedy [16]	Analytical, static. Autoclave in electric furnace	Reaction of CO_2 sample with 1 N NaOH. Na_2CO_3 determined by titration with HCl using phenolphthalein and bromcresol green as indicators.
Coan and King [18]	Analytical, flow method. Technique described in J. Chromatogr., 44, 429 (1969)	Weighing trapped water, measuring the volume of gas.

(Continued)

Table 8.14 Cont.

Source	Experimental method	Water content determination
Kobayashi *et al.*, (1979)	? (Internal ARCO report)	? (Internal ARCO report)
Zawisza and Malesińska [6]	Synthetic, static. Details of the apparatus in J. Chem. Thermodyn. 1977, 9, 153	Measure pressure at first appearance of fog (dew point).
Chrastil [19]	Analytical. High pressure tubing for dissolution of compounds.	Bubbling sample into absolute methyl alcohol and by Fischer's method (titration).
Gillespie and Wilson [20]	Analytical (equilibrium cell)	Passing sample through drying tubes with $Mg(ClO_4)_2$ and weighing the tubes before and after sampling. Passing sample from drying tubes into the wet test meter to determine the volume of the gas sampled.
Gillespie *et al.*, [54]	?	? (AIChE paper 34-b unavailable)
Song *et al.*, [37]	Analytical. Rotating the cylindrical autoclave containing ball bearings.	Not provided (references suggest chromatography).
Song and Kobayashi [21]	Analytical. Rotating the cylindrical autoclave containing ball bearings.	Not provided (references suggest chromatography).
Song and Kobayashi [58]	Analytical. Rotating the cylindrical autoclave containing ball bearings.	Chromatographic analytic technique devised by Bloch and Lifland [55].

(*Continued*)

Table 8.14 Cont.

Source	Experimental method	Water content determination
Briones *et al.*, [42]	Analytical. Flow apparatus with a view cell which functions as a phase separator.	A number of sample traps in series (stainless steel mesh immersed in dry ice-acetone bath). Equilibrium phase compositions calculated from the mass of condensate and/or liquid collected and the measured volume of gas.
Nakayama *et al.*, [22]	Analytical. A new static apparatus, designed and constructed for the measurement of liquid-liquid equilibria up to 177 °C and 20 MPa.	Two types of sampling devices were used, a sliding rod and a capillary tube sampler. Samples were analyzed in a gas chromatograph.
Patel *et al.*, [43]	Synthetic. Burnett isochoric apparatus.	Dew points determined by noting the change in slope of the isochores as they traversed from the single vapor phase to the two-phase region.
Müller *et al.*, [44]	Analytical - equilibrium cell.	Gas chromatograph
d'Souza *et al.*, [45]	Analytical - equilibrium cell with circulating phases.	Difference between final and initial weights of the collector.
Ohgaki *et al.*, [23]	Analytical, flow method.	CO_2 was metered, H_2O was absorbed in magnesium perchlorate and weighed.

(Continued)

Table 8.14 Cont.

Source	**Experimental method**	**Water content determination**
Sako *et al.*, [46]	Analytical – static type apparatus.	Gas chromatograph equipped with a thermal conductive detector.
Sterner and Bodnar [24]	Synthetic fluid inclusion technique (capturing synthetic fluid inclusions in quartz).	CO_2 liquid-vapor homogenization temperatures and total homogenization temperatures were measured using a microscope equipped with a heating / freezing stage.
Mather and Franck [59]	Synthetic - autoclave with variable internal volume.	The transition from two-phase to one (heating) and one to two-phase (cooling) was observed visually.
King *et al.*, [47]	Analytical. High pressure vapor/liquid recirculation apparatus with mechanical stirring.	Samples of the light phase are obtained by recirculating through the light-phase sample bomb and directly back into the light phase in the equilibrium vessel without passing it through the heavy phase. The water in the light phase sample is collected for weighing by first allowing the contents of the sample bomb to expand through receivers cooled to −80 °C and then passing a slow stream of nitrogen through the bomb and a fresh set of receivers (again cooled to −80 °C).

(Continued)

Table 8.14 Cont.

Source	**Experimental method**	**Water content determination**
Dohrn *et al.*, [48]	Analytical. The apparatus is of the vapor phase recirculation type. Its main part, autoclave, serves as a mixing and separating vessel.	Water: Karl-Fischer titration CO_2: PVT measurements
Jackson *et al.*, [25]	Analytical. FT-IR apparatus.	High-pressure near-IR technique to measure water solubility.
Fenghour *et al.*, [26]	Synthetic. Isochoric apparatus that measures the pressure and temperature of the two-phase and single-gas phase region of the mixture.	Mixing known volumes of CO_2 and mass of water.
Frost and Wood [60]	A synthetic fluid inclusion technique – capturing H_2O-CO_2 fluids of known compositions in pre-cracked corundum.	CO_2 homogenization temperatures were determined using a Linkam TH600 heating/cooling stage. The CO_2 liquid and vapor phases generally homogenized to the liquid where an energetic vapor bubble was easy to observe.
Bamberger *et al.*, [27]	Analytical. Apparatus based on the flow technique.	H_2O separated from CO_2 by freezing and weighing. Vapor flow measured by wet-test meter.
Sabirzyanov *et al.*, [28]	Analytical. Experimental circulation facility with the extraction cell and separator.	Weighing method.

(Continued)

Table 8.14 Cont.

Source	Experimental method	Water content determination
Valtz *et al.*, [29]	Static-analytic apparatus with two pneumatic capillary samplers.	Gas chromatograph.
Jarne *et al.*, [30]	Analytical.	Karl Fischer titration for water concentration; chilled mirror for water dew point.
Iwai *et al.*, [31]	Analytical. Static recirculation method.	On-line Fourier transform infrared (FTIR) spectroscopy.
Koglbauer and Wendland [8]	Analytical. View cell in the sample compartment of an FTIR spectrometer.	Determining water dew point, expressed as vapor concentration enhancement factor, by measuring absorbance of IR light by water vapor.
Chapoy *et al.*, [32]	Static-analytic apparatus with fluid phase sampling.	Analysis by gas chromatography.
Youssef *et al.*, (2009)	Analytical - equilibrium cell.	Karl Fischer coulometer.
Tabasinejad *et al.*, [33]	Analytical - an oven consisting of a gas mixing vessel, a demister and a pycnometer.	Gas chromatograph.

(Continued)

Table 8.14 Cont.

Source	Experimental method	Water content determination
Seo *et al.*, [64]	Synthetic, indirect method – equilibrium cell with pre-determined amounts of CO_2 and water.	Visual determination of temperatures at which two-phase L_{CO2}-H mixtures become a single liquid phase at constant pressure.
Chapoy *et al.*, [34]	Analytical - equilibrium cell.	Chilled mirror hygrometer.
Kim *et al.*, [49]	Synthetic, indirect method – equilibrium cell with pre-determined amounts of CO_2 and water.	Visual determination of temperatures at which two-phase L_{CO2}-H mixtures become a single liquid phase at constant pressure.
Hou *et al.*, [5]	Analytical - a circulation-type quasi static analytical apparatus.	Gas chromatograph.
Chapoy *et al.*, [35]	Analytical - equilibrium cell.	TDLAS (Tunable Diode Laser Adsorption Spectroscopy).
Cox *et al.*, [38]	Analytical – saturator.	Moisture analyzer (details not provided).
Wang *et al.*, [65]	Analytical - reaction vessel with optical windows.	NIR (near infrared) spectra recorded with a spectrometer.
Burgass *et al.*, [12]	Analytical - equilibrium cell.	Chilled mirror hygrometer.

(Continued)

Table 8.14 Cont.

Source	**Experimental method**	**Water content determination**
Jasperson *et al.*, [66] (Wiltec Research Company)	Analytical. Water saturated packing material in SS tubing immersed in constant temperature bath (liquid propane).	Water analyzer based on the principle of electrolytic dissociation of water (contained in H_2SO_4) into O_2 and H_2. O_2 converts H_2SO_4 into P_2O_5 which strongly absorbs water from the carrier gas. The electrolytic current is proportional to the water content of the gas.
Jasperson *et al.*, [66] (Korea University)	Synthetic – apparatus similar to the one used by Seo et al. [64].	Temperature at which formed hydrates disappear was visual determined by raising the temperature slowly and keeping the pressure constant.
Foltran *et al.*, [50]	Analytical - infrared equilibrium cell.	FTIR spectrometer.
Meyer and Harvey [67]	Analytical, direct – CO_2 saturation system.	High precision gravimetric hygrometer.
Caumon *et al.*, [36]	Analytical-static: coupling a batch reactor with Raman immersion probes.	Raman spectroscopy.
Comak *et al.*, [51]	Synthetic-dynamic method, using Attenuated Total Reflection (ATR) spectroscopy (ATR-FTIR).	FTIR spectrometer (dew point identified as significant increase in absorbance).

(Continued)

Table 8.14 Cont.

Source	Experimental method	Water content determination
Loring *et al.*, [68]	Analytical-static.	In situ high-pressure titrations with infrared detection (automated fluid-delivery apparatus coupled to a high pressure IR cell with both transmission and ATR IR optics.

Water is more soluble in liquid CO_2 than in vapor CO_2. Water is less soluble in CO_2 than CO_2 in water.

For a given pressure value, the dew point temperature increases when the water amount in CO_2 also increases.

The variation of water solubility in CO_2 with temperature is far greater than the variation with pressure.

There is a minimum in the water content of carbon dioxide.

The H_2O-CO_2 system exists as two liquid phases at T below the critical point of CO_2 at pressures above its vapor pressure.

For a given pressure, the dew point temperature increases when the water amount also increases.

Time required for the H_2O-CO_2 system to attain equilibrium according to different researchers is between a few hours to at least one week.

The system CO_2 – H_2O exhibits the so called density reversal of the equilibrium phases, when the vapour phase becomes denser than the liquid phase with isothermal pressure increase. It takes place at 50 °C and around 80 MPa, at 100 °C and around 125 MPa, and at 250 °C and around 200 MPa.

A small amount of water added to supercritical CO_2 greatly increases the solubility of polar species in CO_2 which can greatly increase the use of supercritical CO_2 in separations and reactions.

References

1. Strebel, Pund., Über den einfluss indifferenter gase auf die sättigungs - Dampfkonzentration von flüssigkeiten. *Zeitschrift für Physikalische Chemie*, 768, 1924.
2. Wiebe, R., Gaddy, V.L., Vaporphase composition of carbon dioxide-water mixtures at varioustemperatures and at pressures to 700 atmospheres. *J. Am. Chem. Soc.*, 63(2), 475–477, 1941.
3. Tödheide, K., Franck, E.U., Das Zweiphasengebiet und die kritische Kurve im System Kohlendioxid–Wasser bis zu Drucken von 3500 bar. *Z. Phys. Chemie Neue. Folge.*, 37(5_6), 387–401, 1963.
4. Scott, R.L., van Konynenburg, P.H., Static properties of solutions. Van der Waals and related models for hydrocarbon mixtures. *Discuss. Faraday Soc.*, 49, 87–97, 1970.
5. Hou, S.-X., Maitland, G.C., Trusler, J.P.M., Martin Trusler, J.P., Measurement and modeling of the phase behavior of the (carbon dioxide+water) mixture at temperatures from 298.15 K to 448.15 K. *J. Supercrit. Fluids*, 73, 87–96, 2013.

6. Zawisza, A., Malesińska, B., Solubility of carbon dioxide in liquid water and of water in gaseous carbon dioxide in the range 0.2-5 MPa and at temperatures up to 473 K. *J. Chem. Eng. Data*, 26(4), 388–391, 1981.
7. Heck, C.K., Hiza, M.J., Liquid-vapor equilibrium in the system helium-methane. *AIChE J.*, 13(3), 593–599, 1967.
8. Koglbauer, G., Wendland, M., Water vapor concentration enhancement in compressed humid nitrogen, argon, and carbondioxide measured by fourier transform infrared spectroscopy
9. Spycher, N., Pruess, K., Ennis-King, J., CO_2-H_{2O} mixtures in the geological sequestration of CO_2. I. Assessment and calculation of mutual solubilities from 12 to 100 °C and up to 600 bar. *Geochim. Cosmochim. Acta*, 67(16), 3015–3031, 2003.
10. Chapoy, A., Mohammadi, A.H., Chareton, A., Tohidi, B., Richon, D., Measurement and modeling of gas solubility and literature review of the properties for the carbon dioxide–water system. *Ind. Eng. Chem. Res.*, 43(7), 1794–1802, 2004.
11. Hu, J., Duan, Z., Zhu, C., Chou, I.-M., PVTx properties of the CO_2–H_2O and CO_2–H_2O–NaCl systems below 647 K: Assessment of experimental data and thermodynamic models. *Chem. Geol.*, 238(3-4), 249–267, 2007.
12. Burgass, R., Chapoy, A., Duchet-Suchaux, P., Tohidi, B., Experimental water content measurements of carbon dioxide in equilibrium with hydrates at (223.15 to 263.15)K and (1.0 to 10.0)MPa. *J. Chem. Thermodyn.*, 69, 1–5, 2014.
13. Aavatsmark, I., Kaufmann, R., A simple function for the solubility of water in dense-phase carbon dioxide. *International Journal of Greenhouse Gas Control*, 32, 47–55, 2015.
14. Aasen, A., Hammer, M., Skaugen, G., Jakobsen, J.P., Wilhelmsen, Øivind., Thermodynamic models to accurately describe the PVTxy-behavior of water / carbon dioxide mixtures. *Fluid Phase Equilib.*, 442, 125–139, 2017.
15. Rowland, D., Boxall, J.A., Hughes, T.J., Al Ghafri, S.Z.S., Jiao, F., Xiao, X., *et al.*, Reliable prediction of aqueous dew points in CO_2 pipelines and new approaches for control during shut-in. *International Journal of Greenhouse Gas Control*, 70, 97–104, 2018.
16. Takenouchi, S., Kennedy, G.C., The binary system H_2O-CO_2 at high temperatures and pressures. *Am. J. Sci.*, 262(9), 1055–1074, 1964.
17. Sidorov, I.P., Kazarnovsky, Y.S., Goldman, A.M., The solubility of water in compressed gases. *Tr. Gosudarst. Nauch-Issled. i Proekt. Inst. Azot. Prom*, 1, 48–67, 1953.
18. Coan, C.R., King, A.D, Jr., Solubility of water in compressed carbon dioxide, nitrous oxide, and ethane. Evidence for hydration of carbon dioxide and nitrous oxide in the gas phase. *J. Am. Chem. Soc.*, 93, 1857, 1971.
19. Chrastil, J., Solubility of solids and liquids in supercritical gases. *J. Phys. Chem.*, 86(15), 3016–3021, 1982.
20. Gillespie, P.C., Wilson, G.M., Association, G.P., Vapor-liquid and liquid-liquid equilibria: water-methane, water-carbon dioxide, water-hydrogen sulfide,

water-n pentane, water-methane-n pentane. *Tulsa: Gas Processors Association, Research report*, 48, 1982.

21. Song, K.Y., Kobayashi, R., Watercontent of co_2-richfluids in equilibrium with liquid water and/or hydrates. 99. Tulsa, OK, GPA, 1986.
22. Nakayama, T., Sagara, H., Arai, K., Saito, S., High-pressure liquid-liquid equilibria for the system of water, ethanol and 1,1-difluoroethane at 323.2 K. *Fluid Phase Equilib.*, 38(1-2), 109–127, 1987.
23. Ohgaki, K., Nishikawa, M., Furuichi, T., Katayama, T., Entrainer effect of H_2O and ethanol on alpha-tocopherol extraction by compressed Co_2. *Kagaku Kogaku Ronbunshu*, 14(3), 342–346, 1988.
24. Sterner, S.M., Bodnar, R.J., Synthetic fluid inclusions; X, Experimental determination of P-V-T-X properties in the CO_2 -H_2O system to 6 kb and 700 degrees C. *Am. J. Sci.*, 291(1), 1–54, 1991.
25. Jackson, K., Bowman, L.E., Fulton, J.L., Water solubility measurements in supercritical fluids and high-pressure liquids using near-infrared spectroscopy. *Anal. Chem.*, 67(14), 2368–2372, 1995.
26. Fenghour, A., Wakeham, W.A., Watson, J.T.R., Densities of (water + carbon dioxide) in the temperature range 304 K to 699 K and pressures up to 24 MPa. *J. Chem. Thermodyn.*, 28(4), 433–446, 1996.
27. Bamberger, A., Sieder, G., Maurer, G., High-pressure (vapor+liquid) equilibrium in binary mixtures of (carbon dioxide+water or acetic acid) at temperatures from 313 to 353 K. *J. Supercrit. Fluids*, 17(2), 97–110, 2000.
28. Sabirzyanov, A.N., Il'in, A.P., Akhunov, A.R., Gumerov, F.M., Solubility of water in critical carbon dioxide. *High Temperature*, 40(2), 203–206, 2002.
29. Valtz, A., Chapoy, A., Coquelet, C., Paricaud, P., Richon, D., Vapour–liquid equilibria in the carbon dioxide–water system, measurement and modelling from 278.2 to 318.2K. *Fluid Phase Equilib.*, 226, 333–344, 2004.
30. Jarne, C., Blanco, S.T., Artal, M., Rauzy, E., Otín, S., Velasco, I., Dew points of binary carbon dioxide + water and ternary carbon dioxide + water + methanol mixtures Measurement and modelling. *Fluid Phase Equilib.*, 216(1), 85–93, 2004.
31. Iwai, Y., Uno, M., Nagano, H., Arai, Y., Measurement of solubilities of palmitic acid in supercritical carbon dioxide and entrainer effect of water by FTIR spectroscopy. *J. Supercrit. Fluids*, 28(2-3), 193–200, 2004.
32. Chapoy, A., Burgass, R., Tohidi, B., Austell, J.M., Eickhoff, C., Effect of common impurities on the phase behaviour of carbon dioxide rich systems: minimizing the risk of hydrate formation and two-phase flow. *SPE Journal*, 16(4), 921–930, 2011.
33. Tabasinejad, F., Moore, R.G., Mehta, S.A., Van Fraassen, K.C., Barzin, Y., Rushing, J.A., *et al.*, Water solubility in supercritical methane, nitrogen, and carbon dioxide: Measurement and modeling from 422 to 483 k and pressures from 3.6 to 134 mpa. *Ind. Eng. Chem. Res.*, 50(7), 4029–4041, 2011.
34. Chapoy, A., Haghighi, H., Burgass, R., Tohidi, B., On the phase behaviour of the (carbon dioxide+water) systems at low temperatures: Experimental and modelling. *J. Chem. Thermodyn.*, 47, 6–12, 2012.

35. Chapoy, A., Nazeri, M., Kapateh, M., Burgass, R., Coquelet, C., Tohidi, B., Effect of impurities on thermophysical properties and phase behaviour of a CO_2-rich system in CCS. *Int. J. Greenh. Gas Control*, 19, 92–100, 2013.
36. Caumon, M.-C., Sterpenich, J., Randi, A., Pironon, J., Measuring mutual solubility in the H_2O–CO_2 system up to 200 bar and 100 °C by in situ Raman spectroscopy. *International Journal of Greenhouse Gas Control*, 47, 63–70, 2016.
37. Song, K.Y., Kobayashi, R., Water content of co_2-rich fluids in equilibrium with liquid water or hydrate. 80. Tulsa, OK, GPA, 1984.
38. Cox, K.R., Chapman, W.G., Song, K.S., Dominik, A., French, R., Water content of liquid CO_2 in equilibrium with liquid water or hydrate. *in: GPA Conference*, 2013.
39. Lowry, H.H., Erickson, W.R., The densities of co ëxisting liquid and gaseous carbon dioxide and the solubility of water in liquid carbon dioxide. *J. Am. Chem. Soc.*, 49(11), 2729–2734, 1927.
40. Malinin, S.D., The system water-carbon dioxide at high temperatures and pressures. *Geochemistry* 3, 292–306, 1959.
41. Li, H., Jakobsen, J.P., Stang, J., Hydrate formation during CO_2 transport: Predicting water content in the fluid phase in equilibrium with the CO_2-hydrate. *International Journal of Greenhouse Gas Control*, 5(3), 549–554, 2011.
42. Briones, J.A., Mullins, J.C., Thies, M.C., Kim, B.-U., Ternary phase equilibria for acetic acid –water mixtures with supercritical carbon dioxide. *Fluid Phase Equilib.*, 36, 235–246, 1987.
43. Patel, M.R., Holste, J.C., Hall, K.R., Eubank, P.T., Thermophysical properties of gaseous carbon dioxide – water mixtures. *Fluid Phase Equilib.*, 36, 279–299, 1987.
44. Müller, G., Bender, E., Maurer, G., *et al.*, Das dampf-flüssigkeits gleichgewicht des ternären systems ammoniak-kohlendioxid-wasser bei hohen wassergehalten im bereichzwischen 373 und 473 kelvin;. *Berichte der Bunsengesellschaft für physikalische Chemie*, 92(2), 148–160, 1988.
45. D'souza, R., Patrick, J.R., Teja, A.S., High pressure phase equilibria i*n* the carbo*n* dioxide - *n* -Hexadecane and carbo*n* dioxide - water systems. *Can. J. Chem. Eng.*, 66(2), 319–323, 1988.
46. Sako, T., Sugeta, T., Nakazawa, N., Okubo, T., Sato, M., Taguchi, T., *et al.*, Phase equilibrium study of extraction and concentration of furfural produced in reactor using supercritical carbon dioxide. *J. Chem. Eng. Japan /. JCEJ.*, 24(4), 449–455, 1991.
47. King, M.B., Mubarak, A., Kim, J.D., Bott, T.R., The mutual solubilities of water with supercritical and liquid carbon dioxides. *J. Supercrit. Fluids*, 5(4), 296–302, 1992.
48. Dohrn, R., Bünz, A.P., Devlieghere, F., Thelen, D., Experimental measurements of phase equilibria for ternary and quaternary systems of glucose, water, CO_2 and ethanol with a novel apparatus. *Fluid Phase Equilib.*, 83(1993), 149–158, 1993.
49. Kim, S., Kim, Y., Park, B.H., Lee, J.H., Kang, J.W., Measurement and correlation of solubility of water in carbon dioxide-rich phase. *Fluid Phase Equilib.*, 328, 9–12, 2012.

50. Foltran, S., Vosper, M.E., Suleiman, N.B., Wriglesworth, A., Ke, J., Drage, T.C., *et al.*, Understanding the solubility of water in carbon capture and storage mixtures: An FTIR spectroscopic study of $H_2O + CO_2 + N_2$ ternary mixtures. *J. Greenhouse Gas Control*, 35, 131–137, 2015.
51. Comak, G., Foltran, S., Ke, J., Pérez, E., Sánchez-Vicente, Y., George, M.W., *et al.*, A synthetic-dynamic method for water solubility measurements in high pressure CO_2 using ATR–FTIR spectroscopy. *J. Chem. Thermodyn.*, 93, 386–391, 2016.
52. Dohrn, R., Peper, S., Fonseca, J.M.S., High-pressure fluid-phase equilibria: Experimental methods and systems investigated (2000–2004. *Fluid Phase Equilib.*, 288(1-2), 1–54, 2010.
53. Shepherd, M., An improved apparatus and method for the analysis of gas mixtures by combustion and absorption. *Bur. Stan. J. Res.*, 6(1), 121, 1931.
54. Gillespie, P.C., Owens, J.L., Wilson, G.M.1984Sour water equilibria extended to high temperature and with inerts present.AIChE Winter National MeetingAtlanta GA.:Paper 34-b;March 11-14.
55. Bloch, M.G., Lifland, P.P., "Catalytic reforming improved by moisture metering,". *Chem. Eng. Prog.*, 69(No. 9), 49–52, 1973.
56. Wiebe, R., Gaddy, V.L., The solubility in water of carbon dioxide at 50, 75 and 100°, at pressures to 700 atmospheres. *J. Am. Chem. Soc.*, 61(2), 315–318, 1939.
57. Stone, H.W., Solubility of water in liquid carbon dioxide. *Ind. Eng. Chem.*, 35(12), 1284–1286, 1943.
58. Song, K.Y., Kobayashi, R., Water content of Co_2 in equilibrium with liquid water and /or hydrates. *SPE Formation Evaluation.* 2 12. pp. 500–508, 1987.
59. Mather, A.E., Franck, E.U, Ulrich Franck, E., Phase equilibria in the system carbon dioxide-water at elevated pressures. *J. Phys. Chem.*, 96(1), 6–8, 1992.
60. Frost, D.J., Wood, B.J., Experimental measurements of the properties of H_2O-CO_2 mixtures at high pressures and temperatures. *Geochim. Cosmochim. Acta*, 61(16), 3301–3309, 1997.
61. Yang, S.O., Yang, I.M., Kim, Y.S., Lee, C.S., Measurement and prediction of phase equilibria for water+ CO_2 in hydrate forming conditions. *Fluid Phase Equilib.*, 175(1-2), 75–89, 2000.
62. Carroll, J.J. 2002.The water content of acid gas and sour gas from 100 to 220°F and pressures to 10,000 psia.81st Annual GPA Convention.
63. Mohammadi, A.H., Chapoy, A., Richon, D., Tohidi, B., Experimental measurement and thermo dynamic modeling of water content in methane and ethane systems. *Ind. Eng. Chem. Res.*, 43(22), 7148–7162, 2004.
64. Seo, M.D., Kang, J.W., Lee, C.S., Water solubility measurements of the co_2 -rich liquid phase in equilibrium with gas hydrates using an indirect method. *J. Chem. Eng. Data*, 56(5), 2626–2629, 2011.
65. Wang, Z., Felmy, A.R., Thompson, C.J., Loring, J.S., Joly, A.G., Rosso, K.M., *et al.*, Near-infrared spectroscopic investigation of water in supercritical CO_2 and the effect of $CaCl_2$. *Fluid Phase Equilib.*, 338, 155–163, 2013.

66. Jasperson, L.V., Kang, J.W., Lee, C.S., Macklin, D., Mathias, P.M., McDougal, R.J., *et al.*, Experimental determination of the equilibrium water content of CO_2 at high pressure and low temperature. *J. Chem. Eng. Data*, 60(9), 2674–2683, 2015.
67. Meyer, C.W., Harvey, A.H., Dew-point measurements for water in compressed carbon dioxide. *AIChE J.*, 61 , (No. 9), 2913–2925, 2015.
68. Loring, J.S., Bacon, D.H., Springer, R.D., Anderko, A., Gopinath, S., Yonkofski, C.M., Water solubility at saturation for CO_2–CH_4 mixtures at $323._2$ K and 9.000 MPa. *J. Chem. Eng. Data*, 62, 1608–1614, 2017.

9

Molecular Simulation of pK Values and CO_2 Reactive Absorption Prediction

Javad Noroozi[1] **and William R Smith**[1,2,3,4,*]

[1]Dept. of Chemical Engineering, University of Waterloo, Waterloo ON, Canada
[2]Dept. of Mathematics and Statistics, University of Guelph, Guelph ON, Canada
[3]Dept. of Chemistry, University of Guelph, Guelph ON, Canada
[4]Faculty of Science, University of Ontario Institute of Technology, Oshawa ON, Canada

Abstract
We propose a molecular simulation methodology to construct an extended Debye-Hückel thermodynamic model for the reactive absorption of CO_2 in aqueous alkanolamine solvents. The methodology is based on the combination of the prediction of the underlying equilibrium constants and the use of the Davies equation for the activity coefficients. The model is purely predictive in nature, requiring no system experimental data.

Keywords: Molecular simulation, alkanolamine, carbon dioxide

9.1 Introduction

Reaction equilibrium constants *K*eq are important parameters required for the thermodynamic modeling of many systems of scientific and industrial interest. Aqueous mixtures involving both neutral and charged solutes are ubiquitous in environmental, geochemical, biological and industrial settings. *K*eq are of fundamental importance for predicting the compositions of such systems based on models of chemical reaction equilibrium, algorithms for the implementation of which in both batch and flow systems are now widely available (for a recent review, see Leal *et al.*, [1]).

**Corresponding author:* bilsmith@uoguelph.ca

Ying Wu, John J. Carroll and Yongle Hu (eds.) The Three Sisters, (185–192) © 2019 Scrivener Publishing LLC

A common experimental paradigm for obtaining values of *K*eq parameters and those in an assumed model for the activity coefficients, $\gamma i(T, P; m)$ (where T is the temperature and P the pressure), is by fitting the parameters to experimental data, which may be an observed equilibrium composition or an measurable property of such a composition. For a system that has not previously been studied, this indirect procedure typically requires expensive experimental data collection over ranges of (T, P) and overall system compositions, and model improvement usually entails the introduction of additional parameters in an ad-hoc manner. In addition, due to the empirical nature of the assumed ln γ model, predictions outside the range of experimental conditions under which the parameters have been obtained is questionable.

In this paper, we propose an approach for directly calculating *K*eq values using molecular simulation methodology. This is based on fundamental scientific principles and on simple mathematical models (force fields, FFs) of the underlying intermolecular forces involving the species, involving a relatively small number of fixed (independent of T and P) parameters for each. These FFs are readily available for many species, and may be constructed for previously unstudied species by means of a well-defined procedure requiring minimal experimental data. Although *K*eq values have been calculated by approximate quantum mechanical (QM) approaches (e.g., [2]), to our knowledge the force-field-based approach used here is new. Furthermore, although the QM-based approaches can only be used to model systems at infinite dilution, the FF approach described herein has the advantage that it can also be extended for use in concentrated solutions.

We consider application of this approach to the reactive absorption of CO_2 in aqueous monoethanol amine (MEA) solvents, an important system for post-combustion CO_2 capture.

9.2 Thermodynamic Background

The chemical potential model most commonly used when analyzing experimental data for modeling aqueous electrolyte solutions is based on Henry-Law ideality, expressed as

$$\mu_i(T,P;m) = \mu_i^{\dagger}(T;P) + RT\,ln\left(\frac{m_i}{m^0}\right) + RT\,ln\,\gamma_i(T,P;m);\, i = 1,2,...,N_S \quad (9.1)$$

where T is the absolute temperature, P the pressure, R the universal gas constant, mi the molality of species i, γi its activity coefficient, Ns the number of solution species, and m^0 is 1 mol kg–1 H_2O. When the $\mu^{\dagger}(T;P)$ values and an analytical expression for the $\gamma i(T, P; \boldsymbol{m})$ values are known, the system equilibrium composition can be calculated by minimizing the system Gibbs free energy subject to the conservation of mass constraints and non-negativity constraints on the species amounts.

If one implements the equilibrium calculation by means of a set of $R = Ns - \text{rank}(A)$ linearly independent stoichiometric reactions, where A is the $M \times Ns$ matrix of chemical formulae and M is the number of chemical elements (it is important to note that the set of reactions need not coincide with a hypothesized reaction mechanism for any process underlying the attainment of the equilibrium composition), one must solve the following equations:

$$\sum_{i=1}^{Ns} v_{ij}\mu_i = \sum_{i=1}^{Ns} v_{ij}\left[\mu_i^{\dagger}(T;P) + RT\,ln\,m_i + RT\,ln\,\gamma_i(T,P;m)\right];j = 1,2,...,R \quad (9.2)$$

$$= \Delta G_j^{\dagger}(T;P) + RT\sum_{i=1}^{Ns} v_{ij}\,ln\,m_i + RT\sum_{i=1}^{Ns} v_{ij}\gamma_i(T,P;m); j = 1,2,...,R \quad (9.3)$$

$$= 0;\ j = 1,2,...,R \quad (9.4)$$

The $\Delta G_j^{\dagger}(T,P)$ values are commonly expressed as equilibrium constants Kj via

$$\Delta G_j^{\dagger}(T,P) = -RT\ln K_j \quad (9.5)$$

and

$$pK_j = \log_{10}K_j \quad (9.6)$$

Knowledge of the pK values alone allows the modelling of system behaviour at low concentrations by setting the activity coefficient terms to zero. To quantitatively model the behaviour at higher concentrations, an activity coefficient model is required. A commonly used model for aqueous solutions is based on the Debye-Hückel model or its extensions [3]. A particularly simple model is that of Davies [4], for which the solute activity coefficients are given by

$$\ln\gamma_i(T,P;m) = -A(T,P)z_i^2\left(\frac{\sqrt{I}}{1+\sqrt{I}} - 0.2I\right)\ln(10) \quad (9.7)$$

where $A(T, P)$ is the dielectric permittivity of the solvent, zi is the valence of species i, and I is the ionic strength of the solution:

$$I = \frac{1}{2}\sum_{i=1}^{N} m_i z_i^2 \quad (9.8)$$

The chemical potential of water is:

$$\mu H_2O(T, P; m) = \mu^*_{H_2O}(T, P) + RT \ln a_{H_2O}(T, P; m) \tag{9.9}$$

where $\mu^*_{H_2O}(T, P)$ is the chemical potential of pure H_2O. Its activity using the Davies model is obtained from application of the Gibbs-Duhem equation to Eqn. (9.7) – (9.9) [5]:

$$\ln a_W = \frac{1000 \ln 10}{M_W} A \left[2\left(\frac{1 + 2\sqrt{I}}{1 + \sqrt{I}} \right) - 4 \ln(1 + \sqrt{I}) - 0.3I^2 \right] - \frac{1 - x_W}{x_W} \tag{9.10}$$

where xw is the mole fraction of H_2O.

9.3 Molecular Simulation Methodology

We use the following expression [6–9] for the reference state chemical potentials, $\mu^\dagger(T, P)$, for species in a solution involving water as the solvent:

$$\mu^\dagger_i(T, P) = \mu^0_i(T; P^0) + \Delta G^{std}(T; P^0) + \Delta G_s^{\rho-m} + \Delta G_i^{intr}(T, P) + \Delta G_i^{corr}(T, P) \tag{9.11}$$

where $\mu^0_i(T, P^0)$ is the ideal gas formation free energy of species i at T and the reference state pressure $P^0 = 1$ bar, ΔG^0 is the standard state correction term, $\Delta G^{\rho-m}$ is a term converting from a density based chemical potential model to a molality-based model for a given solvent (for H_2O at 298.15 K and 1 bar, this term is very small in magnitude). ΔG^{intr} is the intrinsic solvation free energy of species i. The quantities are given by

$$\Delta G^{std} = RT \ln \left(\frac{RT \rho^0}{P^0} \right) \tag{9.12}$$

$$\Delta G_{\rho-m} = RT \ln \left(\frac{\rho_s(T, P) M_{solv} m^0}{1000} \right) \tag{9.13}$$

$$\Delta G^{intr} = \mu^{res,\infty NVT}(T, \rho(T, P)) \tag{9.14}$$

where $\mu^{res,\infty,NVT}(T, \rho(T, P))$ is the residual chemical potential at infinite dilution of the solute, and $\rho s(T, P)$ is the solvent density corresponding to the specified (T, P). This is the usual quantity calculated by inserting/deleting a single particle in the solvent using periodic boundary conditions (PBC) and (for an ion) Particle Mesh Ewald summation (PME) for the electrostatics.

(We note in passing that μ^*_ω, the the chemical potential of pure H_2O in Eqn. (9.9), can also expressed by the sum of the first two terms in Eqn. (9.1) using Eqn. (9.11).

Finally, ΔG^{corr} is a (typically small in magnitude) correction term that includes [6] (1) a finite-size long-range correction term to account for the finite size dependence of the simulations; (2) a correction for the mean effect of electrostatic interactions beyond the simulation cutoff distance, depending on the relative dielectric permittivity of the water model used in the simulation (for SPC/E, this is 71.1); (4) a quantum correction term.

Excluding the terms that are small in magnitude, the reference state chemical potentials are given accurately by the sum of the first four terms in Eqn. (9.11).

9.4 Application to the MEA-H_2O-CO_2 System

We use the following set of reactions to describe the speciation equilibrium in the solution at low CO_2 loadings ($L < 0.5$):

$$2RNH_2 + CO_2 = RNH^+ + RNHCOO^- \tag{9.15}$$

$$RNH_3^+ + H_2O = H_3O^+ + RNH_2 \tag{9.16}$$

$$RNHCOO^- + H_2O = RNH_2 + HCO_3^- \tag{9.17}$$

$$RNH_2 + H_2O = RNH_3^+ + OH^- \tag{9.18}$$

This choice is motivated by the fact that experimental results at low CO_2 loadings indicate that Reaction (9.15) goes essentially to completion with CO_2 as the limiting reactant, and that the equilibrium concentrations of the "minor species" $\{CO_2, HCO_3^-, H_3O^+, OH^-\}$ are very small. This allows an easy and rapid calculation of the equilibrium composition from an initial estimate of the "major species" $\{RNH^+, RNHCOO^-, RNH_2, H_2O\}$ resulting from Reaction (9.15) with CO_2 as the limiting reactant. From the equilibrium composition of CO_2, its vapour phase partial pressure can be calculated assuming ideal-gas behaviour from the following equation:

$$p(CO_2) = exp\left[\frac{\mu(CO_2) - \mu^0(CO_2)}{RT}\right] \tag{9.19}$$

The total vapour phase pressure can be calculated by adding the results for $p(RNH_2)$ and $p(H_2O)$ using a similar equation. This approach incorporates the reasonable assumption that the solution phase composition is independent of pressure.

Similarly, at loadings $L > 0.5$, we use an initial equilibrium estimate from the treatment of MEA as a limiting reactant in Reaction (9.15), and

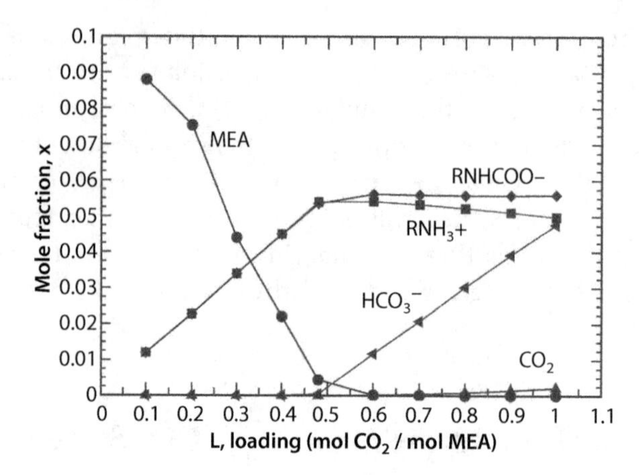

Figure 9.1 Composition of Solution Species as a Function of Loading.

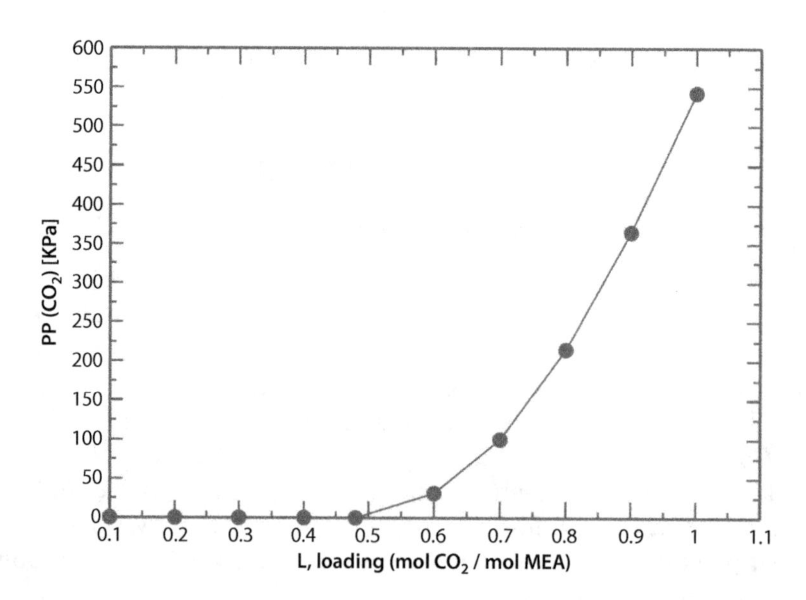

Figure 9.2 CO_2 Partial Pressure as a Function of Loading.

CO_2 as a limiting reactant in Reaction (9.23) in the following set of reactions describing the system:

$$RNHCOO^- + H_2O = RNH_2^+ + HCO_3^- \tag{9.20}$$

$$HCO_3^- + RNH_3^+ = RNHCOO^- + H_3O^+ \tag{9.21}$$

$$RNHCOO^- + 2H_2O = RNH_3^+ + HCO_3^- + OH^- \tag{9.22}$$

$$2HCO_3^- + RNH_3^+ = 2H_2O + RNHCOO^- + CO_2 \tag{9.23}$$

Preliminary results are shown in Figure 9.1 and Figure 9.2 at 298.15 K and 1 bar.

References

1. Leal, A.M.M., Kulik, D.A., Smith, W.R., Saar, M.O., An overview of computational methods for chemical equilibrium and kinetic calculations for geochemical and reactive transport modeling. *Pure Appl. Chem.*, 89(5), 597–643, 2017.
2. Teranishi, K., Ishikawa, A., Sato, H., Nakai, H., Systematic investigation of the thermodynamic properties of amine solvents for CO_2chemical absorption using the cluster-continuum model. *Bull. Chem. Soc. Jpn.*, 90(4), 451–460, 2017.
3. Robinson, R.A., Stokes, R.H., *Electrolyte Solutions*. New York, Dover Publications, 2002.
4. Davies, C.W., The extent of dissociation of salts in water. Part VIII. An equation for the mean ionic activity coefficient of an electrolyte in water, and a revision of the dissociation constants of some sulphates. *J. Chem. Soc.*, 1938, 2093.
5. Hamer, W.J., Wu, Yung-Chi., Osmotic coefficients and mean activity coefficients of uni-univalent electrolytes in water at 25 °C. *J. Phys. Chem. Ref. Data*, 1(4), 1047–1100, 1972.
6. Hofer, T.S., Hünenberger, P.H., Absolute proton hydration free energy, surface potential of water, and redox potential of the hydrogen electrode from first principles: QM/MM MD free-energy simulations of sodium and potassium hydration. *J. Chem. Phys.*, 148(22), 222814, 2018.
7. Lin, F.Y., Lopes, P.E.M., Harder, E., Roux, B., MacKerell, A.D., Polarizable force field for molecular ions based on the classical drude oscillator. *J. Chem. Inf. Model.*, 58(5), 993–1004, 2018.
8. Zhang, H., Jiang, Y., Yan, H., Yin, C., Tan, T., van der Spoel, D., Free-energy calculations of ionic hydration consistent with the experimental hydration free energy of the proton. *J. Phys. Chem. Lett.*, 8(12), 2705–2712, 2017.
9. Nezbeda, I., Moučka, F., Smith, W.R., Recent progress in molecular simulation of aqueous electrolytes: force fields, chemical potentials and solubility. *Mol. Phys.*, 114(11), 1665–1690, 2016.

10

A Dynamic Simulation to Aid Design of Shell's CCS Quest Project's Multi-Stage Compressor Shutdown System

William Acevedo[1,*], Chris Arthur[2] and James van der Lee[2]

[1]*Shell Canada, Calgary, Alberta, Canada*
[2]*Virtual Materials Group, Calgary, Alberta, Canada*

Abstract

Since its official launch in November 2015, Shell's Quest Carbon Capture and Storage (CCS) project located in Alberta, Canada, has successfully captured and sequestered over 1 million tonnes of CO_2, per year, from the Scotford Upgrader hydrogen manufacturing units (HMU). As part of the commissioning practice, the machine was tripped at an intermediate operating pressure of 4.7 MPag to test the performance of the overall system in a shutdown scenario. The tests resulted in reverse rotation of the compressor's bull gear. All further commissioning was halted due to concerns over vibration and potential equipment damage. These observations created the need for further study to prevent a potential escalation in risk associated with the higher normal operating conditions. Initial information provided by the vendors was based on steady-state calculations. These conventional analyses could not be relied upon to replicate the observed behaviour; therefore a dynamic tool was required. The dynamic modelling study was completed using VMGSim v9.5.47 provided by Virtual Materials Group (VMG). This tool allowed us to evaluate the compressor and several strategies to reduce or eliminate reverse rotation in a quick and efficient manner. Dynamic modelling, unlike steady-state modelling, can incorporate additional system details to approach real-life operation (e.g., flow, pressure and temperature operating conditions, torque measurements in the motor shaft, compressor performance curves, existing and new piping configurations, vessel volumes, control valve flow characteristic). As all equations are solved simultaneously, it is possible to predict flow reversal and other abnormal operational issues. This added layer of complexity

**Corresponding author:* William.Acevedo@shell.com

Ying Wu, John J. Carroll and Yongle Hu (eds.) The Three Sisters, (193–218) © 2019 Scrivener Publishing LLC

requires specific knowledge for safe interpretation of results and is dependent on proper data collection. Once all information was collected, data matching was performed prior to using the resulting dynamic model for process predictions. The VMG platform proved to be quite reliable, leading to the final design for the compressor blowdown system at the Scotford Upgrader's Quest CCS project.

Keywords: Quest, carbon capture and storage, compressor dynamic modelling, compressor blowdown, depressurization

10.1 Introduction

Shell's Scotford Complex is a facility that includes heavy bitumen upgrading, oil refinery, chemical plant, and a carbon capture and storage facility (CCS). The upgrader uses a hydrogen-addition process to upgrade bitumen from oil sands' mines north of Fort McMurray. The resulting product is a light synthetic crude oil, as well as other medium to heavy crude products. At Scotford, Hydrogen Manufacturing Units (HMU) produces most of the hydrogen required for the hydro-addition process. The resulting mixture is composed of CO, H_2 and some CO_2. The CO is then reacted with steam in a secondary step to produce H_2 and CO_2, with the raw H_2 obtained containing very small amounts of CH_4 or other hydrocarbons. The CO_2 is then separated from the hydrogen using an amine absorption process. The selected amine for the Scotford Upgrader is Shell's ADIP-X. The ADIP-X solution is suitable for gases with high levels of CO_2 allowing for faster CO_2 and COS removal through enhanced kinetics, higher loading capacity (reducing solvent circulation), lower regeneration heat loads, reduced hydrocarbon co-absorption and lower solvent degradation [1].

Treated H_2 from the top of the contactor is sent to a pressure swing absorption step (PSA) with the H_2 product feeding the hydro-addition process and off gases from the PSA to burners and disposal. The rich in CO_2 amine is regenerated in a still column at high temperature and low pressure. The wet CO_2 gas stream off the top of the regeneration column is sent to the CCS 1st stage compressor suction header. The amine regeneration step operates at a pressure of 131 kPag, while the final compression stage can operate between 8,198 to 12,300 kPag at the rated capacities. At lower gas capacities, the compressor can generate pressures as high as 12,997 to 14,000 kPag. The compressor is an eight-stage, integrally geared centrifugal machine with a 16,550kW (22,118 HP) electric motor drive with a rated speed of 1,200 r.p.m. After each stage, inter-stage coolers and scrubbers cool the gas and remove condensed water. Residual water is then removed via a triethylene glycol (TEG) dehydrator absorber between the sixth and seventh stages. The dry CO_2 rich gas is then fed to the seventh stage. It continues until it reaches its final discharge pressure at the eighth stage of compression. This final pressure is required for the CO_2 rich gas to flow into the pipeline and ultimately the storage reservoir [2].

10.2 Centrifugal Compressor Reversal

Compressors are classified into two main categories: positive displacement and kinetic/dynamic type. Centrifugal compressors are included in the dynamic type when the processed gas flow exceeds 850 to 1,700 m^3/h (500–1000 acfm) range. These machines have a better power-volume per unit weight ratio as well as less expensive for power required. These features come at a penalty in efficiency and errors in required specifications for a given application as well as changes in operating conditions [3].

Centrifugal compressor performance curves are similar to those for centrifugal pumps. This compressor type can be operated between two stability limits, surge and choke conditions [3]. Surge is the operating point where a centrifugal compressor peak head capability and minimum flow is reached; choke/stone wall occurs when the machine is operated at low discharge pressure and high flow rates. Both operational cases can lead to damage of the machine and for these reasons controls are implemented to ensure all operation scenarios stay away from these limits. In the case of surge, anti-surge controls include a controller with pressure, temperature compensated flow metering indication connected to a recycle line-control valve that will ensure the machine sees enough flow to keep away from the surge line. Choking, on the other hand, requires anti-choke valves used to restrict the flow on the outlet line at low pressures as velocity exceeds the speed of sound in the impeller.

The Quest centrifugal compressor is an eight-stage MAN Diesel & Turbo SE/RG 90-8 integrally geared multi-shaft machine. Each stage contains an impeller and every pair of impellers, operating with the same speed, is arranged opposite to each other and share a pinion shaft. Inlet guide vanes are mounted in front of the first stage of each section (MAN Diesel & Turbo). The compressor was selected to process 94.98 mole% of wet CO_2 and 99.23 mole% of dehydrated CO_2. The suction pressure of 131 kPag from the amine regeneration step is raised to a discharge pressure of up to 14 MPag.

The original Quest's vendor design was modified to meet Shell's standards resulting in additional piping and equipment volumes at each compressor stage [2]. During normal commissioning start-up sequence the

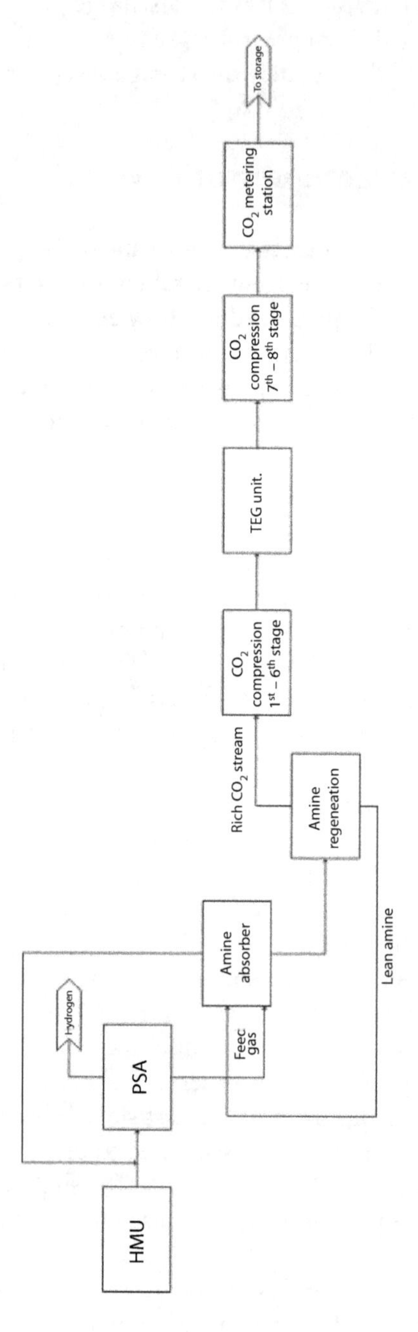

Figure 10.1 Block Flow Diagram for Quest Unit.

Figure 10.2 Shell's Quest Compressor by MAN Diesel & Turbo.

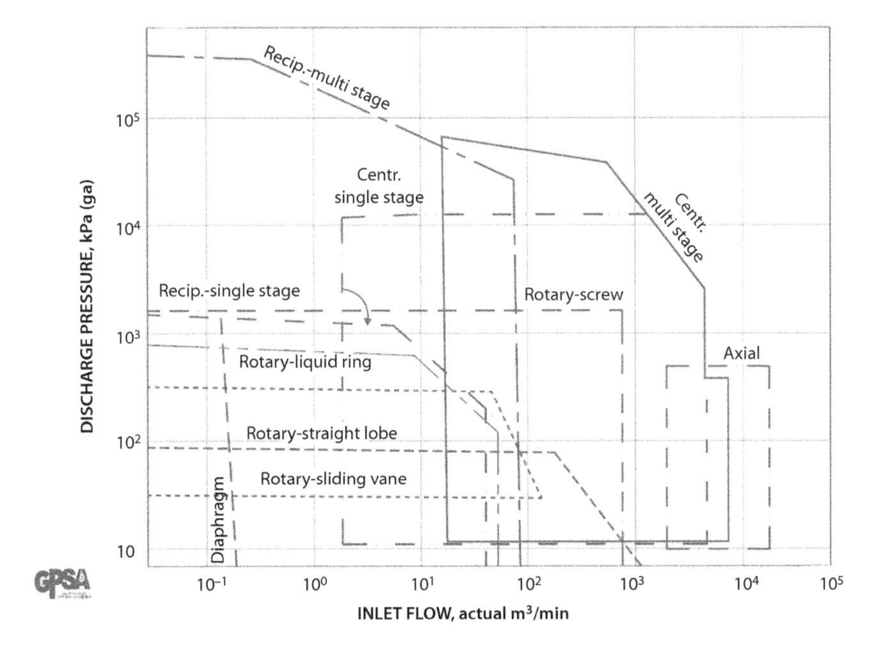

Figure 10.3 Compressor Coverage Chart [from 13th Edition of the GPSA Engineering Data Book (2012)].

machine was tripped at two different operating conditions and allowed to de-pressure. During these relief events, reversal of the bull gear occurred. This flow lead reversal could have resulted in severe equipment damage

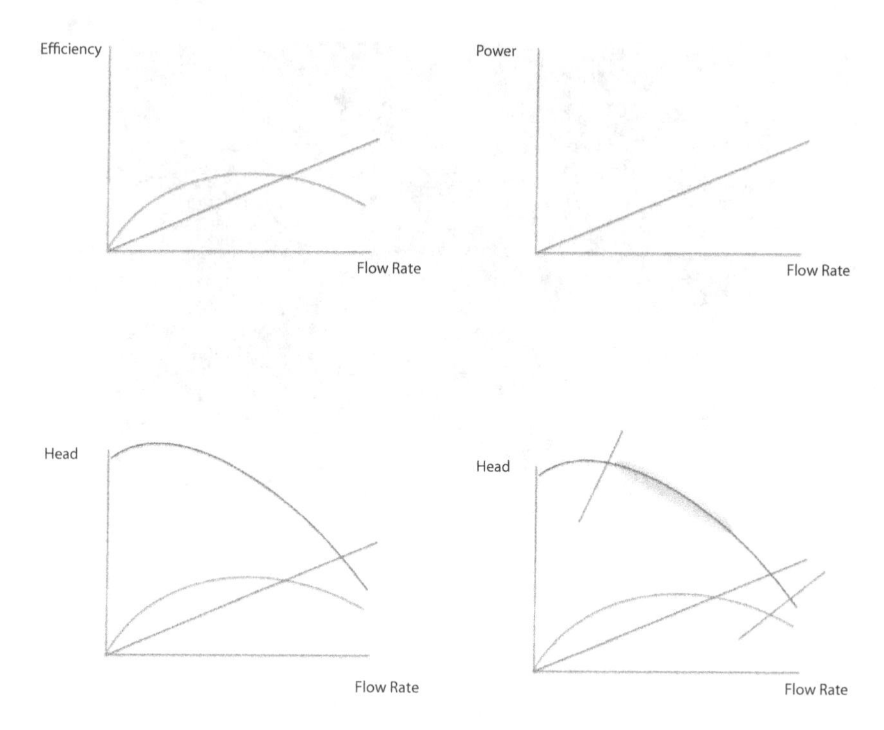

Figure 10.4 Generic Centrifugal Compressor Performance Curves, Courtesy of VMGSim.

Table 10.1 Quest compressor stage speeds, MAN diesel & turbo.

Location	Speed, r.p.m.
Bull Gear	1,200
Pinion 1	6,806
Pinion 2	11,620
Pinion 3	15,880
Pinion 4	22,686

had the lube oil system not been designed to operate during such a scenario. Additional mechanical concerns included thrust bearing damage, which could not be inspected visually without dismantling the top half of the machine [2].

The design included an anti-surge valve downstream of the eighth stage back to the suction of the first stage, blowdown valves downstream of the

Table 10.2 Compressor commissioning bull gear reverse rotation after trips.

Discharge pressure, kPa	Reverse rotation, r.p.m.
4,700	(-) 514
3,700	(-) 371

sixth and eighth stages going to a common atmospheric vent stack and a check valve between the sixth stage and the TEG contactor. Because of the piping modifications the original vent system was unable to relieve pressure in sufficient time leading to backflow as the fluid in the piping and vessels exerted enough momentum to cause reversal of the machine.

The total kinetic energy of the compressor assembly once the motor has been stopped, with no friction and no pressurized gas effects in the system, can be approximated by the following equation:

$$K = \sum_{i=1}^{n} \frac{1}{2} m_i . v_i^2 \tag{10.1}$$

where; K = total kinetic energy, m_i = mass of the ith particle, v_i = speed of the ith particle, and n = total number of particles the rigid body has been subdivided into.

The velocity term in Equation 10.1 can also be expressed as:

$$v_i = \omega \cdot r_i \tag{10.2}$$

where: ω = rotational speed and r_i = radial distance.

Re-arranging Equation (10.1) we have:

$$K = \frac{1}{2} \sum_{i=1}^{n} m_i . \omega^2 \cdot r_i^2 \tag{10.3}$$

$$K = \frac{1}{2} \left(\sum_{i=1}^{n} m_i \cdot r_i^2 \right) \cdot \omega^2 \tag{10.4}$$

The term in the parenthesis is the momentum of inertia and Equation 10.4 can be written as:

$$K = \frac{1}{2} \cdot I \cdot \omega^2 \tag{10.5}$$

The inertia of the compressor bull gear can then be integrated to:

$$I = \frac{1}{2} \left(\rho \cdot \pi \cdot t \cdot R^2 \right) \cdot R^2 \tag{10.6}$$

where: t = thickness of the bull gear, R = radius, and ρ = density of the solid. And the volume, V, and mass, M, are given by:

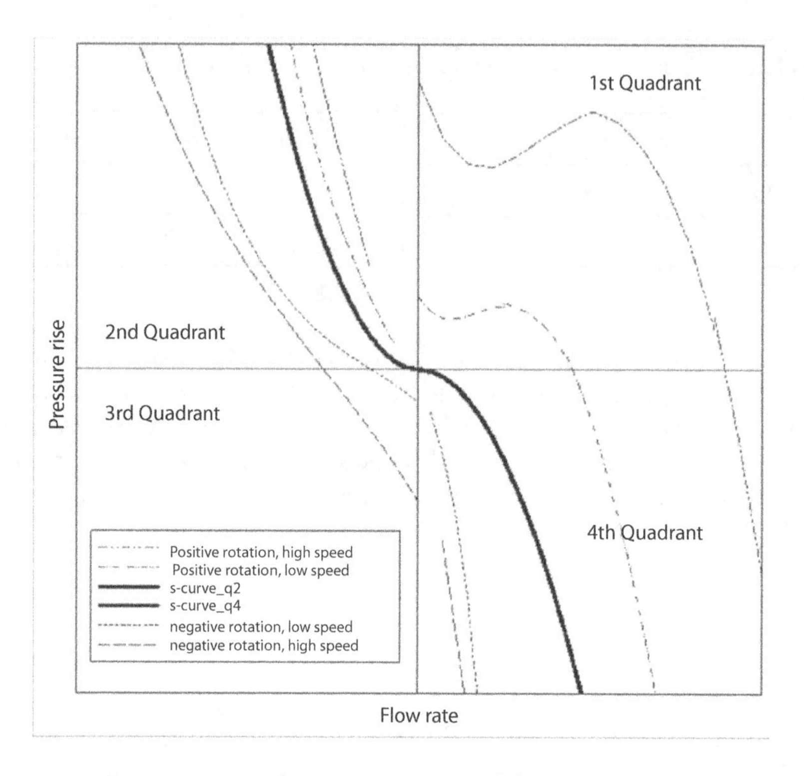

Figure 10.5 Generic Four-Quadrant Compressor Map [4].

$$V = \left(\pi \cdot t \cdot R^2\right) \tag{10.7}$$

$$M = \left(\rho \cdot V\right) \tag{10.8}$$

Substituting yields:

$$I = \frac{M}{2} \cdot R^2 \tag{10.9}$$

To simulate the compressor stages, a four-quadrant model was applied allowing the estimation of energy transfer to and from the fluid under all possible conditions of operation. Predicting compressor operation outside of the positive head, positive flow condition required a different approach extrapolating previous work for centrifugal pumps by A. J. Stepanoff. The following figure shows a generic Four-Quadrant Compressor Map.

In a four-quadrant model, pressure rise is plotted against flow rate and used to describe flow, pressure change, rotation and torque and their respective placement in the the compressor operational modes. The S-curve divide two quadrants into two parts resulting in six operational modes:

Quadrant 1:

- Normal operation – positive flow, pressure change, rotation and torque (positive shaft power)

Quadrant 2:

- Compressor Rotating in positive direction, positive pressure rise but backflow occurs – negative flow, positive pressure change, rotational speed, and torque (positive shaft power)
- Compressor rotates backwards as a turbine under backflow conditions- negative flow, positive pressure change, negative rotational speed, positive torque (negative shaft power)

Quadrant 3:

- Compressor rotates backwards as compressor under backflow conditions – negative flow, pressure change, rotational speed and torque (positive shaft power)

Quadrant 4:

- Compressor rotating backwards but under positive flow – positive flow, negative pressure change, rotational speed and torque (positive shaft power)
- Compressor rotating forwards as turbine – positive flow, negative pressure change, positive rotational speed, negative torque (negative shaft power) [4]

10.3 Dynamic Modelling

A study was undertaken to determine how to mitigate the potential for increased flow reversal when operating at normal discharge pressures. Four parties where tasked with resolving the issue and determining the path forward with support from site operations: MAN Diesel & Turbo, Shell Projects & Technology (P&T), Fluor as the EPC, and Shell Technology Centre Calgary (STCC) Process Engineering. As a reversal of shaft speed was experienced at the reduced start-up pressure of 3.7 MPag, operations halted the ramp up to design pressures to ensure no damage occurred to the machine at the proposed normal operating pressure of 12 MPag.

During extensive commissioning, the machine was tripped at an intermediate operating pressure of 4.7 MPag to test the performance of the overall system in a shutdown scenario. The test resulted in reverse rotation of the compressor bull gear. Concerns around vibration and potential equipment damage led to a study to evaluate and mitigate the potential for a similar event at normal operating conditions. Conventional steady-state analysis could not be relied upon to replicate the dynamic shutdown scenario that was experienced. Therefore, a dynamic modeling study was commissioned with P&T and STCC teams using two different simulation platforms. The STCC team selected VMGSim for the evaluation based on previous successful experience with dynamic modelling for brittle fracture prediction under blowdown scenarios.

Compressor shutdown is an inherently transient process. Designing the shutdown of a multi-stage compressor involves predicting flow rates and energy transfer through the compressor that can change significantly over a small time period. The configuration and size of the surrounding systems and the mechanics of the compressor stages and motor affect this dynamic mass and energy balance.

Given the complexity of a multi-stage compressor shutdown a steady-state or a simplified dynamic approach would not be sufficient to analyse where to locate venting and recycle systems, when to activate them, and design their peak capacities. The quantity and rate of material to be removed from the compressor during a shutdown depends on the fluid inventory stored in vessels and piping, the energy transfer to and from the impellers, the inertia of the rotating machinery, the intermediate resistances to flow and changing stage pressures. The main differences between steady-state vs. dynamic modelling are:

To model this compressor system a state-of-the-art simulation platform, capable of accounting for all of these aspects including a four-quadrant compressor model under all possible conditions of operation, was required. This is necessary to predict compressor operation outside of the positive head. Positive flow condition allowed the replication of undesirable start-up and shutdown scenarios such as the reverse rotation experienced during the Quest commissioning test. The data required for any given compressor dynamic model can be classified based on its importance as shown in the following Table 10.4:

The Quest data used for the dynamic model was:

- Test pressure, temperature and flow for the compressor suction, intermediate stages at discharge conditions of 4.7 MPag

Table 10.3 Steady-state vs. Dynamic modelling comparison.

Steady-State simulation	**Dynamic simulation**
No time domain Mas In – Out = 0	Time domain inserted Mas In – Out = Accumulation
Equipment sizing not considered	Detailed equipment sizing is considered
No piping & equipment volumes	Rigorous piping & equipment volumes
No control scheme	Detailed control scheme implementation
No static head contribution	Rigorous static head contribution is modelled
No IPF function modelling	IPF function can be rigorously modelled
No elevation/nozzle location	Elevation profile/nozzle locations are modelled
Engine solves for a single steady-state point at a time and converges recycle streams	Solves a series of dynamic states, each state is separated in time by a predefined step size
It solves unit operations one at a time and converges recycle streams	It solves pressures and flows using a network (simultaneous) solver and energy/composition balances on a per unit operation basis
A steady-state simulation is built and used to feed the dynamic model initial estimates	All unit operations not ignored must have a consistent set of pressure/flow specifications before running the integrator

Instrumented Protective Function

Table 10.4 Compressor dynamic simulation data requirements [5].

Required data	Critical	Influential	Minor
Compressor vendor performance data/curves	x		
Compressor vendor gas turbine data/curves	x		
Vendor motor data/curves	x		
Compressor-driver rotational inertia	x		
Steam turbine data	x		
P&IDs	x		
Gas Composition/HMB	x		
Equipment volumes	x		
Piping volumes (isometrics)	x		
Piping/equipment resistance (pressure drops)		x	
Mechanical constraints (P, T)		x	
Heat Exchanger data		x	
Vessel data		x	
Anti-surge valve data		x	

(Continued)

Table 10.4 Cont.

Required data	Critical	Influential	Minor
Control valve data		x	
Isolation/check valve data			x
Process/control narrative			x
Cause and effect charts			x
Alarms and trip settings			x
Instrument ranges/vessel taps			x

- Gas composition
- Compressor data sheet e.g., bull gear weight, r.p.m., etc.
- Compressor guide vanes performance curves
- Piping isometrics including pipe run lengths, schedule and fittings
- Recycle, blowdown and surge control valve data sheets
- Inter-stage cooler data sheets to account for volumes
- Compressor inter-stage scrubbers, TEG contactor mechanical drawings for vessel volume calculations (included normal liquid levels)
- Strain gage telemetry used to measure compressor shaft torque
- P&IDs
- Compressor control logic
- Control valve opening times
- CO_2 stack drawings

During the data quality review, it was discovered the measured flow rates had not been properly compensated for pressure and temperature in the DCS and for this reason flow was not used during the model validation.

The thermodynamic property package used for the simulation was the Advance Peng Robinson for Natural Gas, a modification to the original Peng Robinson EOS best suited for hydrocarbon solubility in water. This package is specifically design for the calculation of water content in acid gas, hydrate inhibition, glycols and hydrocarbon solubility. It is also appropriate for polar components. Once the process and mechanical data was collected and reviewed a steady-state model was used to obtain the initialization values for the dynamic model. In order to reduce computation time and complexity the following simplifications to the dynamic model were applied:

- Outlet temperatures from inter-stage coolers were set manually and their volume accounted for
- TEG contactor was set up as a vertical separator to account for its volume inventory
- Water content from the TEG contactor was set up manually using a Component Splitter block
- Pipe fittings and valve equivalent lengths were added to the straight pipe runs

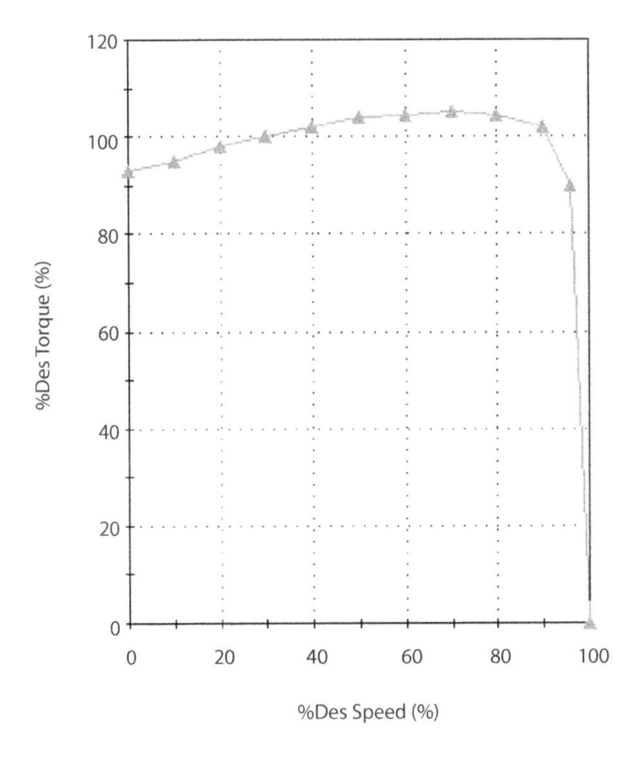

Figure 10.6 Compressor speed vs. torque input to dynamic model in VMGSim.

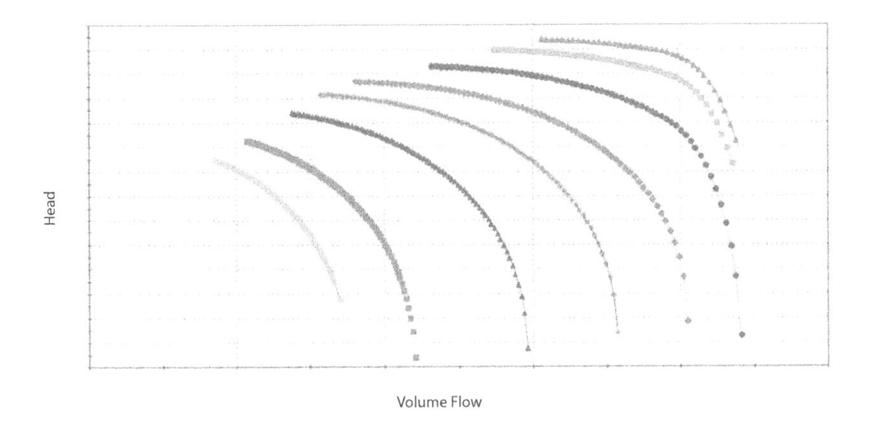

Figure 10.7 Compressor Performance Curves in VMGSim.

Table 10.5 Equivalent length of valves and fittings in feet (FIG. 17.4) [6].

Nominal Pipe size in.	Globe Valve or ball check valve	Angel valve	Swing check valve	Plug cock	Gate or ball valve	45 ell		Short rad. Ell		Long rad. Ell		Hard T		Soft T		90 miter bends			Enlargement					Contraction				
																			Sudden			Std. red.		Sudden			Std. red.	
																			Equiv. L in terms of small d									
						Welded	Threaded	Welded	Threaded	Welded	Threaded	Welded	Threaded	Welded	Threaded	2 miter	3 miter	4 miter	d/D = 1/4	d/D = 1/2	d/D = 3/4	d/D = 1/2	d/D = 3/4	d/D = 1/4	d/D = 1/2	d/D = 3/4	d/D = 1/2	d/D = 3/4
$1^{1/2}$	55	26	13	7	1	1	2	3	5	2	3	8	9	2	3				5	3	1	4	1	3	2	1	1	-
2	70	33	17	14	2	2	3	4	5	3	4	10	11	3	4				7	4	1	5	1	3	3	1	1	-
$2^{1/2}$	80	40	20	11	2	2	-	5	-	3	-	12	-	3	-				8	5	2	6	2	4	3	2	2	-
3	100	50	25	17	2	2		6		4		14		4					10	6	2	8	2	5	4	2	2	-
4	130	65	32	30	3	3		7		5		19		5					12	8	3	10	3	6	5	3	3	-
6	200	100	48	70	4	4		11		8		28		8					18	12	4	14	4	9	7	4	4	1
8	260	125	64	120	6	6		15		9		37		9					25	16	5	19	5	12	9	5	5	2
10	330	160	80	170	7	7		18		12		47		12					31	20	7	24	7	15	12	6	6	2
12	400	190	95	170	9	9		22		14		55		14		28	21	20	37	24	8	28	2	18	14	7	7	2
14	450	210	105	80	10	10		26		16		62		16		32	24	22	42	26	9	-	-	20	16	8	-	-

(Continued)

Table 10.5 Cont.

Nominal Pipe size in.	Globe Valve or ball check valve	Angel valve	Swing check valve	Plug cock	Gate or ball valve	45 ell		Short rad. Ell		Long rad. Ell		Hard T		Soft T		90 miter bends			Enlargement					Contraction				
																			Sudden			Std. red.		Sudden			Std. red.	
																			Equiv. L in terms of small d									
						Welded	Threaded	Welded	Threaded	Welded	Threaded	Welded	Threaded	Welded	Threaded	2 miter	3 miter	4 miter	d/D =1/4	d/D = 1/2	d/D = 3/4	d/D = 1/2	d/D = 3/4	d/D = 1/4	d/D = 1/2	d/D = 3/4	d/D = 1/2	d/D = 3/4
16	500	240	120	145	11	11		29		18		72		18		38	27	24	47	30	10	-	-	24	18	9	-	-
18	550	280	140	160	12	12		33		20		82		20		42	30	28	53	35	11	-	-	26	20	10	-	-
20	650	300	155	210	14	14		36		23		90		23		46	33	32	60	38	13	-	-	30	23	11	-	-
22	688	335	170	225	15	15		40		25		100		25		52	36	34	65	42	14	-	-	32	25	12	-	-
24	750	370	185	254	16	16		44		27		110		27		56	39	36	70	46	15	-	-	35	27	13	-	-
30	-	-	-	312	21	21		55		40		140		40		70	51	44										
36	-	-	-		25	25		66		47		170		47		84	60	52										
42	-	-	-		30	30		77		55		200		55		98	69	64										
48	-	-	-		35	35		88		65		220		65		112	81	72										
54	-	-	-		40	40		99		70		250		70		126	90	80										
60	-	-	-		45	45		110		80		260		80		190	99	92										

The work process used to build any model including dynamic models followed the steps illustrated below:

The compressor model developed was validated with the available plant data and used to aid in the design of the necessary blowdown systems and equipment sizing, based on transient flow to reduce risk to personnel and equipment. Given the transient behavior of this system it was found that items such as valve trim characteristics and actuator stroke time could significantly impact the shutdown event. The timing, magnitude and location of peak flows through recycle and venting systems were evaluated in the simulation during shutdown scenarios under varied operating conditions and configurations. These results were used to allow an assessment of the reverse rotation, noise, velocities, temperatures and forces expected; the result was a robust depressurization system allowing for safe compressor operation.

10.4 Simulation Results

Once the data collection and validation was completed the model was built in small increments using the converged steady-state model as starting point to initialize the dynamic model. Every new block added (e.g., a valve, vessel, compressor stage) was first checked with the Spec Analysis tool prior to running the Integrator. It is best practice once a new process block has been converged that the user should save that case before adding a new block and repeating the process. The compressor impellers and volume size decreases as the gas moves to the final stage; therefore, the inventory stored in the last 2 stages has a very small effect in the reversal of the bull gear. Stages 1 to 6 with their corresponding scrubbers, coolers, and piping have the greatest effect. The gas inventory from the TEG unit followed by the 7th and 8th stages is prevented from flowing backwards via a check valve upstream the TEG contactor. Blowdown valves in the original design were installed downstream of stages 6 and 8 with a surge control valve also downstream of stage 8. The blowdown valve downstream of stage 6 was not able to relieve the gas stored in the compressor and associated equipment fast enough to avoid the bull gear reversal allowing for the gas to flow backward from the higher-pressure side to the suction. The low pressures selected for the compressor test sought to minimize any adverse effect on the machine; however, this resulted in an inability, in the final stage, to generate head, which required the anti-surge valve to be operated during the test.

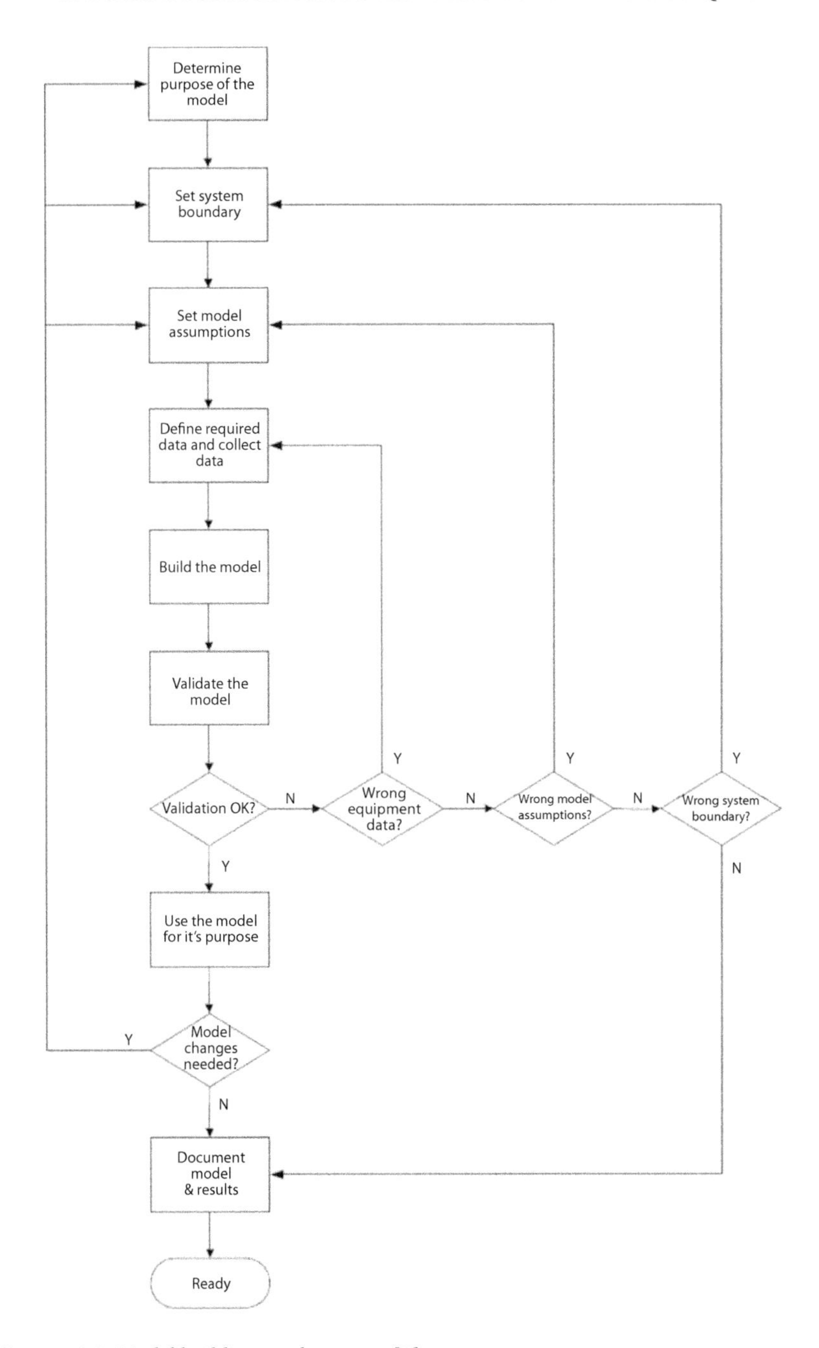

Figure 10.8 Model building work process [7].

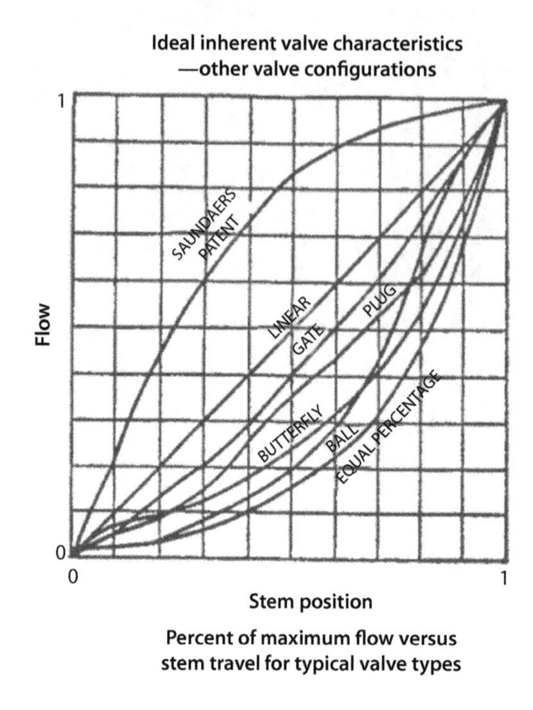

Figure 10.9 Valve Characteristics (Controlling Flow with Ball Valves).

The dynamic model was able to approximate the pressures and reversal of the machine during the test as shown below. Additional tuning of the model and program coding was performed for better data matching:

In order to ensure safe operation a minimum r.p.m. margin above zero was set as a target to account for non-linearity in the model. With confidence that the model was able to approximate the actual operation several different locations for new blowdown valves and check valves were tested with the simulation. The check valve options, although effective and simple in controlling flow reversal, were discarded due to procurement and logistical challenges with the given project deadline.

Once the most favorable piping design from an installation and cost perspective was selected, sensitivities including valve characteristics, valve opening time, and valve with downstream restrictive orifice arrangements were tested as part of the study. The final design included new blowdown valves located upstream the 5th, 6th compression stages and the TEG contactor. In addition, data from the dynamic model was used to estimate the momentum (kinetic energy of the moving fluid) values for the new vent piping. This information was, in turn, used by the EPC to design the required piping supports for the entire system.

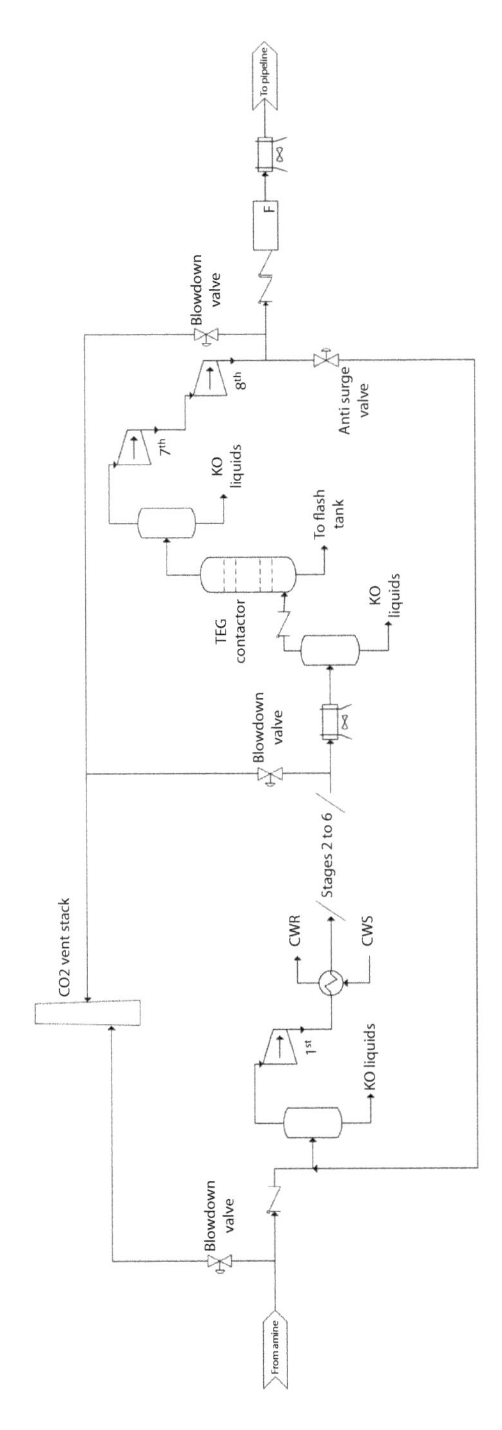

Figure 10.10 Quest compressor original design flow diagram.

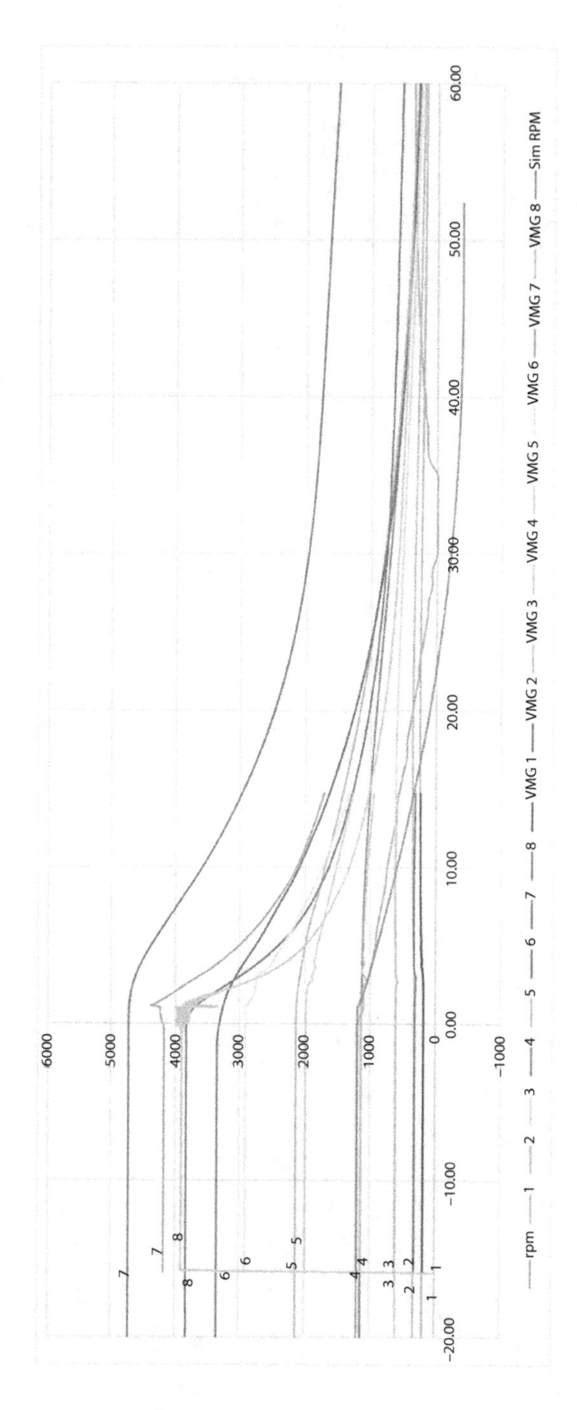

Figure 10.11 Compressor Operating vs. Simulated Discharge Pressures per Stage.

Table 10.6 Bull gear r.p.m. with different blowdown valve combinations.

6/25/2015		Max Rev	Time to Zero	Time to Zero
No.	**Cases**	**Rpm**	**Sec**	**Valve Cv**
1	Base case*	N =(-) 533	17	N/A
2	5 (8")	N =(-) 224	21	1*640
3	4+5 (8")	N =(-) 30	28	2*640
4	4+5+6 (8")	N =(+) 200	29	3*640
5	4+5+6+6 to 1 (8")	N =(+) 200	48	4*640
6	Base +6 to 1 (8")	N =(-) 236	22	1*640
7	5+6+6 to 1 (8")	N =(+) 200	22	3*640
8	6+6 to 1 Recycle (8")	N =(-) 149	24	2*640
9	4+5 (10")	N =(+) 200	25	2*1000
10	5+6 (8")	N =(-) 2	16	2*640
11	5+6 (10")	N =(+) 200	23	2*1000

* Base Case :1*300+1*130+1*155+1*19600

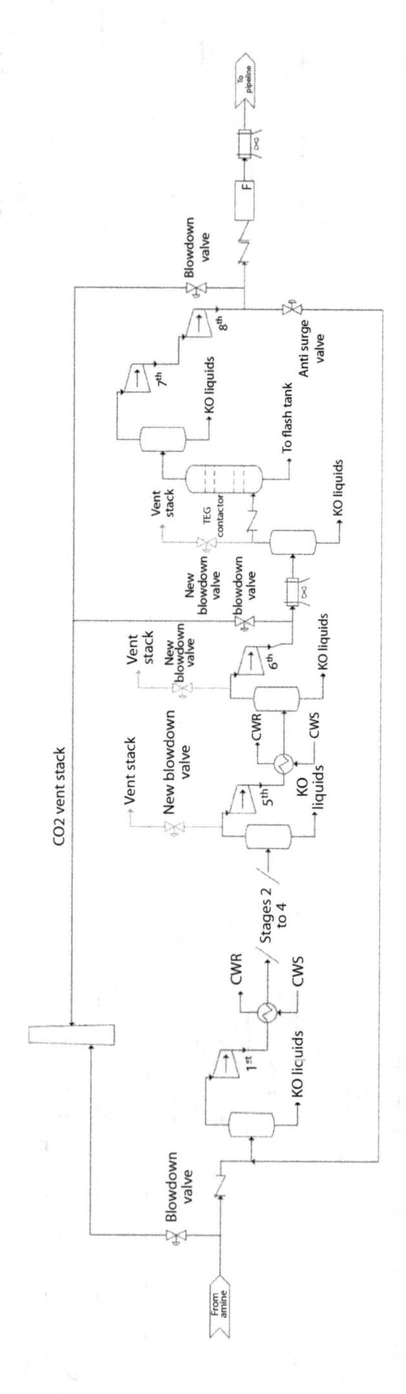

Figure 10.12 Quest compressor original design flow diagram.

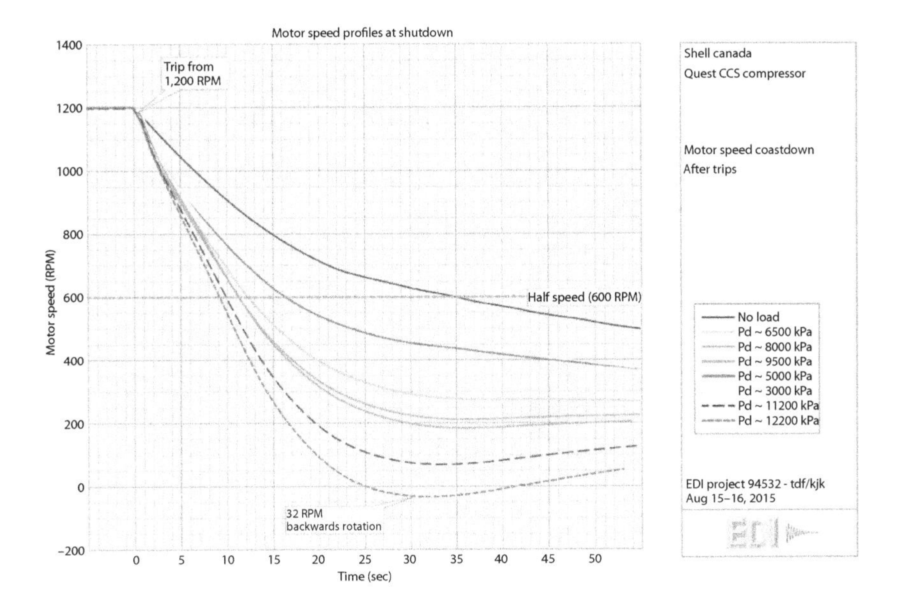

Figure 10.13 Actual Compressor Speed Profile after Blowdown System modifications.

$$\rho V^2 = \text{Momentum, Pa} \tag{10.10}$$

where: ρ = Density of the fluid, kg/m^3, and V = Velocity of the fluid, m/s,

10.5 Modified Blowdown System

After the final implementation of the new blowdown system the compressor was tested at different discharge pressures as illustrated below:

The above results showed that even with the new blowdown valves a −32 r.p.m. occurred when operating at 12,200 kPag, the project accepted the results and implemented a compressor trip at 12,000 kPag. The Quest compressor has since successfully operated continuously storing over 2 MM tonnes of CO_2 from the Scotford upgrader.

10.6 Conclusions

Dynamic modelling requires additional information to account for volume inventory in piping and vessels, machine performance data, complete datasheets, etc., to ensure the model can match reality as closely as possible. It is critical that compression systems be modeled in dynamics to determine if the system will perform as expected; in addition, complete machine performance and equipment data must be provided as part of the handover from projects to operations for future troubleshooting. Dynamic modelling can, in turn, help predict expected performance of the machine after any system modifications, abnormal conditions, as well as a powerful tool for operations & engineering training.

References

1. Shell Global Solutions, S., *ADIP-X For cost-effective, enhanced removal of carbon dioxide (CO2)*. Retrieved from Shell.com:, (2018, February 7). Available from: https://www.shell.com/business-customers/global-solutions/gas-processing-licensing/licensed-technologies/acid-gas-removal/adip-process/_jcr_content/par/textimage.stream/1487623032063/e1036bff9415d92b3c102ab4c57dee386e45468064fca3a58585d8d894ecab48/adip-x.
2. Beausoleil, P., Isley, J., Heinl, B., Kuecker, K., Reverse rotation crops up during quest project testing. *Compressor Tech2*, 34–43, 2017.
3. Campbell, J.M., Hubbard, R., McGregor, K., *Gas Conditioning and Processing V2*. Norman, Oklahoma, Petroskills, 2014.
4. Gill, A., *Four Quadrant Axial Flow Compressor Performance*. South Africa, Department of Mechanical and Mechatronics Engineering University of Stellenbosch, 2011.
5. Hussain, T., Compressor dynamic simulation guideline. *BG Group*, 2015.
6. Tulsa: Gas processors suppliers association. *Engineering Data Book*, 2012.
7. de Wolf, S., Dynamic process modeling guidelines. *Shell Global Solutions International B.V.*, 2011.
8. *Controlling Flow With Ball Valves*. (n.d.). Retrieved 2015, from Industrial Controls. Available from: http://www.industrialcontrolsonline.com/training/online/controlling-flow-ball-valves.
9. Stepanoff, A.J., *Centrifugal and Axial Flow Pumps Theory, Design and Aplication*. Phillipsburg, John Wiley & Sons, Inc, 1957.

11

Benefits of Diaphragm Pumps for the Compression of Acid Gas

Anke-Dorothee Wöhr, Cornelia Beddies and Rüdiger Bullert*

LEWA GmbH, Leonberg, Germany

Abstract

The compression of acid gas or pure CO_2 for acid gas injection (AGI) or carbon capture and storage (CCS) is an energy-hungry process step. In most AGI processes, only compressors are used. The substitution of one or more compressor stages by a high pressure pump can lead to significant energy savings.

First, an overview of reciprocating diaphragm pumps is given, with focus on the safety aspects regarding acid gas. Further, the compression schemes of compressors and the combination of compressor and pump are compared. Examples of current AGI projects where diaphragm pumps are used will be given. Varying process parameters such as changes in flow rate or varying compositions and increasing pressures are discussed. Finally the issue of pressure pulsation is introduced shortly, since this has to be taken into account for critical processes.

Keywords: Diaphragm pump, hydraulic power, leak free, safety, acid gas, diaphragm condition monitoring, compressor

11.1 Characteristics of Diaphragm Pumps

Hydraulically actuated diaphragm pumps are suitable for conveying different fluids. Their most important advantage is that the fluid can be pumped leak free, due to the separation of the fluid from the hydraulic side by a diaphragm. Thus, diaphragm pumps are most suited to handle toxic, highly acidic or basic, abrasive, flammable and environmentally hazardous fluids. For fluids that have to be protected from contamination, like food or pharmaceuticals, diaphragm pumps are also preferred.

**Corresponding author*: Bullert@lewa.de

Ying Wu, John J. Carroll and Yongle Hu (eds.) The Three Sisters, (219–234) © 2019 Scrivener Publishing LLC

A further advantage is that diaphragm pumps have a pressure-firm characteristic. The main influence factor here is the compressibility of the liquid. Thus, varying discharge pressures have a small effect on the flow rate. For large required turn-down ratios, speed controlled pumps using frequency inverters can be installed, having high efficiencies over the whole flow range.

The main disadvantage of diaphragm pumps is that they cause flow and pressure pulsations. This subject will be addressed later.

In the hydraulic part of the pump head, several safety features are installed to render the pump safe against overload and cavitation, even in case of operating errors. Figure 11.1 shows the main components of a diaphragm pump head.

As standard, LEWA diaphragm pumps are equipped with multilayered diaphragms, usually two layers, and integrated diaphragm condition monitoring. In case of diaphragm damage, the other layer remains hermetically tight and prevents leakage of the process fluid to the atmosphere. The diaphragm condition monitoring system will show the wear of the diaphragm, due to a pressure rise between the two layers. The basic design is a pressure gauge, but pressure switches, contact pressure gauges or pressure

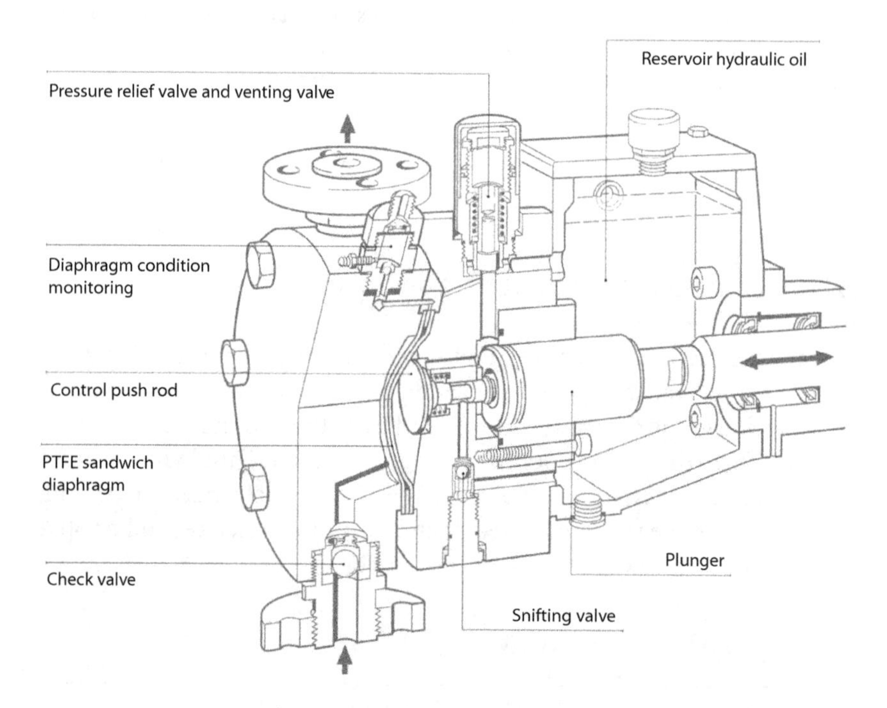

Figure 11.1 Diaphragm Pump Head.

transmitters can also be installed, with either fixed or adjustable trigger values. Some fluids, especially liquefied gases including carbon dioxide and hydrogen sulfide, permeate through PTFE, which is commonly used as diaphragm material. For these fluids, a special diaphragm condition monitoring has to be used, which releases small amounts of diffused gas. If fluids permeate through the diaphragm, the pressure in between the diaphragm layers rises constantly, but slowly. In case of a diaphragm rupture, the pressure rises suddenly and strongly. The high amount of intruding fluid cannot be released. Thus, the rupture signal will be displayed. The different cases are illustrated in Figure 11.2.

a. Standard diaphragm condition monitoring, sound diaphragm: Slow pressure rise due to permeating gas. Rupture signal is shown after some time.
b. Special diaphragm condition monitoring for liquefied gases, sound diaphragm: Small amount of permeating gas can be released. No rupture signal is shown.

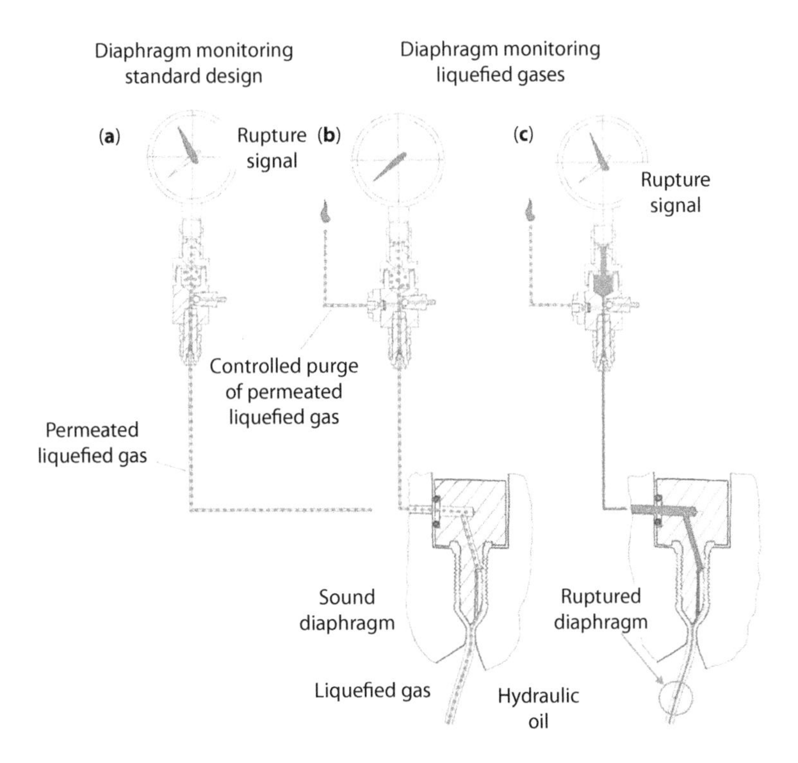

Figure 11.2 Diaphragm Condition Monitoring for Liquefied Gases.

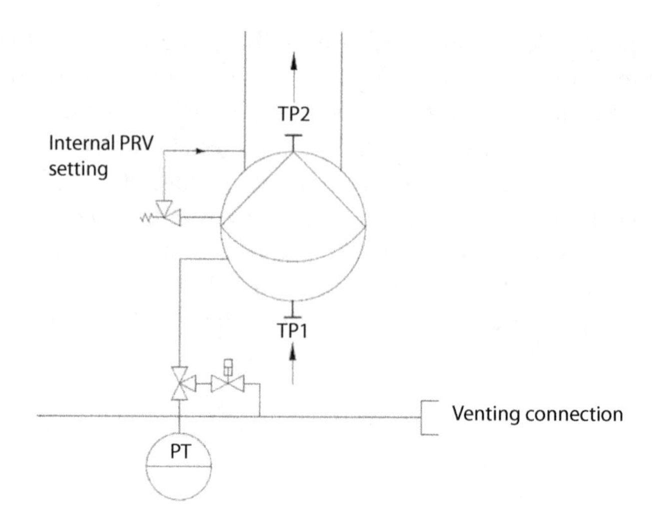

Figure 11.3 Diaphragm Condition Monitoring With Solenoid Valve.

c. Special diaphragm condition monitoring for liquefied gases, ruptured diaphragm: Large amount of intruding gas is not released. Rupture signal is shown.

With critical fluids like acid gas, even small amounts must not be released to the atmosphere. For this case, a special solution with a solenoid valve is available that releases the diffused gas in a controlled manner to a safe area. Figure 11.3 shows the P&ID of this solution for one pump head.

11.2 Current Projects

LEWA has recently completed two orders for acid gas injection projects in the Canadian Wapiti area. Both customers need a high pressure and have decided to use LEWA pumps in the new plants for a significant pressure increase in their processes.

The first project required two triplex diaphragm pumps with 50% service each. The operating conditions ask for compression from 59 bar g to a minimum of 79 bar g and a maximum of 143 bar g. The total required flow rate is 34 m^3/h. This is roughly a hydraulic power of 80 kW which requires an installed power of 150 kW. The stroke frequency is varied between 50 spm and 150 spm. However, a maximum turndown ratio of 1:10 with a stroke frequency from 20 spm to 200 spm would also be possible. The composition varies between 27% CO_2 and 73% H_2S for the low flow rate

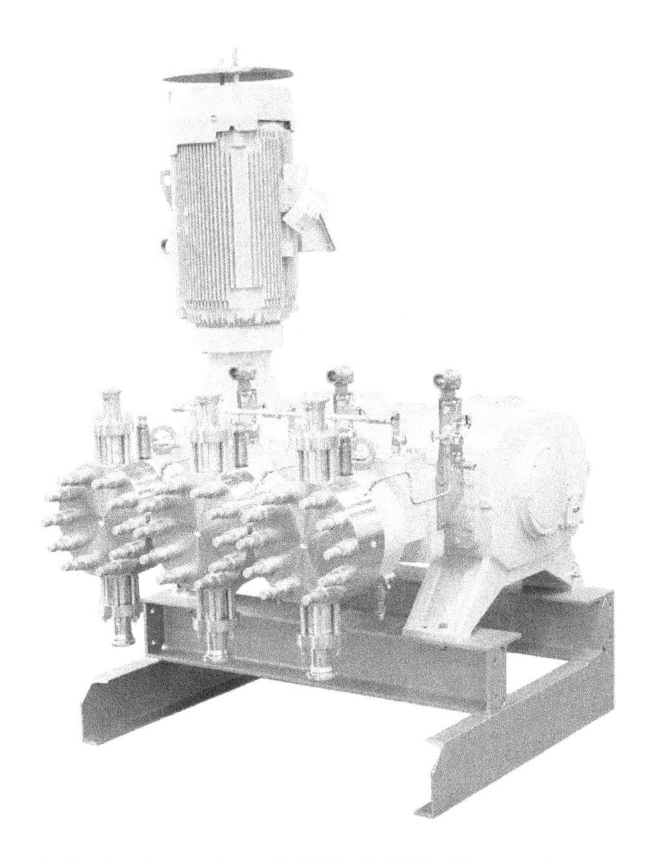

Figure 11.4 Triplex Diaphragm Pump G3M for Acid Gas Compression.

and 3% CO_2 and 97% H_2S for the maximum flow rate. The pumps are sold as bare pumps with only a diaphragm monitoring system. Figure 11.4 shows a picture of one of the pumps.

The other project also requires two triplex pumps which will be installed to compress acid gas from a suction pressure between 49 bar g and 62 bar g to 241 bar g with a flow rate from 2.5 m^3/h to 14 m^3/h. Here we talk about 75 kW hydraulic power with a 185 kW rated motor. The composition is approx. 90% H_2S and 10% CO_2. These pumps are supplied as complete skids and are equipped with baseplates, piping, resonators for dampening, additional instrumentation and a lifting beam. Additionally a pulsation study was performed, considering the pumping process in relation to the customer's piping network. All equipment match the special requirements for Canada/Alberta like the CSA B51 Boiler, pressure vessel and pressure piping code, ABSA registrations or P.Eng. certified lifting lug design. Figure 11.5 shows the skid of this pump.

Figure 11.5 Triplex Diaphragm Pump G3R for Acid Gas Compression.

Acid gas is a challenging fluid. Therefore LEWA has integrated some special solutions into these projects. Firstly, for all fluid wetted parts, material for H_2S containing applications has been used. The wetted parts are NACE MR0175 compliant. For the process valves, acid gas resisting sealing material has been chosen. The aforementioned special diaphragm condition monitoring for liquefied gases has been installed, which bleeds the acid gas in between the diaphragm layers to a safe area. This has also been implemented into the control logic on site. Another special solution has been the installation of additional tubing at the pump head holder for the hydraulic fluid. It is used as a vacuum line for additional safety. Thus potential acid gas in the hydraulic fluid can be led to a safe area, too.

11.3 Improving Efficiency of Acid Gas Compression

If a pump is used in combination with a compressor in the compression process, energy savings can be high, since compression in the liquid state consumes less power than in the gaseous state. In this hybrid approach,

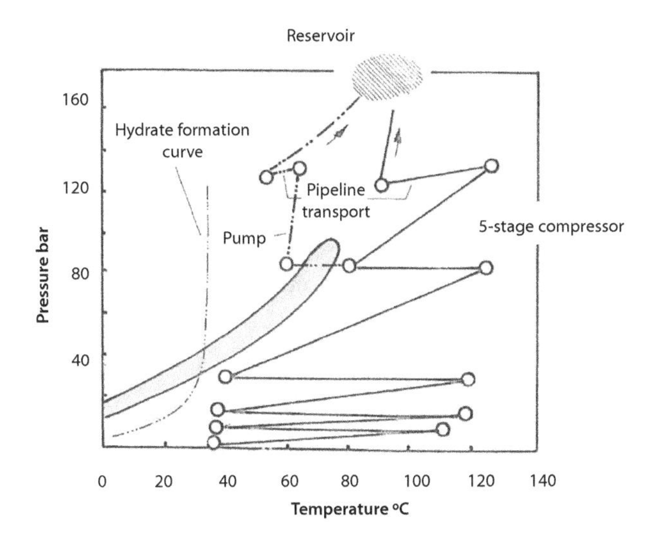

Figure 11.6 Different Approaches for the Compression of Acid Gas.

the fluid at the discharge of the compressor is cooled and liquefied, and the pressure is finally increased in a final single step to the required pressure by the pump. The refrigeration power to liquefy and sub cool the gas has to be also considered in the energy balance. Which compression approach is the best and how much energy can be saved by using a pump has to be checked individually for each project. It depends on several factors like injection pressure, composition of the acid gas and availability of cooling medium.

Figure 11.6 shows these two possible compression paths: The "conventional" way with a multistage compressor, with 5 compressor stages in this example, and the combination of compressor and pump. With this approach, one or more compressor stages can be omitted, depending on the final discharge pressure.

Since H_2S and CO_2 can form hydrates at certain pressures and temperatures, the suction conditions of the pump have to be designed properly. Keeping the temperature above 35 °C will avoid hydrate formation. Alternatively, hydrate inhibitors can be used if it is not possible to obtain a water-free acid gas mixture before it enters the pump.

11.4 Increasing Pressures

One aspect where pumps could play an important role is the retrofit of existing AGI plants. As the reservoir fills up, the reservoir pressure and thus the required discharge pressure rises. Let's imagine a reservoir with an initial injection pressure of 140 bar which increases to 200 bar over time. The compression is done with a compressor only, and the existing compressor can handle a maximum discharge pressure of 160 bar. The challenge is now to evaluate whether upgrading the compressor or adding a diaphragm pump is the more economical approach. It is very unlikely that another cylinder can be added to the existing compressor. There are reciprocating separable compressors available, but even if an extension would be possible, there are still problems remaining. For example, the motor would have to be changed anyway, and the fluid becomes too dense above a certain pressure. Adding a pump to reach the final discharge pressure solves the problem. However, a suction pressure of 160 bar is far too high for typical diaphragm, since it exceeds the permissible rod load during the suction stroke. In this case, it would be necessary to bypass at least one compressor stage.

11.5 Varying Compositions

Since diaphragm pumps work on a volumetric principle, density and compressibility are decisive for the determination of the flow rate. Acid gas streams do not only consist of hydrogen sulfide and carbon dioxide, but can also contain small amounts of volatile components like methane or other short-chained hydrocarbons. If the percentage of these components or the percentage of carbon dioxide increases compared to the percentage of hydrogen sulfide, the compressibility of the mixture also increases and thus the volumetric flow rate decreases. To still meet the required flow rate, a speed-controlled pump is required. Further it has to be ensured that the acid gas mixture remains liquid since the vapor pressure also rises with an increasing percentage of CO_2. It might be necessary to increase the suction pressure provided by the compressor or cool the mixture further down.

To illustrate the influence of the acid gas composition, a case study has been conducted with varying suction temperatures, suction pressures and acid gas compositions. A triplex pump type G3R has been selected as example, with a maximum allowed suction pressure of 65 bar g, a minimum allowed stroke frequency of 30 spm and a maximum allowed stroke frequency of 160 spm. The required discharge pressure is 241 bar g and the required flow rate 12 m^3/h. The H_2S and CO_2 contents vary from 0% to 100%.

The fluid data have been calculated with the NIST Standard Reference Database 23, Version 8.0. Other proven software programs for

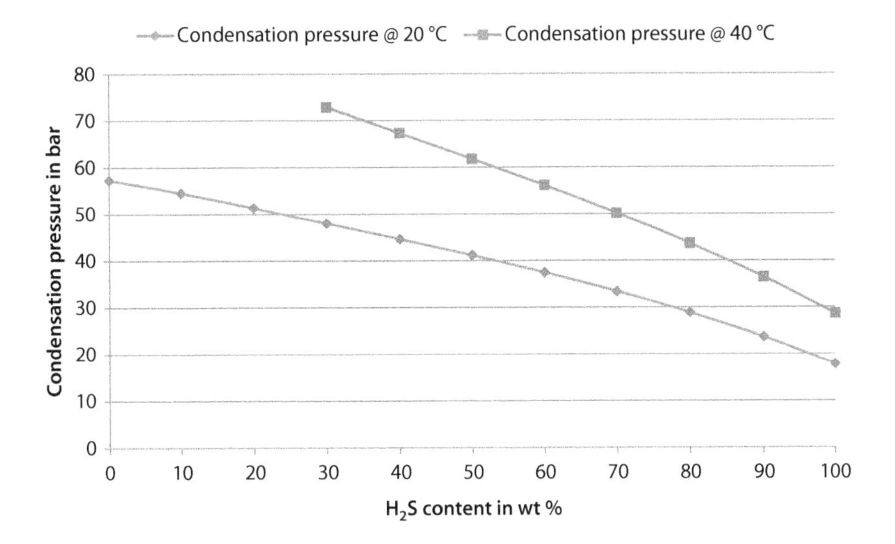

Figure 11.7 Condensation Pressure of Acid Gas at Varying Compositions.

calculating physical fluid properties are Aspen HYSIS, VMG Thermo and AQUAlibrium, which has been especially designed for systems containing acid gas and water. All these programs may differ in their results since different calculation models are implemented.

For determining the temperature rise due to heat of compression and the relative compression, which is defined as the ratio of density change to the suction density in %, an isentropic approach has been assumed. This approximation can be made since the compression in the pump takes place rapidly.

The following three suction conditions have been evaluated:

a. 20 °C, 65 bar g
b. 20 °C, suction pressure 10 bar above condensation pressure
c. 40 °C, suction pressure 15 bar above condensation pressure.

Figure 11.7 shows the condensation pressures at 20 and 40 °C. At 40 °C, the calculation of the condensation pressure was not possible above 70% of CO_2 since the temperature is above or close to the critical temperature. At 40 °C and CO_2 content higher than 50%, the necessary condensation pressure exceeds the allowed suction pressure of the chosen pump.

Figure 11.8 shows the relative compression for the three cases, and Figure 11.9 the resulting discharge temperature. The high differential pressure above condensation pressure had to be chosen to keep the mixture

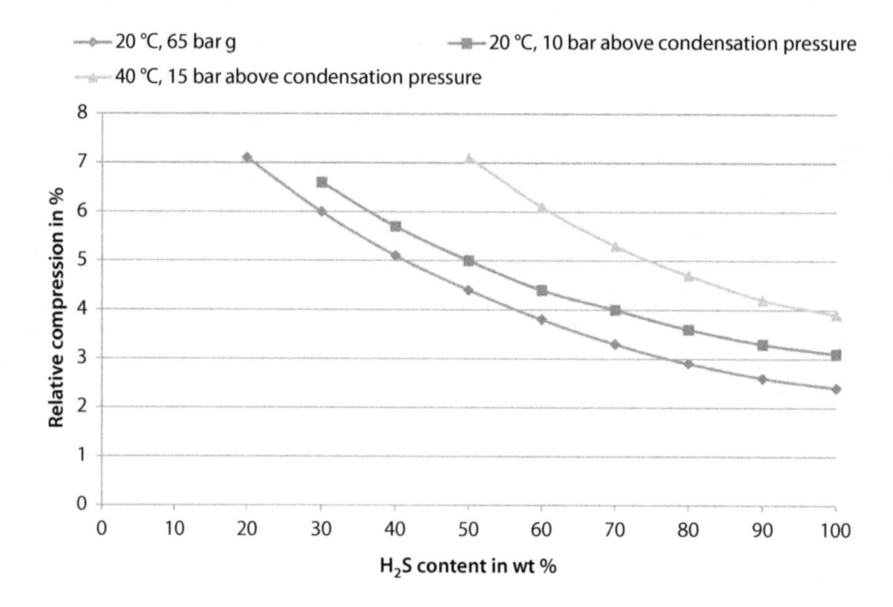

Figure 11.8 Relative Compression of Acid Gas at Varying Compositions and Suction Conditions, at Compression to 241 bar g.

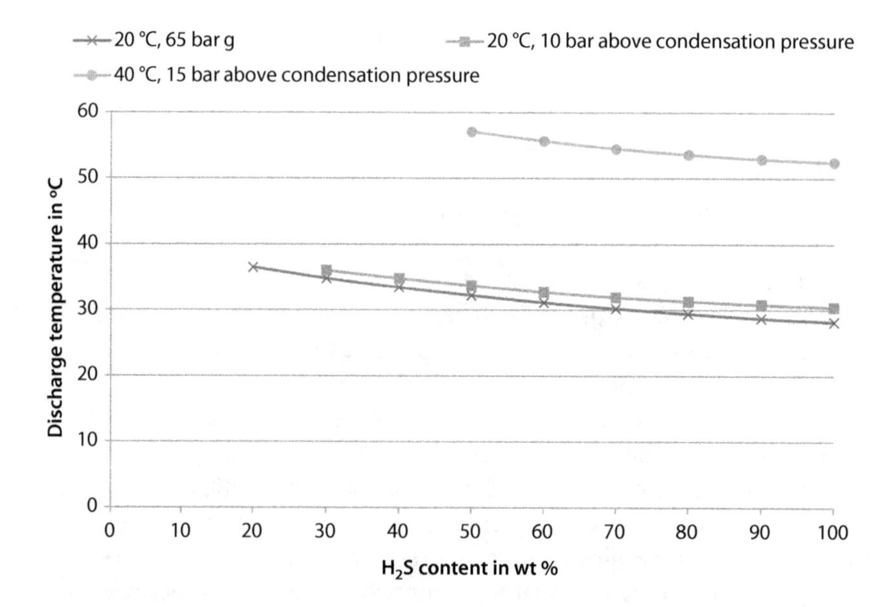

Figure 11.9 Resulting Discharge Temperature of Acid Gas at Varying Compositions and Suction Conditions, at Compression to 241 bar g.

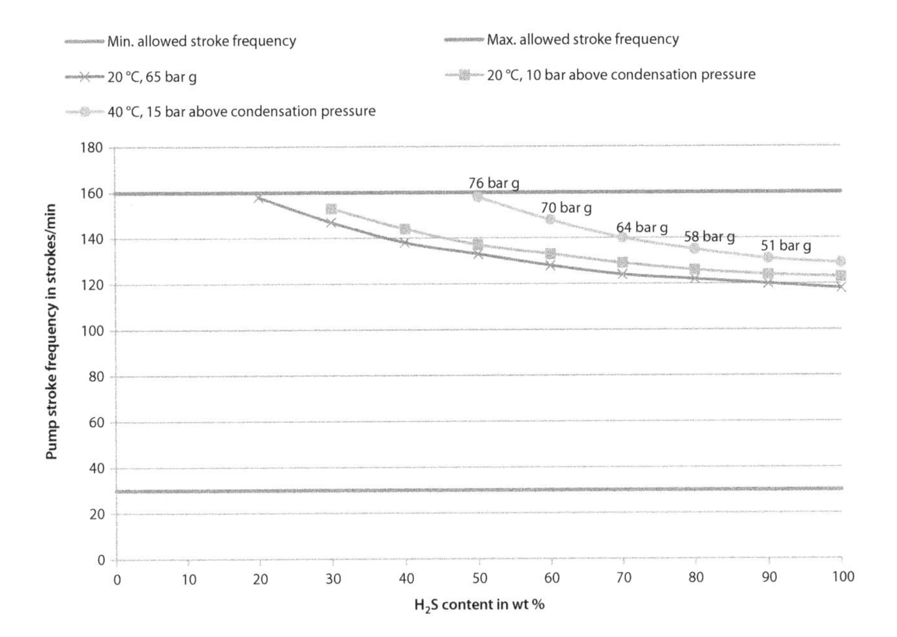

Figure 11.10 Required Stroke Frequency of Selected Pump to Compress Acid Gas at Varying Compositions and Suction Conditions to 241 bar g.

liquid on the suction side, taking into account the heat of compression. However, for high CO_2 contents, not all points could be calculated, since the mixture is not liquid at the chosen conditions. Alternatively, a still higher, unrealistic suction pressure would have to be chosen for this theoretical approach.

With the calculated values of the relative compression, the necessary stroke frequency to meet the required flow rate of 12 m^3/h was determined. The results are shown in Figure 11.10. To see the influence of temperature and composition over a range as large as possible, the stroke frequency was determined also for those points where the maximum allowed suction pressure of the chosen pump would be exceeded. This concerns CO_2 contents above 30% at 40 °C. For better illustration, the values of the necessary suction pressure have been added next to the 40 °C-curve. The min. and max. allowed stroke frequencies are also displayed in the diagram.

The results show that the turn down ratio is not the limiting factor, since the minimum allowed stroke frequency is as low as 30 spm. Instead the limiting factor is the allowed suction pressure, or rather the combination of temperature, suction pressure and CO_2 content with regard to the requirement of keeping the acid gas mixture liquid.

11.6 Pressure Pulsation and Synchronization

Due to the reciprocating movement of the plunger, reciprocating positive displacement pumps cause flow and pressure pulsations in the pipeline. The magnitude of these pulsations depends on many factors like density and speed of sound of the fluid, stroke frequency of the pump and number of plungers, length and diameter of the pipeline, etc. If the system of pump and pipe network is not properly designed, pressure pulsations can become high and cause cavitation, impermissibly high pressure values which can trigger safety valves, and vibrations of the pipeline which can even lead to fatigue failure. To avoid high pressure pulsations, a pulsation study can be carried out already in the planning phase. The API 675 standard proposes criteria for maximum allowed pressure amplitudes and safety margins with regard to vapor pressure and maximum allowed pressure. This standard also describes two modelling methods to calculate the resulting pressure pulsations which are called approach I and approach II. For approach I, analytical calculation models are permitted, whereas approach II requires a tried and tested acoustic simulation model. There are no explicit guidelines that specify where approach I or II is required. LEWA recommends carrying out a pulsation study according to approach II if at least one of the following criteria is fulfilled:

- More than 50 kW of hydraulic power
- Pump stroke frequency greater than 200 rpm
- Complex pipeline network
- Critical process
- Critical fluid.

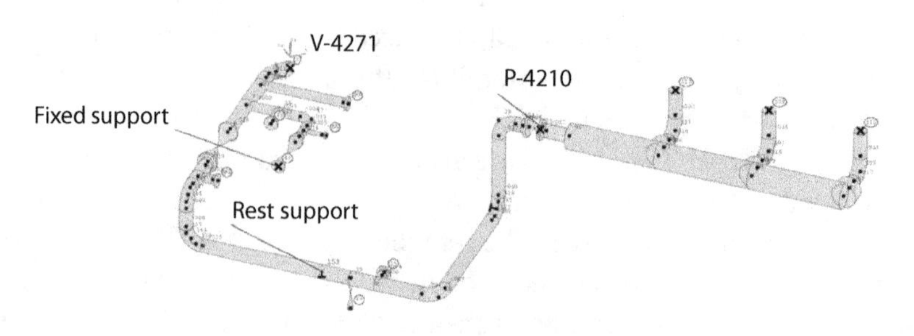

Figure 11.11 Model of a Suction Side Pipeline for a Pulsation Study.

For one of the projects mentioned above, a pulsation study has been carried out by LEWA. Figure 11.11 shows the model of the suction side as example.

If two or more speed-controlled pumps operate in parallel in the same process, there is also the option of synchronizing these pumps. Synchronization means that the pumps are operated at a defined phase shift. The basic objective of synchronization is to reduce pressure pulsations, but this configuration also allows for flexible operation and maintenance concepts. Instead of using a true stand-by pump, all pumps operate during normal operation at reduced speed. If maintenance of one pump is necessary, this pump is excluded from the configuration and the speed of the other pumps is increased. If increasing flow rates have to be met or if the volumetric efficiency of the pumps decreases, e.g., due to higher suction temperatures or higher CO_2 content, the pump speed can be increased or a stand-by pump can be included in this configuration for a limited period.

Figure 11.12 shows the theoretical volume flow pulsation of a three-headed pump and of two synchronized three-headed pumps with a phase shift of 60°, meeting the same average flow rate. For better understanding, high frequency pulsations, which are always present, are not shown. It can be clearly seen that the flow pulsations are significantly lower with two synchronized pumps, which also results in lower pressure pulsations in the pipe system.

Figure 11.13 shows the maximum pressure amplitude at one location in the pipe network for two synchronized pumps at a phase shift of 0 and 60°. At a phase shift of 0° there is a resonance peak at approximately 130 strokes per minute. At a phase shift of 60°, the pressure amplitude is significantly reduced. With selective variations in the phase shift, resonance points can be bypassed with a synchronization control system.

11.7 Conclusion

Diaphragm pumps are most suited to handle highly toxic acid gas. To prevent leakages to the atmosphere even in the case of diaphragm damage, additional safety features can be implemented, such as a special solution for diaphragm condition monitoring and coverage of the holder of the hydraulic oil. To avoid too high pressure pulsations, in the worst case resulting in fatigue failure of the pipeline, it is recommended – as for critical fluids in general – to carry out an acoustical pulsation analysis.

Using a pump in combination with a compressor for acid gas compression can lead to significant energy saving, the amount of which depends

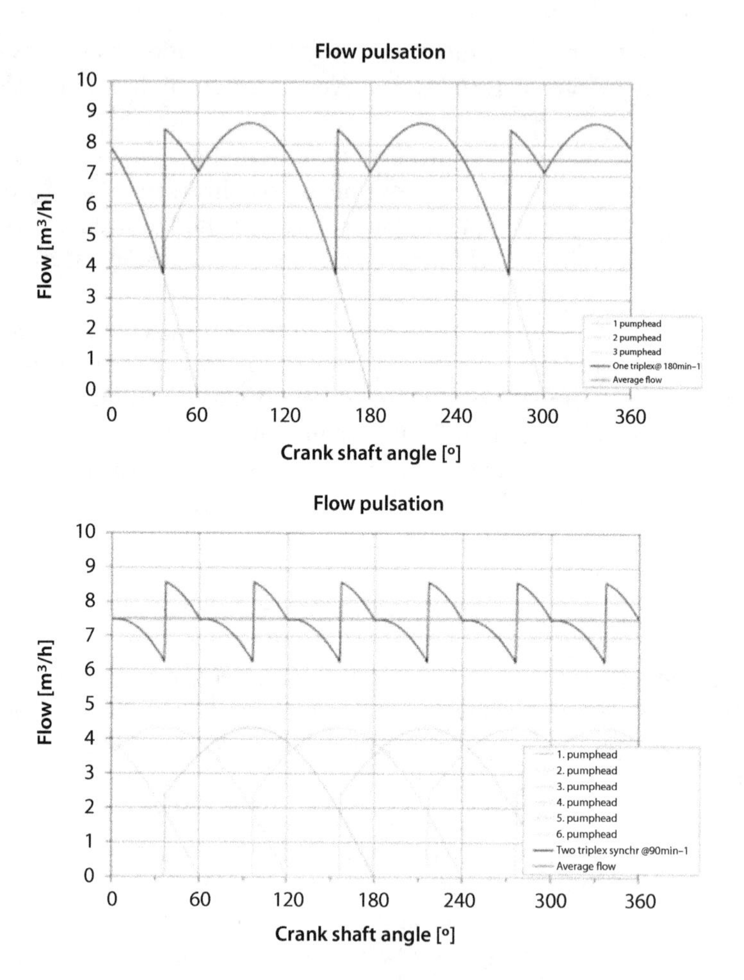

Figure 11.12 Comparison of Theoretical Volume Flow Pulsations; Left: Three-Headed Pump; Right: Two Synchronized Three-Headed Pumps with a Phase Shift of 60°.

on various factors. When the reservoir pressure increases, the extension of an existing compressor poses problems. Also here, the implementation of a pump provides an adequate solution which avoids retrofitting the compressor.

Varying percentages of CO_2 and H_2S in the acid gas mixture influence the compressibility and thus the volumetric efficiency of the pump. This can be compensated by using a speed-controlled pump. However, the acid gas composition is also decisive for the condensation pressure. This means that the suction conditions have to be checked carefully, since the acid gas has always to be kept in the liquid state when using a pump.

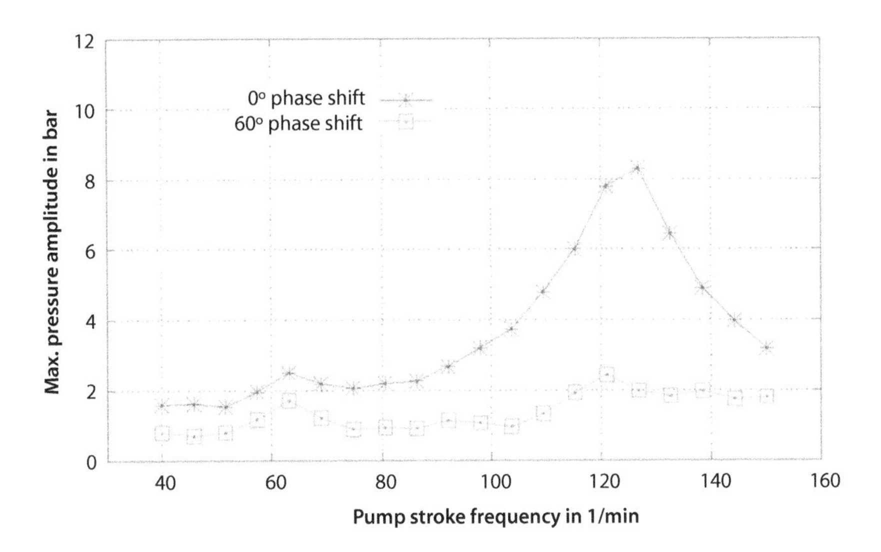

Figure 11.13 Maximum Pressure Amplitudes at One Location in a Pipe Network for Two Synchronized Three-Headed Pumps with a Phase Shift of 0 and 60°.

References

1. J, D., Moore, J., Ms. Marybeth Nored, D., Klaus brun: Novel concepts for the compression of large volumes of carbon dioxide. *Southwest Research Institute, Oil & Gas Journal*, 2007.
2. Simone bertolo: Four post-combustion CO_2 compression strategies compared. *Carbon Capture Journal, Sept*, 2009.
3. Baldwin, P., Williams, J., Capturing CO_2: Gas compression vs. liquefaction. *Power Magazine*, 2009.
4. Bachu, S., William, D., Gunter: Overview of acid-gas injection operations in western Canada. *Alberta Energy and Utilities Board, Edmonton, AB, T6B 2X3, Canada.*
5. Bachu, S., William, D., Gunter: Characteristics of acid-gas injection operations in Western Canada. *Report to international Energy Agency – Greenhouse Gas R&D Programme*, 2003.
6. Positive displacement pumps – controlled volume for petroleum, chemical, and gas industry services; API Standard 675, third edition, November 2012; Errata, June, 2014.

12

Dynamic Solubility of Acid Gases in a Deep Brine Aquifer

Liaqat Ali[1,*] and Russell E Bentley[2]

[1]*XHorizons, Houston, Texas, USA*
[2]*WSP USA, Houston, Texas, USA*

Abstract

A great deal of literature exists regarding the injection of acid gases (H_2S and CO_2) into deep brine aquifers. However, very little is known about the dynamic solubility of these gases under reservoir conditions.

We present the results of simulations depicting the dynamic solubility of H_2S and CO_2 in a deep brine aquifer. We compared the cases where there is solubility with cases of no solubility. We explain the effect of simultaneous injection of acid gases and water.

The results of the comparison of the cases with solubility to the cases with no solubility show that the acid gases dissolve laterally, as well as vertically. The solubility of H_2S or CO_2 in formation water (especially of H_2S) has more effect than any other variable on the distribution of H_2S and CO_2 concentration in their respective plumes. Wherever the H_2S concentration decreases due to dissolution of H_2S into the formation water, the CO_2 concentration correspondingly increases in the gas phase. This keeps the total gas phase concentration constant. The concentration distribution is different in different layers of the formation depending upon the pressure, temperature, porosity and permeability of the formation. Furthermore, simultaneous injection of water with acid gases makes the plume larger, in some high permeability layers. The concentrations of H_2S and CO_2 in the plumes created when there is simultaneous injection of acid gases and water are different from the concentration in the plumes where only acid gases are injected.

The results show how acid gas concentrations change with increasing injection rates. We also present the concentration rate changes with time, after the injection is stopped. The study results shed light on the fate of acid gases in deep brine aquifers, which, in turn, have broader implications for CO_2 sequestration.

**Corresponding author:* liaqat@xhorizonsllc.com

Ying Wu, John J. Carroll and Yongle Hu (eds.) The Three Sisters, (235–254) © 2019 Scrivener Publishing LLC

Keywords: Dynamic solubility, acid gas injection, reservoir modelling

12.1 Introduction

Much research has been conducted on the injection of acid gases into underground geological formation to advance carbon sequestration efforts [1–5, 7]. However, very little is known about the dynamic solubility of acid gases under reservoir conditions.

Recently, Li and Jiang [9] investigated the partitioning behavior of CO_2 with three impurities, N_2, CH_4 and H_2S in the formation brine by numerical simulation. They found that in a mixture of 98% CO_2 and 2% H_2S, the less soluble N_2 and CH_4 components breakthrough earlier than H_2S which has more time to come in contact with brine, hence, giving rise to more efficient stripping of H_2S. This phenomenon causes late breakthrough of H_2S. Zhang *et al.*, [6] investigated preferential dissolution of H_2S gas compared with CO_2 gas. They also found out that dissolution of H_2S into brine caused delayed breakthrough of H_2S. They concluded that iron bearing sandstone and carbonate formations are favorable for H_2S mineral trapping. Bachu *et al.*, [1] and Darvish *et al.*, [6] investigated the chromatographic partitioning of CO_2 and H_2S injected into a water saturated porous medium. They found out through lab experiments and numerical simulations that the preferential solubility of H_2S in brine over CO_2 leads to H_2S being stripped off at the leading edge of the gas displacement front, resulting in its delayed breakthrough. Ghaderi *et al.*, [7] studied the solubility of H_2S in a sour saline aquifer vs. a sweet aquifer (with no H_2S). They concluded that based on the results of their simulations, when CO_2 is injected into a sour saline aquifer, the H_2S initially dissolved in the brine comes out of solution and is released into an expanding CO_2 plume. In contrast to sour saline aquifers, brine in a sweet aquifer strips away the H_2S from the CO_2 stream and consequently the mole fraction of H_2S decreases at the leading edge of the plume. Ozah *et al.*, [13] performed detailed investigation of the storage of pure CO_2 and CO_2-H_2S mixtures in deep saline aquifers. In addition to other mechanisms for storage, they also studied the preferential solubility of the H_2S in brine by simulating the injection of a mixture of 70% CO_2 and 30% H_2S for 50 years followed by natural gradient flow for 10,000 years. They concluded that the leading edge of the gas mixture was depleted with respect to H_2S due its preferential solubility to that of CO_2.

Most of the research is focused on studying and confirming why the H_2S breaks through earlier than CO_2 in porous media. Many researchers have confirmed through laboratory experiments and simulations that when a mixture of CO_2 and H_2S is injected into a porous medium, the H_2S does break through earlier than CO_2. A few studies (two-dimensional simulations) have shown images of what the H_2S and CO_2 plumes look like during injection and many years after the injection has stopped [13].

We present images from our 3D reservoir simulations showing how the H_2S and CO_2 behave in a deep brine aquifer. We show that due to its preferential solubility, the concentration of H_2S deceases and correspondingly the CO_2 concentration increases at the leading edge of the plume.

We describe the static and dynamic model construction and history matching with the available data. We show how the solubility of H_2S in a porous brine aquifer affects the concentrations of H_2S and CO_2 in the gas zones near the wellbore as well as at a considerable distance away from the wellbore. We present results that concur with the results of the above-cited studies that the preferential solubility of H_2S over CO_2 causes the H_2S to breakthrough early.

12.2 Reservoir Simulation Modeling

Reservoir simulations were performed for a proposed well to inject a mixture of acid gases (CO_2 and H_2S) into the Glen Rose Formation. The proposed well (AGI-Y2) is located in the state of Texas, USA. There are two active disposal wells nearby. One is a salt water disposal well (SWD-X), which is about 4.5 miles away. The other active disposal well is an acid gas injection (AGI-Y1) well, which is located about 19 miles from the planned well. A model was built including these wells. To save computing time, a smaller model was extracted from this bigger model to history match the injection of the existing wells.

12.3 3D Static Model

The structure map of the Edwards Top was used as the top surface and the thicknesses of the Glen Rose Top and Base were added to build the 3D static model. The 3D model is shown in Figure 12.1. The proposed well is located at a relatively higher structural position than the other disposal wells in the area.

A porosity log was available from one well, located at a distance of about 2.8 miles north-east of the planned well (Figure 12.2). As can be

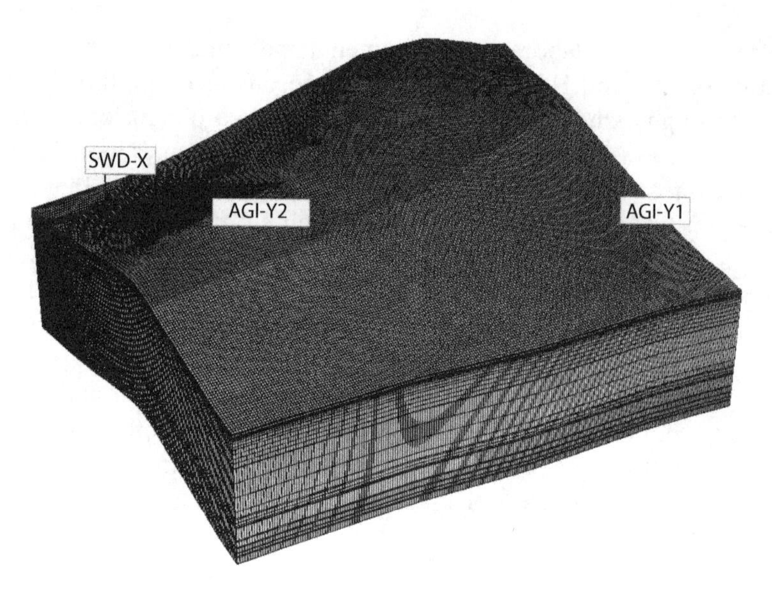

Figure 12.1 The 3D Structure of the Model.

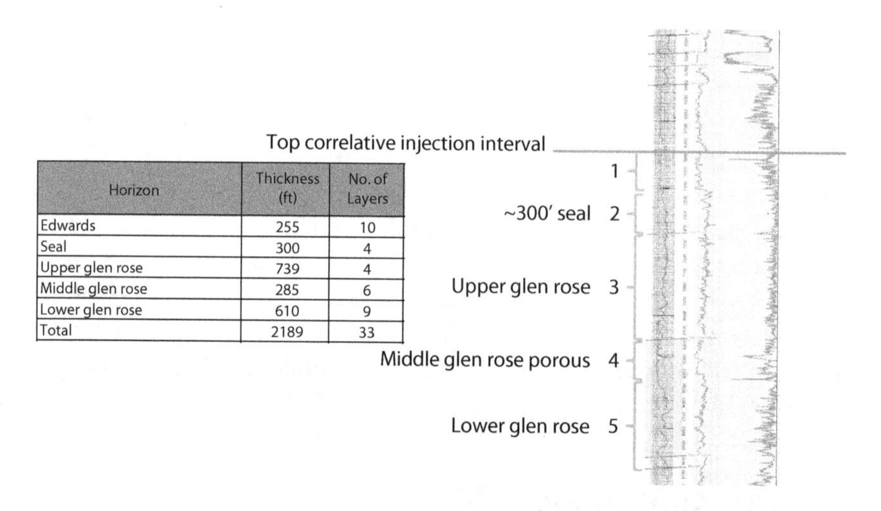

Horizon	Thickness (ft)	No. of Layers
Edwards	255	10
Seal	300	4
Upper glen rose	739	4
Middle glen rose	285	6
Lower glen rose	610	9
Total	2189	33

Figure 12.2 The Correlative Intervals Depicted on a Well Log in the Area of Interest.

seen, the porosity log shows that the porosity was variable. The model was divided into 5 zones corresponding to the five geological layers of the formation. Zone 1 consisted of the Edwards Formation, and was about 255 ft thick. It was divided into 10 simulation layers. There is a seal, about 300 ft. thick, of low porosity and permeability which was designated as Zone 2. Zone 2 was divided into 4 simulation layers. Below

Zone 2 is a 739 ft thick Upper Glen Rose section (Zone 3). The Upper Glen Rose is a zone of low porosity and permeability. Zone 3 is divided into 4 simulation layers. Zone 4 is a 285 ft thick, relatively porous and permeable zone, also called the Middle Glen Rose. Zone 4 is divided into 6 simulation layers. The lowest zone, Zone 5, is a 610 ft thick section called the Lower Glen Rose. Zone 5 is divided into 9 simulation layers. The thickness of the simulation layers (33 in total) in each zone varies depending upon the porosity. The total formation thickness of all layers is about 2200 ft.

The porosity, as indicated from the well log, is variable over the entire interval. No core data is available in Glen Rose, in this area. The permeability was calculated using Jorgensen's equation (Jorgensen, 1988):

$$k = 84105\frac{\phi^{m+2}}{(1-\phi)^2} \tag{12.1}$$

The maximum permeability is about 19 md in a layer in Zone 4. One small layer of 15 md exists in the Upper Glen Rose (Zone 3), towards the top of the zone (Layer 15). Another small layer in the Middle Glen Rose has a permeability of 18.8 md (Layer 20).

The model's dimensions are 221 × 156 × 33 with 1,137,708 active cells. Each cell represents 600 ft. X 600 ft. In order to forecast the injection rate and pressure, the grids were refined in a 1 square mile area around the well, AGI-Y2. The cells' dimensions in the refined area are 200 ft. X 200 ft. A smaller model was extracted to history match the AGI-Y1 data. The model dimensions were 122 × 96 × 33 with 386, 496 active cells. After history matching the injection rate and pressure, the matched properties were used in another extracted model around AGI-Y2. The dimensions of this model were 130 × 126 × 33 with 523,908 active cells. This is the model that was used to study the solubility of H_2S and CO_2 in the brine.

A Peng-Robinson equation of state model was built based on the acid gas composition from AGI-Y1 well, nearby. The gas composition consisted of 68.2 mole percent of CO_2, 31.6 mole percent of H_2S and 0.2 mole percent of CH_4. The AGI-Y1 well did not inject any water with the acid gas.

An injection of 3.5 MMCFD of acid gas is planned along with 1000–5000 bpd of plant waste water. The co-injection of water and acid gas will represent 90.87 mole % of water, 6.23 mole % of CO_2, 2.89 mole % of H_2S and 0.002 mole % of CH_4.

The injection interval is at a depth of 7500 ft to 9250 ft. The 3.5" (2.992" ID) tubing is set at 7400 ft. The surface injection temperature was 90 degrees F and a reservoir temperature (BHT) of 220 degrees F was

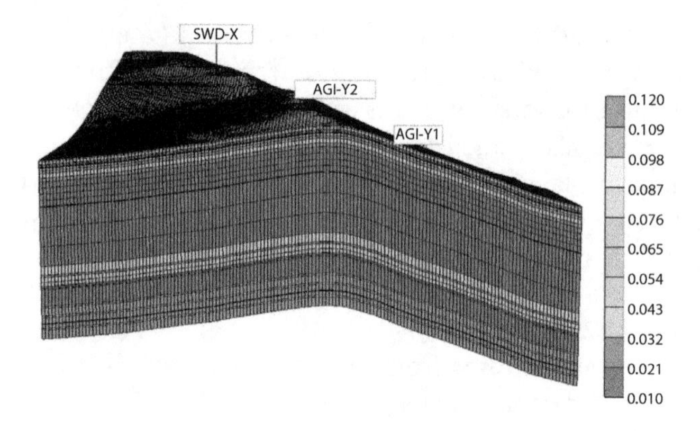

Figure 12.3 Vertical Porosity Distribution in the Model.

assumed. The reservoir pressure, calculated at the top of the zone, with a pressure gradient of 0.465 psi/ft, is 3488 psi, at a reference depth of 7500 ft.

12.4 History Matching

The effort to history match the water disposal data of the SWD-X well was unsuccessful. It seems that the pressure data was not of good enough quality to be reliable; however, a good history match was achieved for the AGI-Y1 well. The matching parameters were: the surface injection temperature and the permeability. A successful history match was achieved at a surface injection temperature of 90 degrees F and by multiplying the permeability by a factor of 10. The maximum permeability, after the history match, was 188 md in one of the layers and the minimum permeability was 0.04 md.

The history match is shown in Figure 12.4. As can be seen, the injection rates, cumulative injection volumes, and the surface injection pressures have reasonably good matches. The history match of the surface injection pressure is especially good in the initial years. Then, the observed injection pressure seems to decrease. This may be due to the reaction of the acid gas with in-situ minerals, since the injection is in the limestone reservoir. In reality, most likely, the permeability of the near wellbore region is enhanced due this reaction, hence reducing the wellhead pressure. Since the reaction of the acid gas with rock minerals was not modeled, the simulation pressure is higher than the observed pressure. At high rates, injection pressures match better.

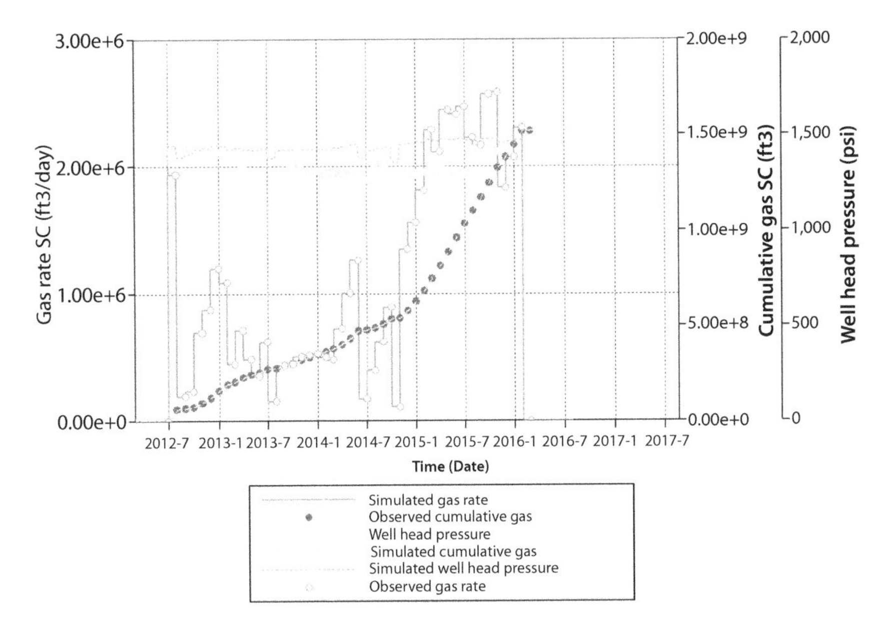

Figure 12.4 History Match of Well Pressures and Injection Rates.

12.5 Results

The objectives of the reservoir simulations were to determine the size and movement of the H_2S and CO_2 plumes in the porous brine reservoir, after 10 years of injection, followed by 30 years of injection shutdown. The results of these investigations are not presented here. The objective of this paper is to demonstrate how H_2S and CO_2 partition and move through a porous media. Therefore, only the solubility of H_2S and CO_2 in brine under dynamic conditions are discussed.

From the history matching results we found out that a surface temperature of 90 degrees F is appropriate. At the beginning of the simulation the surface temperature was an unknown. We choose a maximum surface injection pressure of 3750 psi, which is 0.5 psi/ft; a maximum bottom hole pressure (BHP) of 7400 psi, which is a little less than 1 psi/ft; an acid gas injection rate of 3.5 MMCFD and water disposal rates of 1000 bpd and 5000 bpd.

The solubility of H_2S and CO_2 under reservoir conditions in the following four cases are discussed in detail in this paper.

- **Reference Case.** Continuous injection of 3.5 MMCFD of dry acid gas for 10 years with solubility protocol turned off.

This is followed by 30 years of injection shutdown to let the plumes expand under natural gradient flow. This case was used as a reference to show the contrast between the Base Case and the Reference Case.

- **Base Case.** Continuous injection of 3.5 MMCFD of dry acid gas for 10 years with solubility protocol turned on. This is followed by 30 years of injection shutdown to let the plumes expand under natural gradient flow.
- **Case 1.** Co-injection of 3.5 MMCFD of acid gas and 1000 bpd of water continuously for 10 years with solubility protocol on. This is followed by 30 years of injection shutdown to let the plumes expand under natural gradient flow.
- **Case 2.** Co-injection of 3.5 MMCFD of acid gas and 5000 bpd of water continuously for 10 years with solubility protocol on. This is followed by 30 years of injection shutdown to let the plumes expand under natural gradient flow.

Table 12.1 presents the initial conditions of the acid gas injection at the well. The acid gas injection rate is 3.5 million cubic feet per day (MMCFD) and the water rates are 1000 and 5000 barrel per day (bpd). The injected gas composition at the wellhead is 88.2 mole% CO_2 and 31.6 mole% H_2S. The wellhead pressure (WHP) ranges from 1677 psi to 1839 psi and the bottom hole pressure (BHP) varies between 3768 and 4060 psi.

Figure 12.5 presents a reference case where H_2S and CO_2 are not soluble in brine. Figure 12.5 shows a cross-section of the plumes of H_2S and CO_2 after 10 years of injection and 30 years after the injection was stopped. The red color in the picture indicates the injected gas mole fractions at the well. It decreases towards the leading edge of the plumes of H_2S and CO_2 differently, in each layer, depending upon the permeability of the layer. In relatively low permeability layers the distribution of H_2S and CO_2 concentration decreases towards the leading edge of the plumes. Thirty years after the injection is stopped, no change in H_2S and CO_2 concentration occur.

Figure 12.6 presents a comparison of the Reference Case (no solubility) and the Base Case (with solubility) for H_2S and CO_2. The H_2S and CO_2 plumes in the Base Case, at the end of 10 years of injection, appear smaller than those of the Reference Case. Also, the distribution of H_2S and CO_2 is quite different. The full injected concentration of H_2S (31.6 mole%) is shown only in few layers (Fig. 12.6a). The concentration of CO_2 in the Base Case is more than the injected CO_2 concentration (68.1%) as shown in Fig. 12.6b. The maximum amounts of CO_2 and H_2S dissolved in water

Table 12.1 Initial conditions of the acid gas injection at the well.

	Acid gas injuction rate, MMCFD	Water injuction rate, BPD	Acid gas composition (mole%) at the well head				
			CO_2	H_2s	Total	WHP(psi)	WHP(psi)
Base Case	3.5	0	68.2	31.6	99.8	1674	3768
Case 1	3.5	1000	68.2	31.6	99.5	1741	3862
Case 2	3.5	5000	68.2	31.6	99.8	1886	4060

are about 1.51 mole% and 2.84 mole%, respectively. Since, in each cell and in every layer, the H_2S concentration is decreasing due to dissolution in water, the CO_2 concentration is increasing. Some CO_2 also dissolves in the

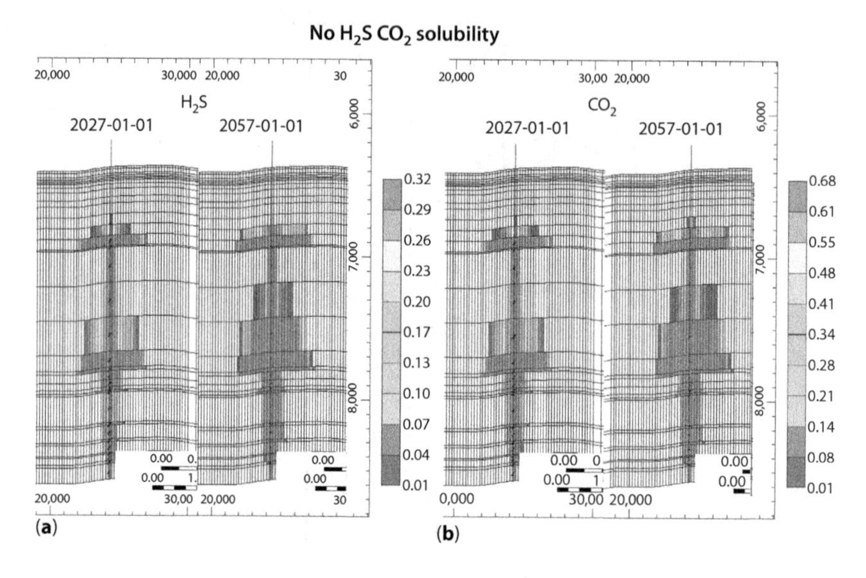

Figure 12.5 Comparison of H_2S and CO_2 Plumes at the End of 10 Years of Injection for no Solubility Case.

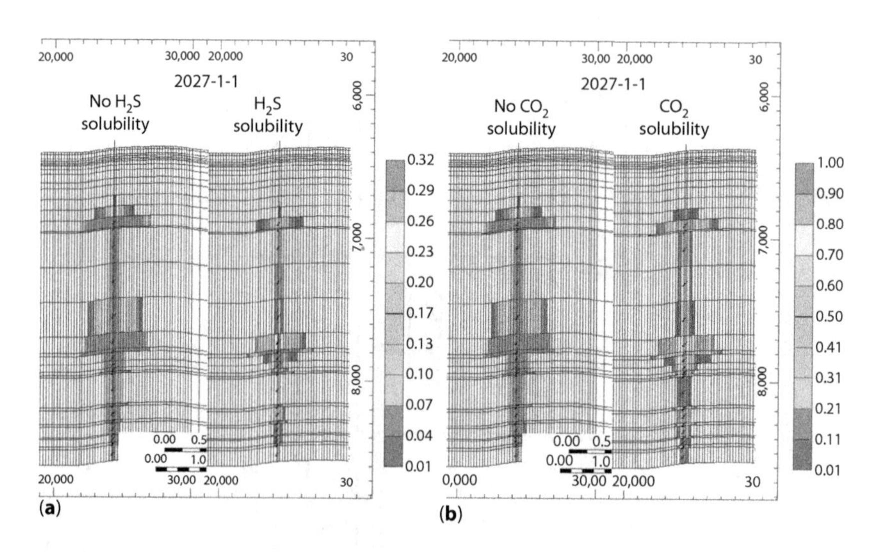

Figure 12.6 Comparison of H_2S and CO_2 Plumes in Layers 15 and 20 at the End of 10 Years of Injection for the Reference Case.

water. However, in each cell the total gas concentration remains constant. This behavior is obvious in the results published by Ozah *et al.*, (2005); however, they did not explain this behavior in their paper. The H_2S concentration decreases from 30% (injected concentration) to 10% and correspondingly the CO_2 concentration increases to 90% which is above the injected concentration of 70% (Figures. 12.12 and 12.13 of their paper). This happens at the leading edge of the respective plumes. However, the

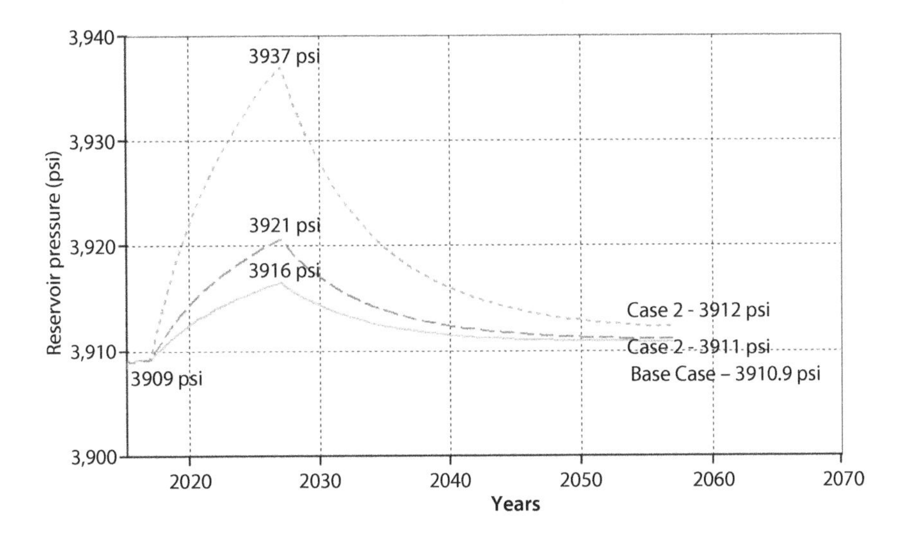

Figure 12.7 Reservoir Pressure Profiles During the Periods of Injection and Injection Shutdown.

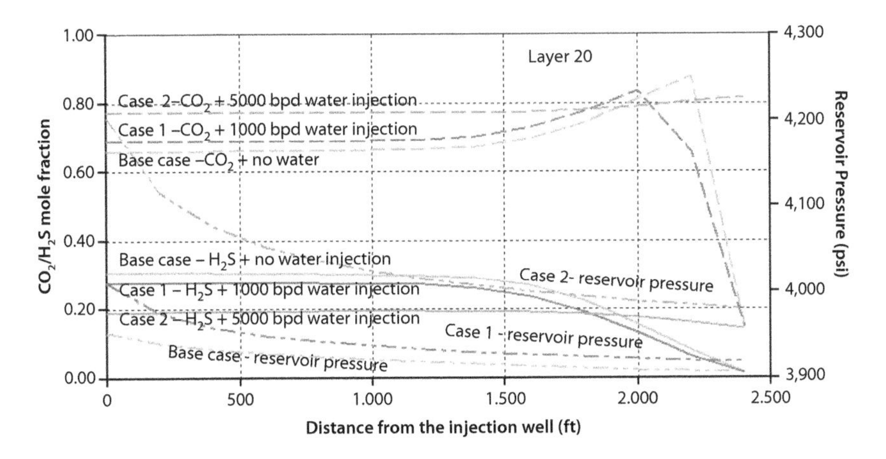

Figure 12.8 An Evolution of the Concentration of H_2S and CO_2 Away from the Injection Well for Base Case, Case 1 and Case 2 for Layer 20.

total concentration remains constant. This solubility behavior will be explained later when we discuss the solubility of acid gases in the high permeability Layers 15 and 20.

In the rest of the paper only the Base Case, Case 1 and Case 2 are discussed since they consider the solubility of H_2S and CO_2 in the reservoir brine. It is worthwhile to observe the reservoir pressure profile.

Figure 12.7 shows the reservoir pressure profile during the injection period and the injection shutdown period. The initial reservoir pressure is 3909 psi. The Base Case has the lowest reservoir pressure (3916 psi), just before the injection was stopped. Since water was being injected along with H_2S and CO_2 mixture, the reservoir pressure is higher in Cases 1 and 2 than in the Base Case. However, the difference in reservoir pressure is not very significant. After 30 years of injection shutdown the reservoir pressure comes back close to the initial reservoir pressure. The reservoir pressure increased only 1-3 psi.

Figure 12.8 presents an evolution of the concentration of H_2S and CO_2 away from the injection well in Layer 20 for the Base Case, Case 1 and Case 2. Several observations are very noteworthy. For the Base Case, where no water is being injected with the H_2S and CO_2 mixture, the initial injected composition (31.6 mole% H_2S and 68.2 mole% CO_2, as given in Table 12.1) is a little less at the wellbore (30.5 mole% H_2S and 65.8 mole% CO_2). As soon as the acid gas is injected, some of it dissolves in brine and reduces the H_2S concentration in the gas phase. Since water is being injected along with H_2S and CO_2 in Cases 1 and 2, the initial composition is significantly different than the injected composition at the injection well. The higher the injection rate, the lower the H_2S concentration and the higher the CO_2 concentration. However, the total concentration of the mixture remains the same most of the time (~96.5 mole%), as shown in Table 12.2 and 12.3. At about 1000 ft. from the injection well, the reservoir pressure dropped considerably. The concentration of H_2S starts decreasing, whereas the concentration of CO_2 starts increasing, as the H_2S and CO_2 plumes move further into the reservoir. Probably, H_2S is being stripped away from this point onward. In the Figure 12.8 for Case 2, the change in the concentration of H_2S and CO_2 occurs until the plumes are about 1500 ft. away from the injection well. This may be due to the high volume of water injection, in this case. However, the change in concentrations of H_2S and CO_2 is not as pronounced as in Base Case and Case 1.

Figure 12.9 shows the evolution of the concentration of H_2S and CO_2 away from the injection well for Base Case, Case 1 and Case 2, 30 years after the injection shutdown. After 30 years of injection shutdown, the

Table 12.2 Acid gas composition and pressure at end of 10 years of injection for layer 15.

Layer 15		Acid gas injuction rate, MMCFD	Water injuction rate, BPD	Acid gas composition (mole%) never Wellbore				Acid gas composition at 2000 ft.			
				CO_2	H_2S	Total	BHP(psi)	CO_2	H_2S	Total	Pressure(psi)
1/1/2027	**Base Case**	3.5	0	65.9	30.5	96.4	3952	77.3	19.4	96.7	3909
	Case 1	3.5	1000	68.1	31.6	99.7	4011	69.5	27	96.5	3922
	Case 1	3.5	5000	70.8	25.6	96.4	4203	84	12.5	96.5	3989
1/1/2057	**Base Case**	3.5	0	67.0	27.1	94.1	3952	76	24	100.0	3909
	Case 1	3.5	1000	59.6	29.5	89.1	4011	69.8	20.5	90.3	3922
	Case 1	3.5	5000	70.8	25.7	96.5	4203	83.9	12.6	96.5	3989

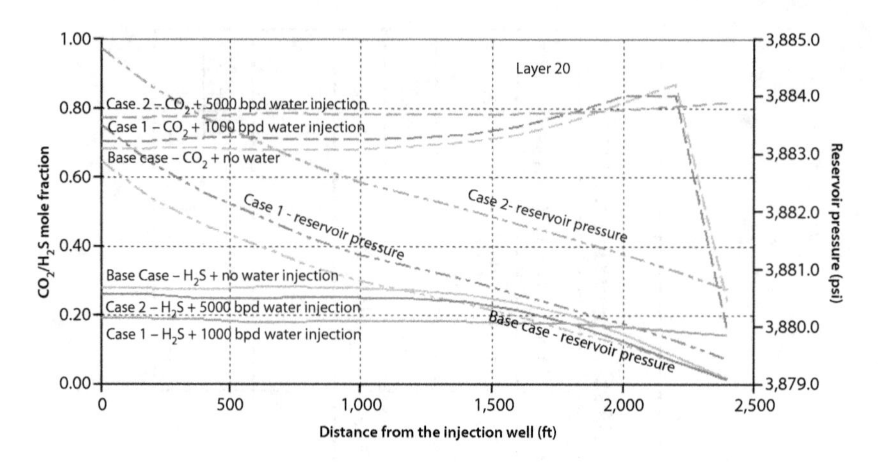

Figure 12.9 Evolution of Mole Fraction of H_2S and CO_2 Away from the Injection Well for Base Case, Case 1 and Case 2, 30 Years after the Injection Shutdown.

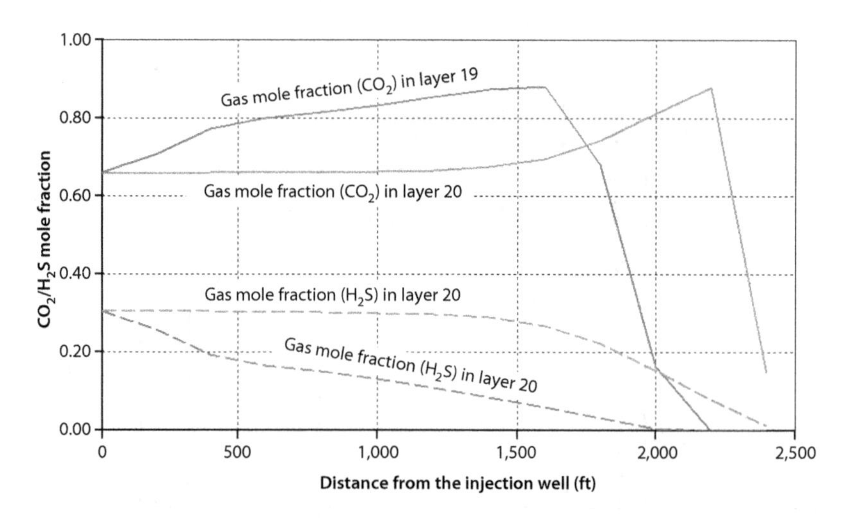

Figure 12.10 Comparison of H_2S and CO_2 Solubility at a Distance Away from the Wellbore for Case 1 in Layers 19 and 20 at the End of 10 Years of Injection.

change in concentration of H_2S and CO_2 is very similar to that of the cases before injection shutdown. Here again, at about 1000 ft. away from the wellbore, the H_2S concentration in Base Case and Case 1 starts decreasing with a corresponding increase in the concentration of CO_2. However, for Case 2, a slight change in the concentration of H_2S and CO_2 starts to occur at about 1500 ft. away from the injection well. It is interesting to note that the drop in pressure is only a few psi in 30 years. It seems like most of the H_2S has already been stripped away during the injection phase. The change in H_2S concentration at the end of 10 years of injection and 30 after the injection shutdown is very small (Table 12.2 and 12.3).

In Figure 12.10, a comparison of the solubility of H_2S and CO_2 is shown at distance away from the wellbore at the end of 10 years of injection in Layers 19 and 20 for Case 1. We show here only Case 1. Similar H_2S and CO_2 behavior is observed in other cases. Layer 19 is a lower permeability layer whereas Layer 20 is the highest permeability layer. As can be seen, the H_2S concentration at the wellbore is 28 mole% and the CO_2 concentration is 68 mole%. Some of the H_2S is already dissolved in the brine since brine is being injected along with acid gases at the rate of 1000 bpd in this case. Away from the wellbore, the concentrations of H_2S and CO_2 decrease significantly more in the lower permeability layer than that in the higher permeability layer. In Layer 19, the concentration of H_2S is zero just before the leading edge and the concentration of CO_2 is zero at the leading edge. In Layer 20, the concentration of H_2S is zero at the leading edge whereas the concentration of CO_2 is 14.6 mole% at the leading edge of the plume.

Figure 12.11 shows the preferential solubility of H_2S in Layer 20 for Case 1. To show the preferential solubility of H_2S, we selected a cell at a distance away from the wellbore. As you can see, the CO_2 breakthrough is earlier in that cell, earlier than that of H_2S. When the H_2S reached 9 mole%, the CO_2 concentration reached a maximum of 86.7 mole%. The injected CO_2 concentration was only 68.2 mole%. Once the concentration of CO_2 reaches a maximum value, it starts decreasing while the concentration of H_2S is still increasing. A maximum concentration of 19.1 mole% of H_2S is reached in the H_2S plume much later than the maximum concentration of CO_2 is reached in the CO_2 plume (about 3 years later).

Figure 12.12 presents a comparison of H_2S and CO_2 plumes for Base Case. The four images on the left are of H_2S plumes and the four images on the right are CO_2 plumes. In each set of images, H_2S and CO_2, the top images are for Layer 15 and the bottom images are for Layer 20. As you can see, the size of the plumes for H_2S and CO_2 remain unchanged before and after 30 years of injection stoppage. The concentration of H_2S and CO_2 change slightly in Layer 15, whereas in Layer 20 a significant change in concentration of H_2S and CO_2 is observed after 30 years of injection stoppage.

Figure 12.13 shows a comparison of H_2S and CO_2 plumes for Case 1. As can be seen, the size of the H_2S plume is the same. The H_2S concentration, on the other hand, changes slightly in Layer 15 and a marked

Table 12.3 Acid gas composition and pressure at the end of 10 years of injection for layer 20.

Layer 20		Acid gas injuction rate, MMCFD	Water injuction rate, BPD	Acid gas composition (mole%) never Wellbore				Acid gas composition at 2000 ft.			
				CO_2	H_2S	Total	BHP(psi)	CO_2	H_2S	Total	Pressure(psi)
1/1/2027	**Base Case**	3.5	0	65.8	30.5	96.3	3952	81.2	15.4	96.6	3909
	Case 1	3.5	1000	68.7	27.8	96.5	4011	83.3	13.7	97.0	3922
	Case 1	3.5	5000	77.5	18.9	96.4	4203	78.9	17.5	96.4	3989
1/1/2057	**Base Case**	3.5	0	68.1	28.2	96.3	3882.9	81.6	14.6	96.2	3879.7
	Case 1	3.5	1000	70.1	26.1	96.2	3883.5	83.5	12.6	96.1	3880
	Case 1	3.5	5000	77.2	19.2	96.4	3884.8	79.4	16.7	96.1	3881

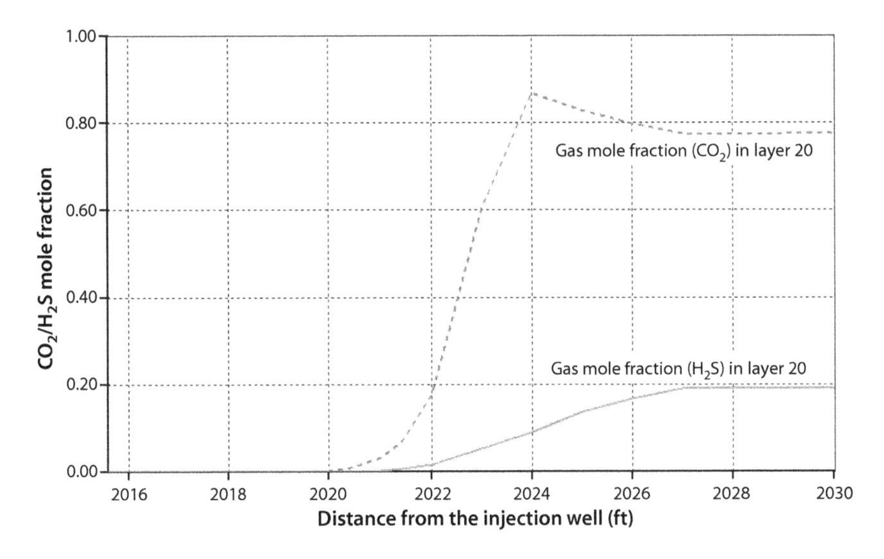

Figure 12.11 Preferential Solubility of H_2S in Layer 20 for Case 1.

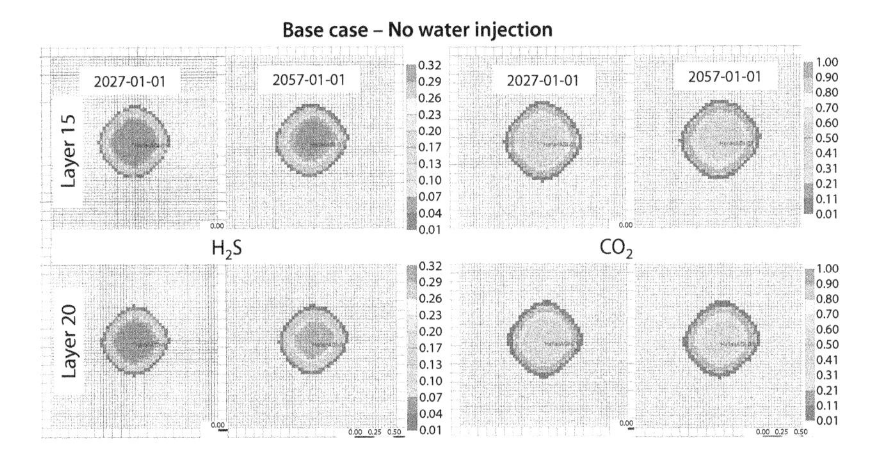

Figure 12.12 Comparison of H_2S and CO_2 Plumes for Base Case.

change is observed in Layer 20, after 30 years of injection shutdown. The CO_2 plume size remains the same in both layers. A significant change in CO_2 concentration is observed in Layer 20 after 30 years of injection shutdown.

Figure 12.14 presents a comparison of H_2S and CO_2 plumes for Case 2. A significant change in H_2S concentration is observed in Layer 15, after 30 years of injection shutdown. The plume size remains unchanged. In Layer

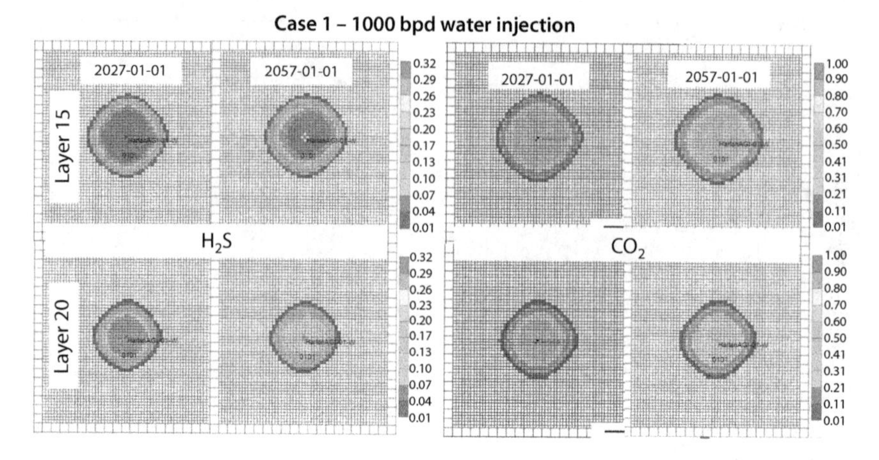

Figure 12.13 Comparison of H_2S and CO_2 Plumes for Case 1.

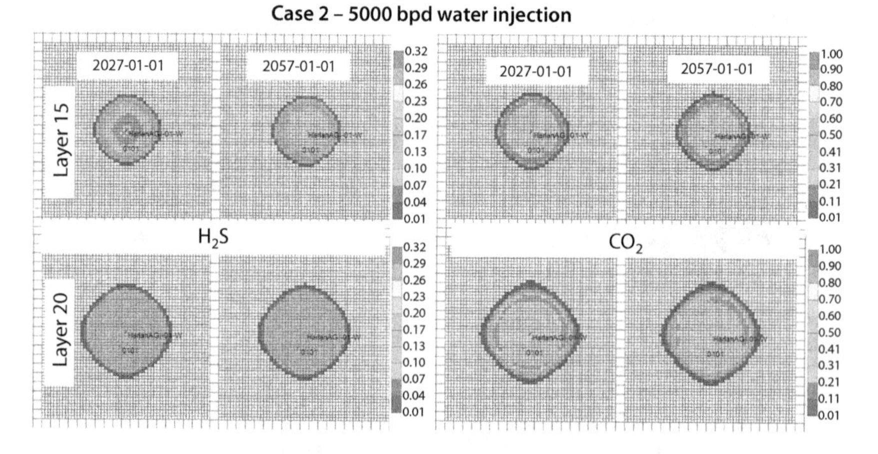

Figure 12.14 Comparison of H_2S and CO_2 Plumes for Case 2.

20, the plume size and the concentration of H_2S remains the same. No significant change is observed either in plume size or CO_2 concentration in both layers.

12.6 Summary and Conclusions

The simulation model is based on very limited data. The model is based on well log data from three wells penetrated in Glen Rose formation of which one well had a porosity log. The maximum permeability, after history matching the AGI-Y1 well's injection and pressure data, is 188 md in Layer 20. The simulation layering in the model is based on the porosity profile of a nearby well.

We successfully history matched the acid gas injection/pressure data of AGI-Y1 well. The observed acid gas injection pressure data compared with that of the simulated data indicated that the injection pressure into the formation decreases after about a year. This could be due to reaction of the acid gas with the in-situ minerals, since Glen Rose is a limestone formation.

The solubility of H_2S and CO_2, especially the H_2S, has the greatest effect on the distribution of H_2S and CO_2 concentrations in their respective plumes. Wherever H_2S concentration decreases due to dissolution of H_2S into the formation water, the CO_2 concentration correspondingly increases in the gas phase, keeping the total gas phase concentration constant. This happens at the leading edge of the plume.

The concentration distribution is different in different layers depending upon the pressure, temperature, porosity and permeability of the layer. Co-injection of water with acid gas makes the plume larger, in some high permeability layers. Also, the concentrations of H_2S and CO_2 are different in the plumes with water, compared with those with acid gas injection only.

In general, in all the cases run, whether acid gas only or co-injection of acid gas with water, the injection pressure increased over the years.

The rate of acid gas injection was set constant in all the cases at 3.5 MMCFD. A maximum surface injection pressure of 1674 psi and a bottom hole pressure (BHP) of 3768 psi were observed in the case of continuous injection of acid gas only. With continuous injection, a maximum surface injection pressure of 1741 psi and a BHP of 3862 psi were observed for the case of co-injection of acid gas with 1000 bb/d of water. With continuous co-injection with water of 5000 bpd, a maximum surface injection pressure of 1886 psi and a BHP of 4060 psi were observed.

The H_2S and CO_2 plumes were larger in the cases where there was co-injection of acid gas and water. The H_2S and CO_2 concentration in the plumes is different for each case and for each layer.

References

1. Hosseini, S.A., Lashgari, H., Choi, J.W., Nicot, J.-P., Lu, J., Hovorka, S.D., Static and dynamic reservoir modeling for geological CO_2 sequestration at

Cranfield, Mississippi, U.S.A. *International Journal of Greenhouse Gas Control,* 18, 449–462, 2013.
2. Michael, K., Golab, A., Shulakova, V., Ennis-King, J., Allinson, G., Sharma, S., *et al.*, Geological storage of CO_2 in saline aquifers—A review of the experience from existing storage operations. *International Journal of Greenhouse Gas Control,* 4(4), 659–667, 2010.
3. Wigand, M., Carey, J.W., Schütt, H., Spangenberg, E., Erzinger, J., Geochemical effects of CO_2 sequestration in sandstones under simulated in situ conditions of deep saline aquifers. *Applied Geochemistry*, 23(9), 2735–2745, 2008.
4. Leonenko, Y., Keith, D.W., Reservoir engineering to accelerate the dissolution of CO_2 stored in aquifers. *Environ. Sci. Technol.*, 42(8), 2742–2747, 2008.
5. Nordbotten, J.M., Celia, M.A., Bachu, S., Injection and Storage of CO_2 in Deep Saline Aquifers: Analytical Solution for CO_2 Plume Evolution During Injection. *Transp. Porous Med.*, 58, , (3), 339–360, 2005.
6. Bachu, S., Nordbotten, J.M., A.:, C.M., Evaluation of the spread of acid gas plumes injected in deep saline aquifers in western Canada as an analogue to CO2 injection in continental sedimentary basins. *Proceedings of the 7th International Conference on Greenhouse Gas Control Technologies*, 5–9, 2004.
7. Bachu, S., Gunter, W.D., Acid-gas injection in the Alberta basin, Canada: a CO_2-storage experience, Geological Society, London, Special Publications, 233, 225–234, 2004.
8. Saripalli, P., McGrail, P., Semi-analytical approaches to modeling deep well injection of CO_2 for geological sequestration. *Energy Conversion and Management*, 43(2), 185–198, 2002.
9. Li, D., Jiang, X., Numerical investigation of the partitioning phenomenon of carbon dioxide and multiple impurities in deep saline aquifers, 7th International Conference on Applied Energy (ICAE2015), March 28–31, Abu Dhabi, UAE, 2015.
10. Zhang, W., Xu, T., Li, Y., Modeling of fate and transport of coinjection of H_2S with CO_2 in deep saline formations. *J. Geophys. Res.*, 116, 2011.
11. Pooladi-Darvish, M., Hong, H., Stocker, R.K., Bennion, B., Theys, S., Bachu, S., Chromatographic Partitioning of H_2S and CO_2 in Acid Gas Disposal. *Journal of Canadian Petroleum Technology*, 48(10), 52–57, 2009.
12. Ghaderi, S.M., Keith, D.W., Lavoie, R., Leonenko, Y., Evolution of hydrogen sulfide in sour saline aquifers during carbon dioxide sequestration. *International Journal of Greenhouse Gas Control*, 5(2), 347–355, 2011.
13. Ozah, R.C., Pope, G.A., Numerical Simulation of the Storage of Pure CO_2 and CO_2-H_2S Gas Mixtures in Deep Saline Aquifers, 12, 2005.

13

Tomakomai CCS Demonstration Project of Japan, CO_2 Injection in Progress

Yoshihiro Sawada*, Jiro Tanaka, Chiyoko Suzuki, Daiji Tanase and Yutaka Tanaka

Japan CCS, Co., Ltd., Tokyo, Japan

Abstract
A large-scale CCS demonstration project is being undertaken by the Japanese government in the Tomakomai area, Hokkaido Prefecture, Japan. The project will run between JFY 2012 – 2020 (JFY is from April of calendar year to following March) to demonstrate the viability of a full-cycle CCS system, from CO_2 capture to injection and storage. One hundred thousand tonnes per year or more of CO_2 is being captured, injected and stored in sub-seabed saline aquifers in Tomakomai Port. The construction and commissioning of the facilities was completed in March 2016, and CO_2 injection began in April. CO_2 injection will be conducted for three years, and monitoring will be for five years. This paper describes and reviews the progress of the main features of this project: 1) Extensive monitoring system in a seismically active country, 2) Deviated CO_2 injection wells drilled from onshore to offshore, 3) Application of law reflecting London Protocol, 4) Low energy CO_2 capture process utilizing high CO_2 partial pressure gas, 5) Injection of CO_2 near urban area.

Keywords: CCS demonstration project, seismically active country, deviated injection well, London Protocol, low capture energy, two-stage absorption system, low pressure flash tower, injection near urban area

13.1 Introduction

Japan has conducted several CO_2 injection projects in the past, as shown in Figure 13.1. Two CO_2- EOR test projects were carried out in the

**Corresponding author:* yoshihiro.sawada@japanccs.com

Ying Wu, John J. Carroll and Yongle Hu (eds.) The Three Sisters, (255–276) © 2019 Scrivener Publishing LLC

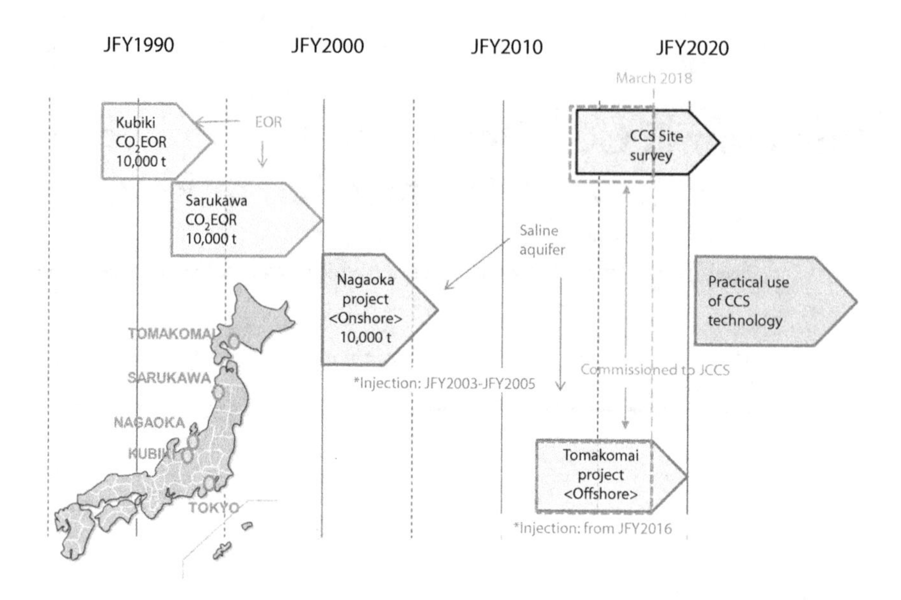

Figure 13.1 CO_2 Injection and Site Survey Projects in Japan.

1990s, injecting approximately 10,000 tonnes of CO_2 in each project. In the early 2000s, approximately 10,000 tonnes of CO_2 was injected into a saline aquifer in the Nagaoka Project, and monitoring technology was developed through this project.

A large-scale CCS demonstration project is being undertaken by the Japanese government in the Tomakomai area, Hokkaido Prefecture, Japan. The objective is to demonstrate the viability of a full CCS system, from CO_2 capture to injection and storage. One hundred thousand tonnes/year or more of CO_2 is being injected and stored in offshore saline aquifers in the Tomakomai port area. The implementation of this project has been commissioned to Japan CCS Co., Ltd. (JCCS) by the Ministry of Economy, Trade and Industry (METI). The investigation of the potential sites for CO_2 storage is also being undertaken by JCCS, under consignment by the Ministry of the Environment (MOE) and METI. These projects are being conducted in accordance with Japan's Strategic Energy Plan [1], which states that "research and development will be conducted with a view to practical use of the carbon capture and storage (CCS) technology around 2020 and a study will be conducted on introducing CCS-ready facilities as early as possible with due consideration given to the possible timing of the commercialization of CCS."

The company profile of JCCS and the project framework are shown in Figure 13.2. JCCS was founded in May 2008 when a group of major

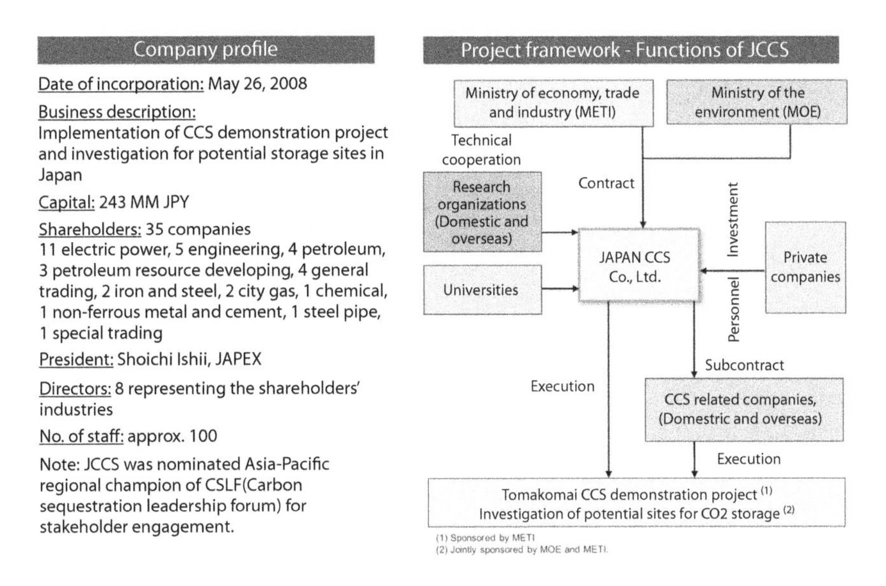

Figure 13.2 Company Profile and Project Framework.

companies with expertise in CCS-related fields, including electric power, petroleum, oil development, and plant engineering, joined forces to answer the Japanese government's call for development of CCS technology as a countermeasure against global warming. JCCS is a unique company in the world in that it was founded and dedicated explicitly to developing integrated CCS technology.

13.2 Overview of Tomakomai Project

The flow scheme of the Tomakomai Project is shown in (Figure 13.3). The CO_2 source is a Pressure Swing Adsorption (PSA) offgas from a hydrogen production unit (HPU) of an oil refinery located in the coastal area of Tomakomai Port. The HPU provides CO_2 rich offgas to the Tomakomai demonstration project CO_2 capturefacility via a 1.4 km pipeline. In the capture facility, gaseous CO_2 of 99% or higher purity is recovered from the offgas by a commercially provenamine scrubbing process with a design capacity of 200,000 tonnes per year.

At the injection facility, the gaseous CO_2 is compressed and injected into two different offshore reservoirs by two dedicated deviated injection wells. The storage points are located 3 to 4 km offshore.

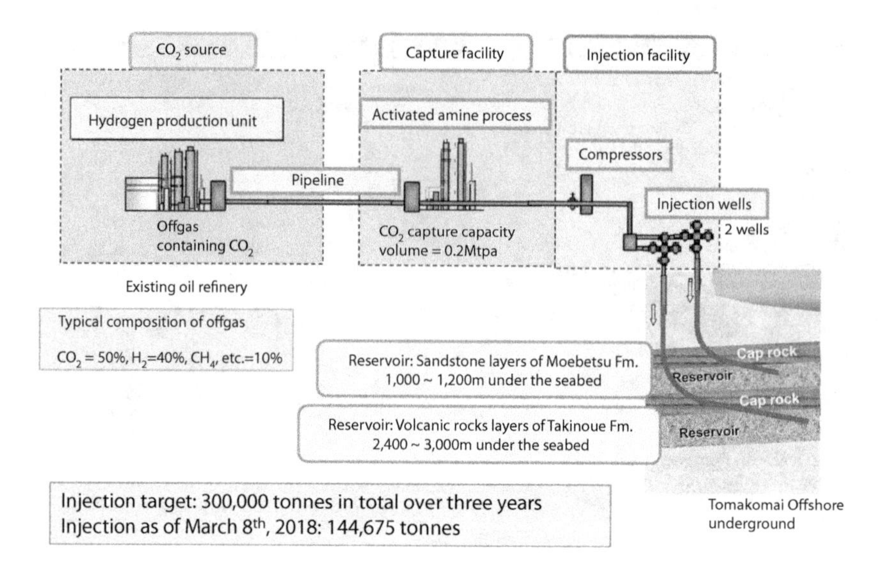

Figure 13.3 Flow Scheme of Tomakomai Project.

Figure 13.4 Aerial Photo of Capture and Injection Facilities.

An aerial photo of the capture and injection facilities, CO_2 capture facilities and compressors, heads of injection wells are shown in Figure 13.4, Figure 13.5, and Figure 13.6, respectively.

Figure 13.5 CO_2 Capture Facilities and Compressors.

Figure 13.6 Heads of Injection Wells.

As the number of dedicated geological storage projects is small, the Tomakomai Project is gaining valuable experience in CO_2 injection into deep saline aquifers, comprising a sandstone layer and a volcanic rock layer.

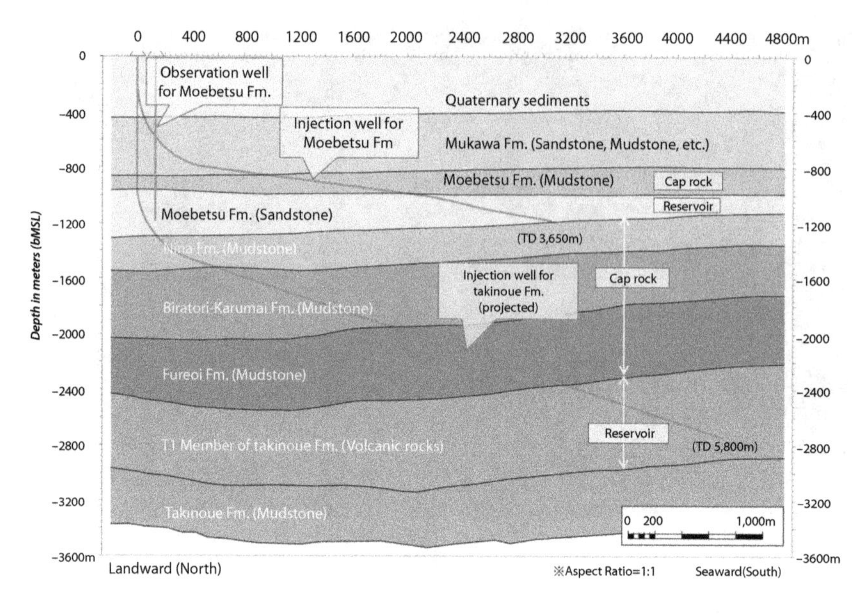

Figure 13.7 Schematic Geological Section.

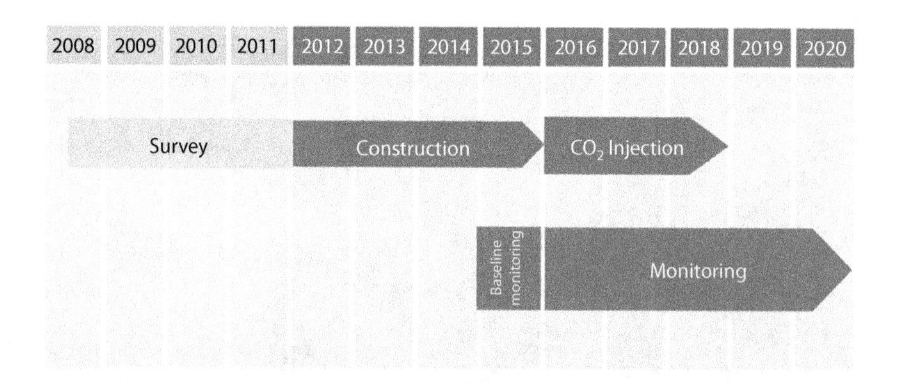

Figure 13.8 Project Schedule.

A schematic geological section is shown in Figure 13.7. The shallow reservoir (Moebetsu Formation) is located at a depth of approximately 1,000 m below the seabed. This reservoir is a Lower Quaternary saline aquifer, mainly composed of sandstone and is approximately 200 m thick. The deep reservoir (Takinoue Formation) is located at a depth of approximately 2,400 m below the seabed. This reservoir is a Miocene saline aquifer composed of volcanic and volcaniclastic rocks and is approximately 600 m thick.

The project schedule is shown in (Figure 13.8). The project is scheduled to run between the period of JFY 2012 - 2020 (JFY is from April of calendar year to following March) to demonstrate the viability of a full-cycle CCS system, from CO_2 capture to injection and storage. The construction and commissioning of the onshore facilities was completed in March 2016, and CO_2 injection began in April 2016. The injection of CO_2 will be conducted for three years and monitoring for five years.

At the annual meeting of the Carbon Sequestration Leadership Forum (CSLF), a ministerial-level international climate change initiative with 26 member governments for the development of improved cost-effective technologies for CCS, in Tokyo in October 2016, the Tomakomai Project was formally certified as a CSLF-recognized project, and JCCS was nominated Asia-Pacific regional champion for stakeholder engagement.

13.3 Injection Record

The injection record of the Moebetsu Formation (sandstone) is shown in Figure 13.9. The cumulative injected volume is 150,038 tonnes as of March 18th, 2018. The injection rate varied between 7.6 ~ 25.3 tonnes/hr (180 ~ 600 tonnes/day), as it depends on the supply of CO_2 containing gas from the oil refinery. The bottom hole pressure was 9.2MpaG before

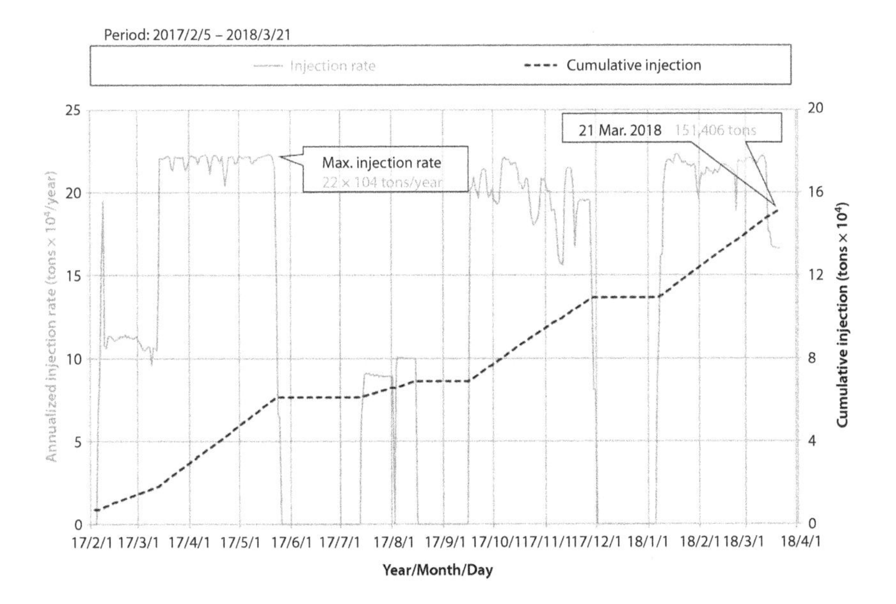

Figure 13.9 Injection Record to Moebetsu Formation (Sandstone).

injection and reached 10MpaG during injection at a maximum injection rate of 600 tonnes/day. The bottom hole pressure at the maximum injection rate is much smaller than the allowable upper limit of the injection pressure (12.6MPaG) which was set at 90% of the breaking strength of the overlying cap rock. The permeability of the Moebetsu Formation is high.

The injection into the Takinoue Formation (volcanic rocks) started in February 2018. The analysis of initial injection results is underway. It is expected that the injectivity of the Takinoue Formation will be very small due to its low permeability.

13.4 Features of Tomakomai Project

13.4.1 Feature 1: Extensive Monitoring System in a Seismically Active Country

To confirm the safety and stability of CO_2 injection, it is necessary to monitor the CO_2 behavior in the reservoirs and conduct observation continuously to detect any CO_2 leakage. This is particularly important in a seismically active country.

Japan is located in a very seismically active zone. Thus, the most important objective of the Tomakomai CCS Demonstration Project is to verify that:

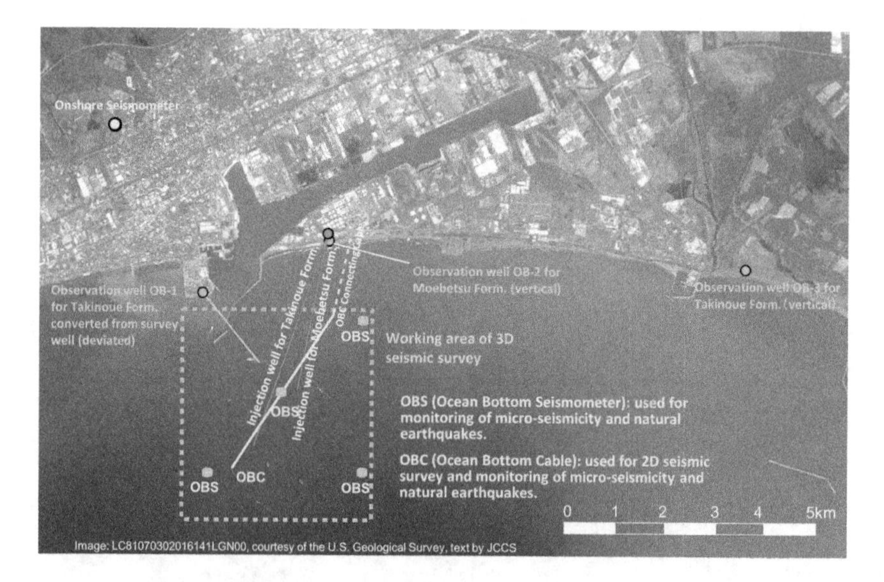

Figure 13.10 Location of Wells and Monitoring Facilities.

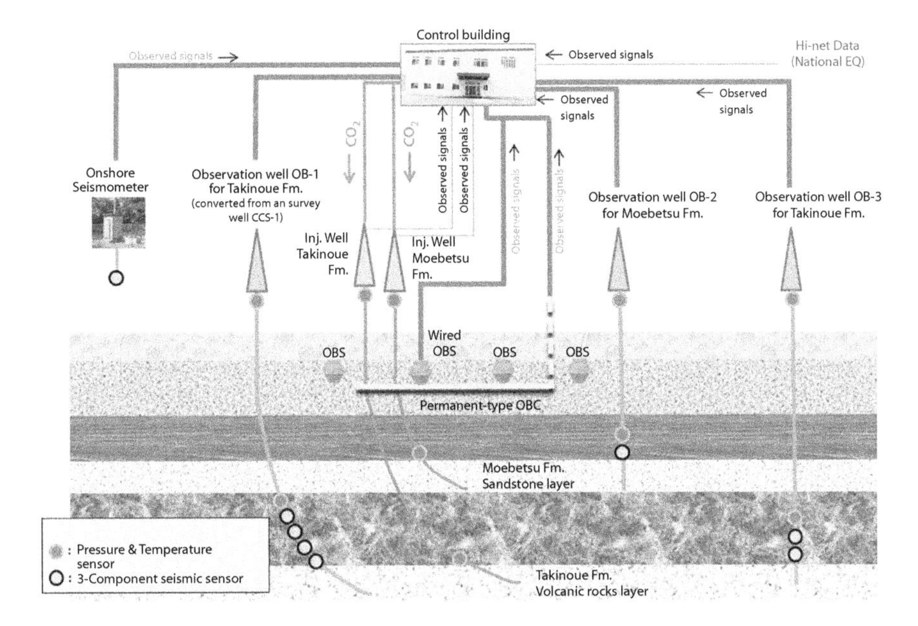

Figure 13.11 Schematic Diagram of Monitoring System.

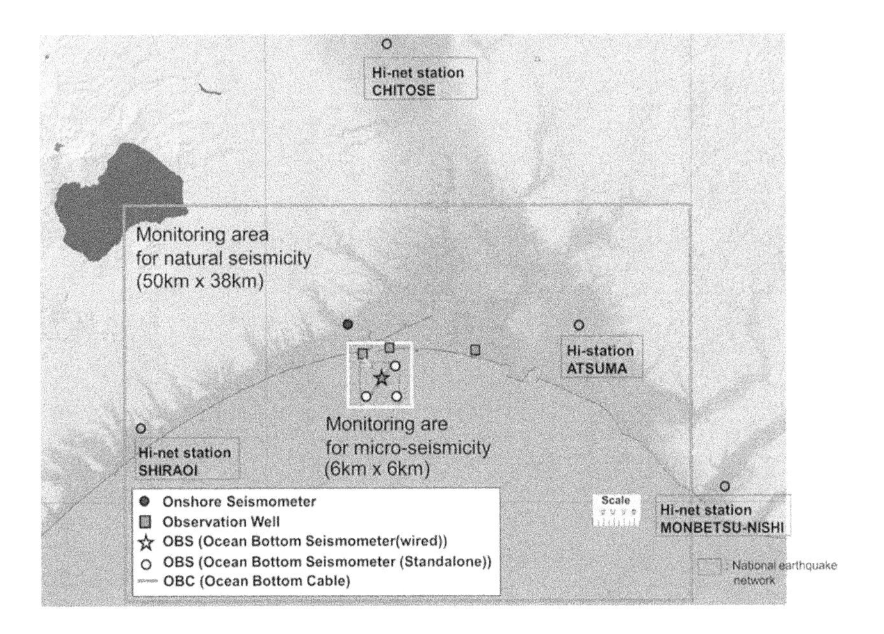

Figure 13.12 Seismic Monitoring Area.

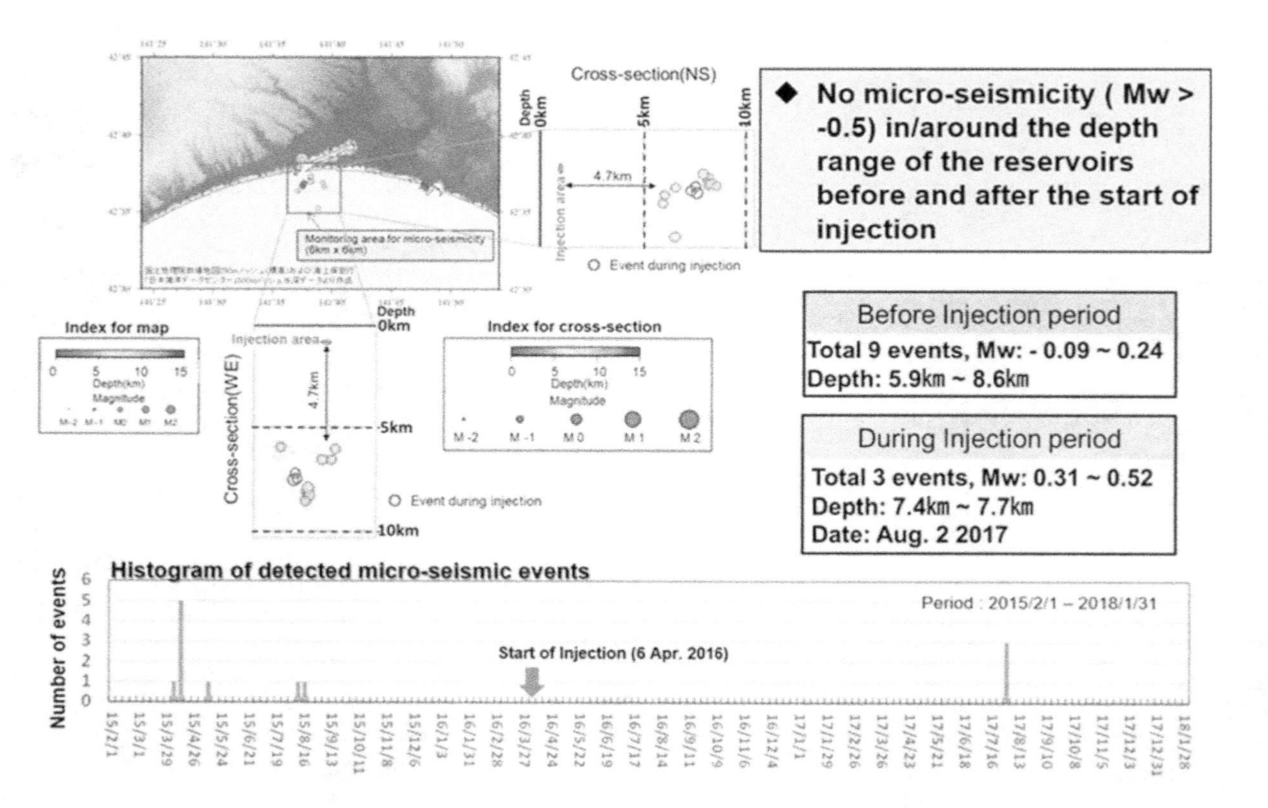

Figure 13.13 Micro-Seismicity Monitoring Results.

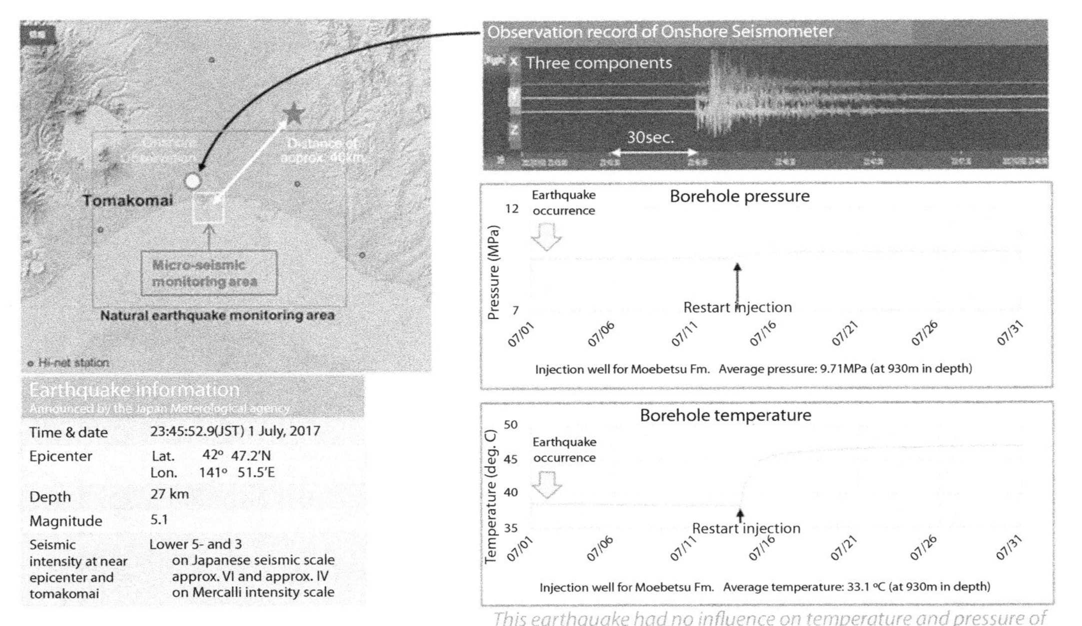

Figure 13.14 Natural Earthquakes Monitoring Results.

- natural earthquakes have no influence on the CO_2 stored
- no perceptible earth tremors are induced by the CO_2 injected thereby removing concerns about CCS regarding earthquakes. To this end, an extensive monitoring system has been installed.

The location of the wells and the monitoring facilities are shown in Figure 13.10, and a schematic diagram of the monitoring system is shown in Figure 13.11. The monitoring system is comprised of three observation wells equipped with pressure and temperature (PT) sensors, seismic sensors, comprehensive seismic instrumentation of one onshore seismic station, four ocean bottom seismometers (OBSs) and an ocean bottom cable (OBC) with a total of 72 seismometers. Baseline data acquisition of the monitoring system started from February 2016. A baseline 3D seismic survey was conducted in 2009 during the investigation period prior to the project, and a 2D seismic survey was conducted in 2013. The first time-lapse 2D seismic survey after the start of CO_2 injection was conducted from January to February 2017 when the cumulative injection was at 7,200 tonnes, but did not show any anomaly. The first time-lapse 3D seismic survey was conducted from July to August 2017 at cumulative injection of 61,000 to 69,000 tonnes, and a clear anomaly was detected along the injection interval. Further data processing is currently underway.

The seismic monitoring area is shown in Figure 13.12. The micro-seismicity monitoring results are shown in Figure 13.13. Before injection of CO_2, a total of 9 events with a moment magnitude (Mw) of −0.09 ~ 0.24 were detected at depths of 5.9 km ~8.6 km, whereas a total of 3 events with a Mw of 0.31 ~ 0.52 were detected at depths of 7.4 km ~7.7 km after commencement of injection. As no micro-seismicity (Mw >−0.5) in/around the depth range of the reservoirs before and after the start of injection has been detected, it is deemed that the injection of CO_2 has not induced micro-seismic activity.

On July 1, 2017, a natural earthquake of magnitude 5.1 occurred at a distance of about 40 km from the CO_2 injection area. The seismic intensity near the epicenter and in Tomakomai were approximately VI and IV respectively. At that time, injection was not being conducted due to annual maintenance of the facilities. As shown in Figure 13.14, the borehole pressure and the borehole temperature maintained constant, and it was concluded that the large natural earthquake did not have any adverse effect on the stored CO_2. On July 14, 2017, injection

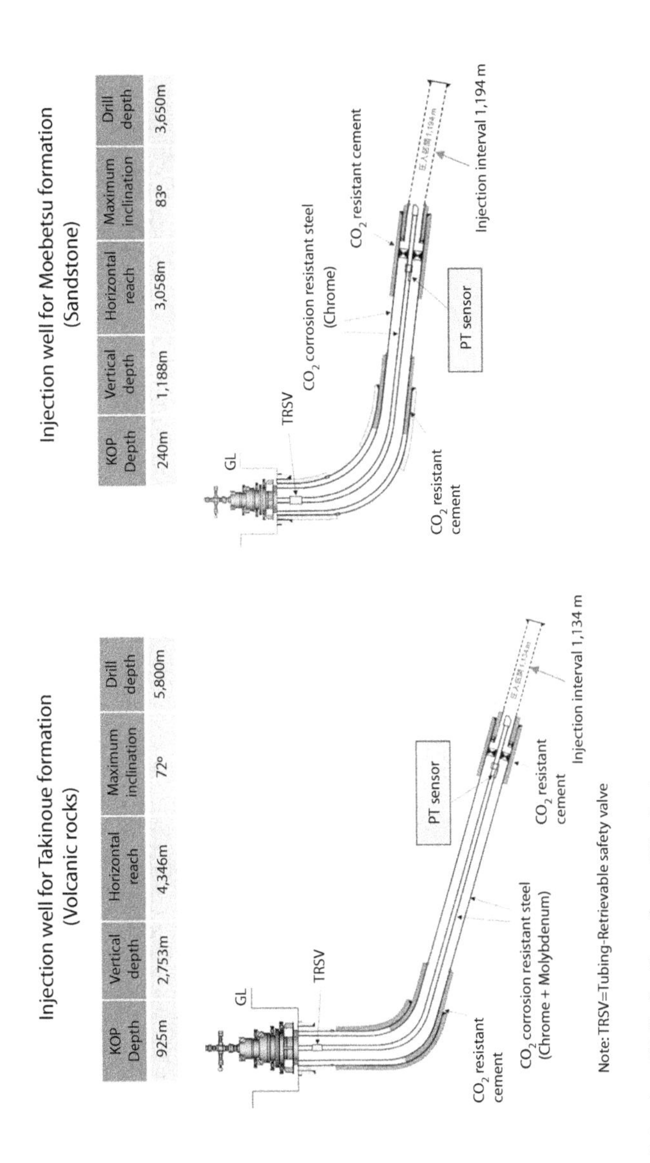

Figure 13.15 Injection Wells for Tomakomai Project.

was restarted, and the borehole pressure and temperature increased in response.

13.4.2 Feature 2: Deviated CO_2 Injection Wells Drilled From Onshore to Offshore

The injection wells are shown in Figure 13.15. The injection well for the Moebetsu Formation is an extended reach drilling (ERD) well with a maximum inclination of 83 degrees, a drill depth of 3,650 m, a vertical depth of 1,188 m and a horizontal reach of 3,058 m. The injection well for the Takinoue Formation has a maximum inclination of 72 degrees, a drill depth of 5,800 m, a vertical depth of 2,753 m and a horizontal reach of 4,346 m. The injection intervals of both injection wells exceed 1,100 m, and are completed with slotted or perforated liners.

The injection wells were drilled from onshore to offshore, resulting in significant reduction of drilling, operation and maintenance costs, and the long injection interval length is intended to enhance injection efficiency.

A tubing-retrievable safety value is installed in the upper portion of the tubing to prevent the blow out of CO_2 in the extreme case of the wellhead being destroyed by a tsunami.

The tubing and casing are made of CO_2 corrosion resistant steel (chrome steel), and CO_2 resistant cement is used to prevent CO_2 corrosion.

13.4.3 Feature 3: Application of Law Reflecting London Protocol

In Japan sub-seabed CO_2 storage is governed by the Act for the Prevention of Marine Pollution and Maritime Disaster reflecting the London 1996 Protocol, and regulated by the Ministry of the Environment (MOE). On February 22, 2016, METI submitted a permit application for offshore CO_2 storage to MOE attaching a "Marine Environmental Survey Plan". On March 31, 2016, METI received a 5 year permit from MOE. CO_2 injection into Moebetsu Formation was launched on April 6, 2016. Since the start of CO_2 injection, seasonal marine environmental surveys in compliance with the "Marine Environmental Survey Plan" have been conducted as shown in Figure 13.16.

The monitoring items are as follows:

- Seabed survey by side-scan sonar and sub-bottom profiler
- Current direction and speed survey by current meter

- Sampling of seawater by water sampler to obtain partial pressure of CO_2 (pCO_2), dissolved oxygen (DO) etc [2] and plankton observation
- Seabed mud survey by bottom sampler
- Collection of benthos by net or dredge unit
- Observation of benthos by divers or ROV

Baseline seasonal marine environmental surveys prior to CO_2 injection were conducted from August 2013 to May 2014.

13.4.4 Feature 4: Low Energy CO_2 Capture Process Utilizing High CO_2 Partial Pressure Gas

The CO_2 source is a hydrogen production unit (HPU) of an adjacent oil refinery, which supplies off gas containing approximately 50% CO_2 from a Pressure Swing Adsorption (PSA) hydrogen purification unit. In the capture facility, gaseous CO_2 of 99% purity is recovered by a commercially proven amine scrubbing process (OASE® by BASF). A two-stage absorption system with a low pressure flash tower (LPFT) shown in (Figure 13.17) reduces the amine reboiler duty in the capture system, and an energy consumption of approximately 1.16 GJ/tonne-CO_2 for CO_2 capture is achieved at almost maximum operation load as shown in Table 13.1. This is significantly lower than a conventional capture process for hydrogen production, which is typically around 2.5 GJ/tonne-CO_2. The Tomakomai capture process utilizes the high pressure of the feed gas to the CO_2 absorption tower

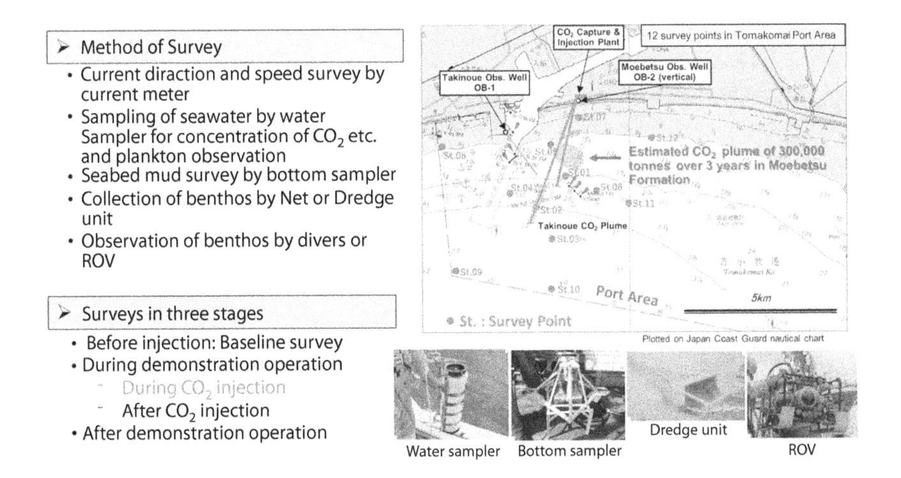

Figure 13.16 Monitoring of Marine Environment.

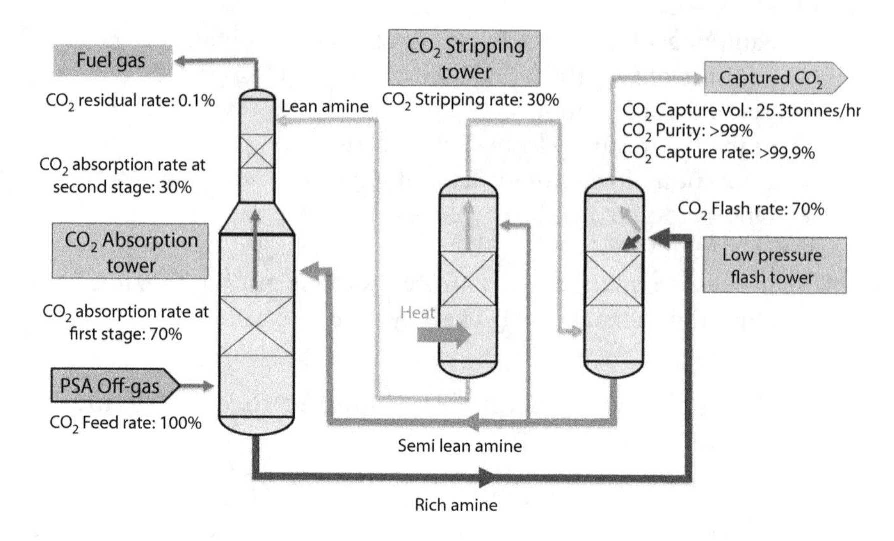

Figure 13.17 CO_2 Capture Process.

Table 13.1 CO_2 Capture energy.

Loading factor	98%
Heat energy(GJ/t-CO_2)	0.98
Electric energy (GJ/t-CO_2)	0.18
CO_2 capture energy (GJ/t-CO_2)	1.16

Note 1: Loading factor of 100% corresponds to 25.3t-CO_2/h of PSA offgas with standard composition
Note 2: CO_2 caputre energy (GJ/t-CO_2) = heat energy + electric energy
where: heat energy = reboiler heat(steam) consumption(GJ/t-CO_2)/steam boiler efficiency
electric energy = pump electricity consumption(kWh/t-CO_2) x electricity-heat conversion factor/power generation efficiency
where: steam boiler efficiency = 0.9
electricity-heat conversion factor = 0.0036(GJ/kWh)
power generation efficiency = 0.42(LHV)

and the high partial pressure of the CO_2 in the feed gas. Around 70% of the CO_2 is stripped at the LPFT and around 30% of the CO_2 is stripped at the CO_2 stripping tower. In the LPFT, CO_2 is stripped by depressurization. The thermal energy of water vapor of entrained from the CO_2 stripping tower is also utilized to strip CO_2 in the LPFT. The greater part of semi-lean amine from the LPFT is returned to the CO_2 absorption tower. As only the remaining smaller portion of semi-lean amine from the LPFT is sent to the CO_2 stripping tower, the reboiler heat required is reduced.

13.4.5 Feature 5: Injection of CO_2 Near Urban Area

Tomakomai City has a population of 172,000, and as the operation is taking place in the port area (Figure 13.18), intensive stakeholder engagement has been implemented since JFY2011.

Prior to the selection of Tomakomai as the project site, JCCS started its public outreach activities by contacting the local government. The mayor of Tomakomai expressed an interest towards the project, for the city was committed to environment conservation to address climate change, and he saw this as an opportunity to promote the legacy of Tomakomai City. In 2010, the city launched the Tomakomai CCS Promotion Association with the mayor as chairman, and secured membership from all the major local industries including the fishery cooperatives who were acutely aware of the effect of climate change from the changing conditions of their fishing. From this point, METI and JCCS started their collaboration with the local government in connecting with the local industries, especially the fishery cooperatives, keeping them involved in the process.

In 2011, when the city was still a candidate site, METI, JCCS and the local government hosted the first CCS forum in Tomakomai City, and shared information about CCS and the potential of this technology with

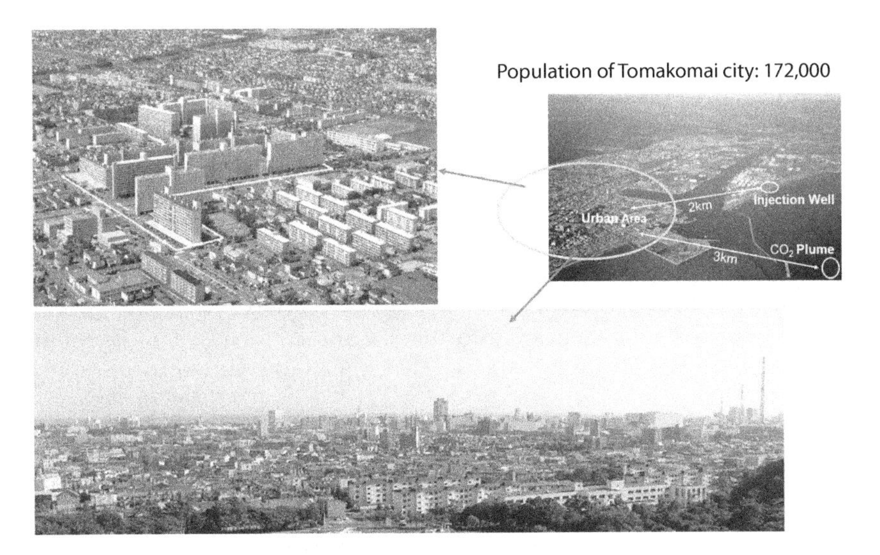

Figure 13.18 Location of CCS site and Tomakomai City (Source of photo [3]).

the local citizens. A survey conducted in conjunction with the forum found that they were concerned due to their unfamiliarity with the technology, and requested open disclosure of information, confirmation of safety and reliability, and the dissemination of information to the younger generation as a technology for the future.

From this point, JCCS started to follow a higher intensity educational approach to a variety of generations in Tomakomai City: young people, adults, and senior citizens, by holding science classes, onsite summer schools and university lectures for young people, panel exhibitions, lectures and women's workshops, annual CCS forums and site tours for adults, excursions with CCS tours and lectures for senior citizens. For these activities, in the beginning when the public did not have much knowledge about CCS, JCCS focused on providing basic information, starting with climate change, and the role and potential of CCS. After the completion of the CCS facilities, JCCS hosted site tours and onsite lectures for the local communities and received visitors from within and outside Japan.

For the people with concerns about CCS, we addressed them individually, carefully explaining the technology without denying their viewpoints. This led in some cases to persons that initially had concerns about CCS becoming leading proponents of CCS in their communities.

In addition to these activities in Tomakomai City, JCCS has been collaborating with universities in Tokyo and outside Tomakomai to hold lectures on CCS for students, and has been participating in environmental exhibitions in order to enhance the understanding of the general public.

JCCS's main approach to public outreach activities was first, sharing correct information based on current CCS technology; second, maintaining constant, thorough cooperation with the local government; third, avoiding a one-way flow of information by encouraging conversation with all parties; fourth, designing its activities to create a personal connection with the audience; fifth and last, planning all of the above mentioned approaches in consideration of the benefits to the local communities.

Another key focus of our activities is creating highly visual, easy to understand materials that would draw public attention and help us provide people with clear and accurate information and context. JCCS constantly updated and customized these materials from community to community, and place to place.

The Tomakomai project is recognized as a case of the central government, local community, and private sector working together in carrying out a CCS project. We have received a total of over 6,000 visitors to the project site from JFY 2014 to the present, of which about 400 are from overseas.

One of the most significant lessons we have learned from our public outreach experience is that there is no single formula that will work in all cases. Our core policy is the building of trust with each community we interact with, which we believe will broaden the opportunities for growing our audiences.

Our public outreach activities, examples of the information disclosure on the website, examples of the public outreach activities are shown in Figure 13.19, Figure 13.20 and Figure 13.21.

13.5 Conclusion

A large-scale CCS demonstration project is being undertaken in the Tomakomai area, Hokkaido Prefecture, Japan to demonstrate the viability of a full-cycle CCS system, from CO_2 capture to injection and storage. One hundred thousand tonnes per year or more of CO_2 is being captured, injected and stored in sub-seabed saline aquifers in Tomakomai Port. The construction and commissioning of the facilities was completed in March 2016, and CO_2 injection began in April. CO_2 injection will be conducted for three years, and monitoring for five years.

An extensive monitoring system has been installed to confirm the safety and stability of CO_2 injection. To date, it is deemed that the injection of CO_2 has not induced micro-seismicity activity. Also, a large natural

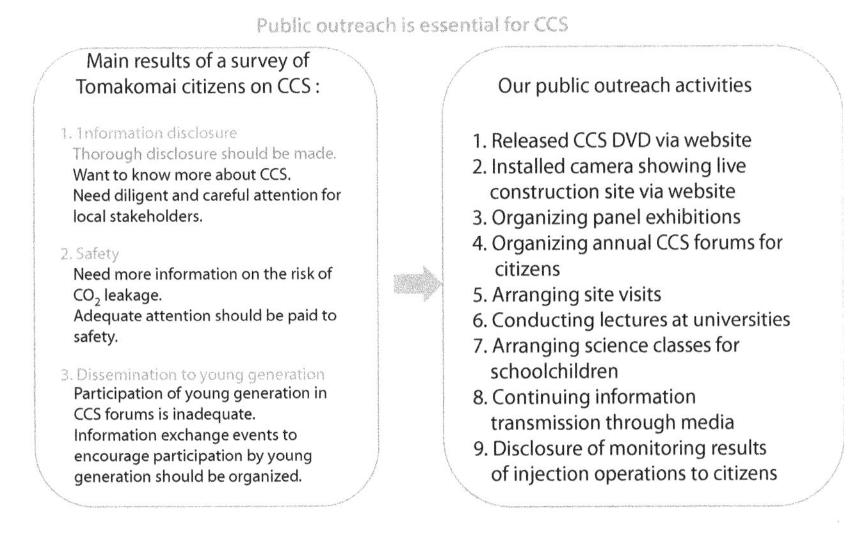

Figure 13.19 Public Outreach Activities.

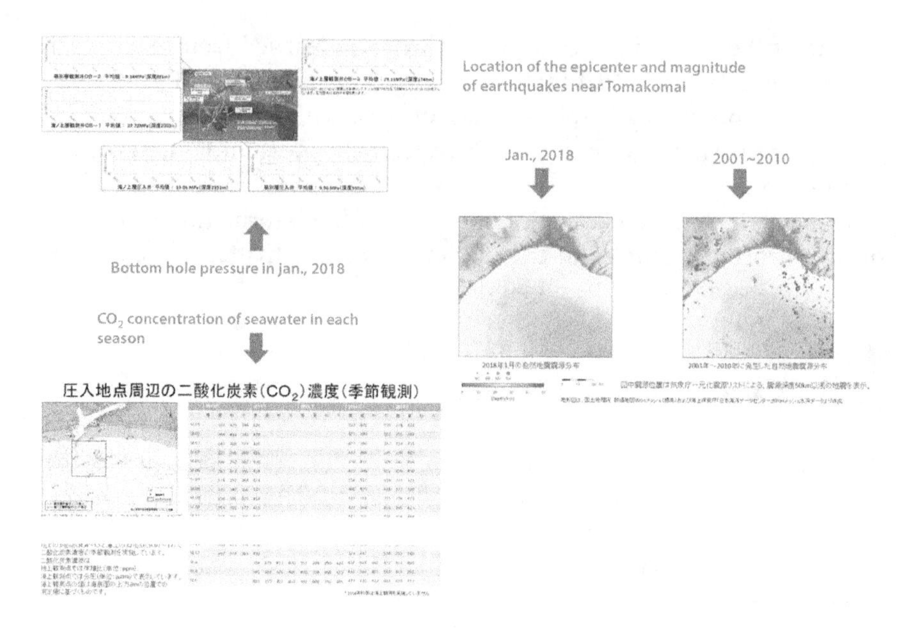

Figure 13.20 Example of Information Disclosure on Website.

Figure 13.21 Examples of Public Outreach Activities.

earthquake occurring some 40 km from the project site did not have any effect on the stored CO_2.

The CO_2 injection wells are deviated wells drilled from onshore to offshore which has resulted in a significant reduction of drilling, operation and maintenance costs, and a long injection interval length is intended to enhance injection efficiency.

Japan has amended a domestic law reflecting the London Protocol, and the monitoring of the marine environment is being conducted in accordance with this law.

A low energy CO_2 capture process utilizing high CO_2 partial pressure gas has been applied, and an energy consumption of approximately 1.16 GJ/tonne-CO_2 for CO_2 capture was achieved at almost maximum operation load.

As the injection of CO_2 is conducted near a large urban area, intensive stakeholder engagement is being implemented.

CO_2 injection into the Moebetsu Formation (sandstone layers) is proceeding smoothly. As of March 18th, 2018, 150,038 tonnes of CO_2 has been injected. The target is to inject a total of 300,000 tonnes over three years.

Acknowledgments

The authors would like to express thanks to METI for its kind permission to disclose information on the Tomakomai CCS Demonstration Project.

References

1. Japanese Government, "Strategic Energy Plan". April, 2014.
2. Kita, J., "Long-term sea water monitoring in coastal Japanese waters". *Combined Meeting of the IEAGHG Modeling and Monitoring Networks, Edinburgh Centre for Carbon Innovation*. Scotland, Edinburgh, 2016.
3. Available from: https://www.google.co.jp/search?q=%E8%8B%AB%E5%B0%8F%E7%89%A7%E5%B8%82+%E5%86%99%E7%9C%9F&tbm=isch&tbo=u&source=univ&sa=X&ved=0ahUKEwiTn_PjpPnXAhWFHpQKHRteB3AQsAQIVw&biw=1536&bih=771

14

The Development Features and Cost Analysis of CCUS Industry in China

Mingqiang Hao[1], Yongle Hu[1,*], Shiyu Wang[2] and Lina Song[3]

[1]*Research Institute of Petroleum Exploration and Development*
[2]*China University of Geosciences (Beijing)*
[3]*China University of Petroleum (Beijing)*

Abstract

Nowadays the CCS industry is developing rapidly worldwide, and projects are gradually turning from single-section items to whole-industry ones. The target of capture has expanded from power plants and natural gas processing to steel, cement, kerosene, fertilizers, and hydrogen production. Among China's CCS demonstration programs, use of CCUS takes a main role. The scale of projects in operation and under construction is relatively small, but the scale of planned projects is larger. High emission reduction cost has become the bottleneck in applications of CCS. Since the industry will develop from demonstration stage to industrial application in the future, national policies and carbon trading will be the main driving factors in the operation of CCS industrial chain, so for developing countries, in order to exploit the CCS techniques, it is more urgent to obtain technology and funds through market mechanism. Costs of CO_2 sources are comprised of three main parts: capture, compression and transportation, all of which are affected by the scale of capture. The cost of capture is also related to the concentration of emission source. For the type of high CO_2 concentration, the expense of compression takes the lead in accounting, and for the low CO_2 concentration type is capture cost. Considering the tolerance of CO_2 is lower than source cost for most oil fields, it is necessary to seek ways including technology, policies, markets, and so on to fill the gap and promote sustainable development.

**Corresponding author:* hyl@petrochina.com.cn

Ying Wu, John J. Carroll and Yongle Hu (eds.) The Three Sisters, (277–294) © 2019 Scrivener Publishing LLC

14.1 Introduction

As one of the technical approaches in the global fight against climate change, Carbon capture and sequestration (CCS) has received extensive attention all over the world. According to the IEA, by 2050, among all the emission reduction technologies used to limit the greenhouse gas concentration to 450ppm, carbon capture, utilization and sequestration (CCUS) will account for 20%. Nowadays, several main energy research institutes in the world, including the IEA, IEF and OPEC, as well as some organizations which actively advocate carbon reduction, have all agreed that CCUS technology will be taken as the main carbon reduction technology in the future [1, 2]. Among the numerous carbon reduction technologies, CCUS has the following features: (1) It has a huge potential for emission reduction; (2) It will be well combined with fossil fuels; (3) CO_2 could be utilized as a resource; (4) Costs for long-term emission reduction are relatively low. As a new industry, CCUS is still in the stage of research and demonstration in terms of the whole industry chain [3, 4].

14.2 Characteristics of CCUS Project

14.2.1 Distribution and Characteristics of CCUS Project

According to the statistics of GCCUSI, by May 2012, there are more than 300 CCUS projects in the world, of which 74 are massive integrated items, 14 are in operation and 52 are in planning. Figure 14.1 shows the distribution of the world's large-scale integrated CCUS projects. It shows that most large-scale integrated items and the largest amount (62%) of captured CO_2 are mainly concentrated in North America and Europe, followed by Canada, Australia and China [5, 6].

14.2.2 Types and Scales of CCUS Emission Sources

Emission sources of large-scale integrated projects in the world cover several categories, including power plants, natural gas treatment, syngas, coal liquefaction, fertilizer, hydrogen production, steel, refining and chemical industries. Among them, the power plant has the largest amount of carbon capture (52%), followed by natural gas treatment (20%) and syngas (14%) [7].

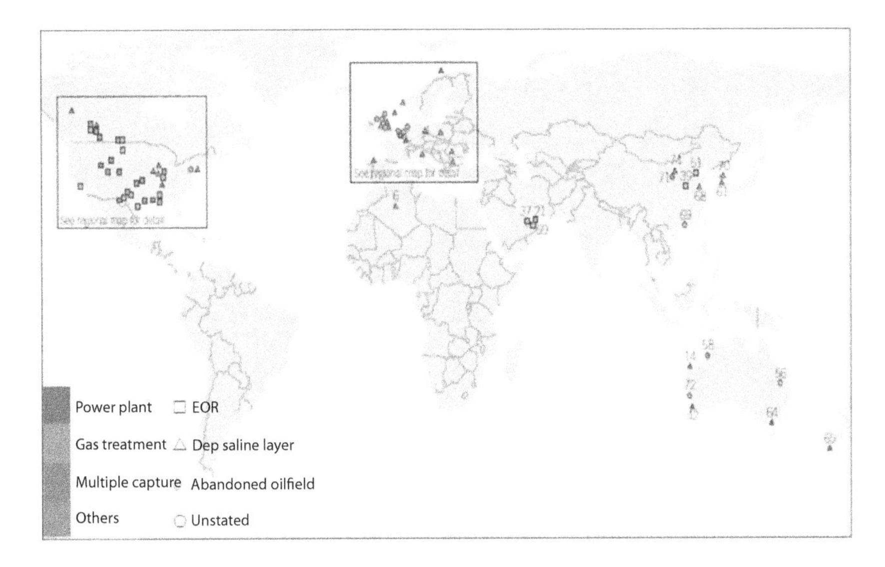

Figure 14.1 Distribution of the world's large-scale integrated CCUS projects.

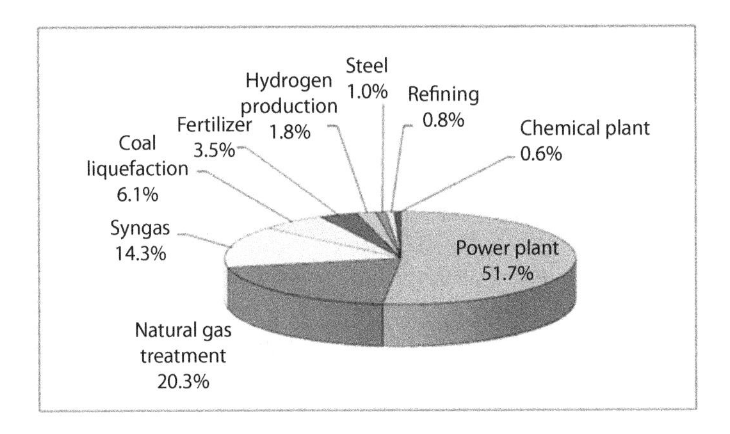

Figure 14.2 Pie chart of the carbon capture amount constitution of the world's large-scale integrated CCUS projects.

Considering the capture amount of an item on average, CO_2 capture scale in the natural gas treatment, syngas, coal liquefaction and power plants is very large, up to 5.0 ~ 8.5 million tons per year, and the average capture amount of a single project is 2.0 ~ 3.7 million tons per year. The scale of fertilizer, hydrogen production, steel, oil refining and chemical industries is relatively small, ranging from 0.6 ~ 2.5 million tons per year, with an average of 0.9 ~ 1.2 million tons per year [8].

14.2.3 Emission Scales and Composition of CO_2 Emission Enterprises in China

Main types of emission sources in China are power plants, cement, steel, and coal chemical industry, accounting for 92% of the total, and the other 4 categories account for only 8%.

Among the emission sources collected, CO_2 emissions from coal and electric power are about 10 million tons per year, while the CO_2 emissions from companies like calcium carbide, refining, synthetic ammonia and polystyrene are relatively small, ranging from hundreds of thousands of tons to several million tons, mostly within 5 million tons per year. CO_2 emissions from coal chemical industry, steel and cement vary considerably, mostly from 1 ~ 30 million tons per year [9].

14.2.4 Distributions of CO_2 Sources in China

The distribution of carbon emission enterprises of the eight major industries is consistent with the population and economic development of China, mainly in eastern China, and relatively few in the west. Most of them are power plants, cement, steel and chemical industries of low concentration. The sources of high concentration, including coal chemical industries, synthetic ammonia, calcium carbide and moderate concentration like polyethylene, are relatively small. Overall, abundant CO_2 emission sources are all founded near several main oilfields of PetroChina, such as

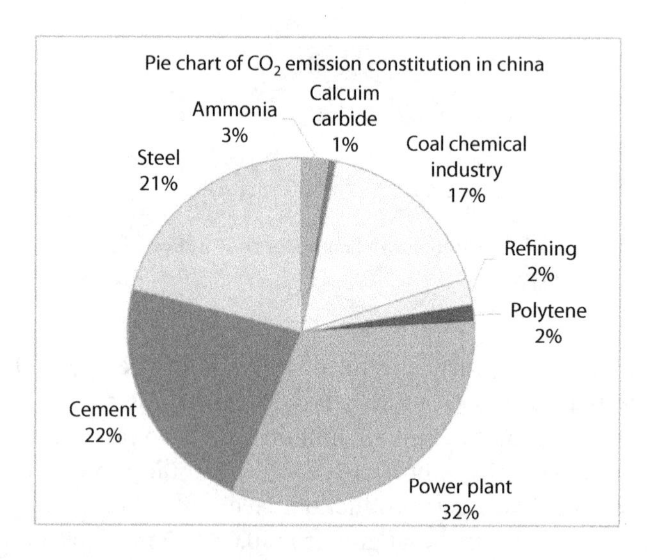

Figure 14.3 Pie chart of CO_2 emission constitution in China.

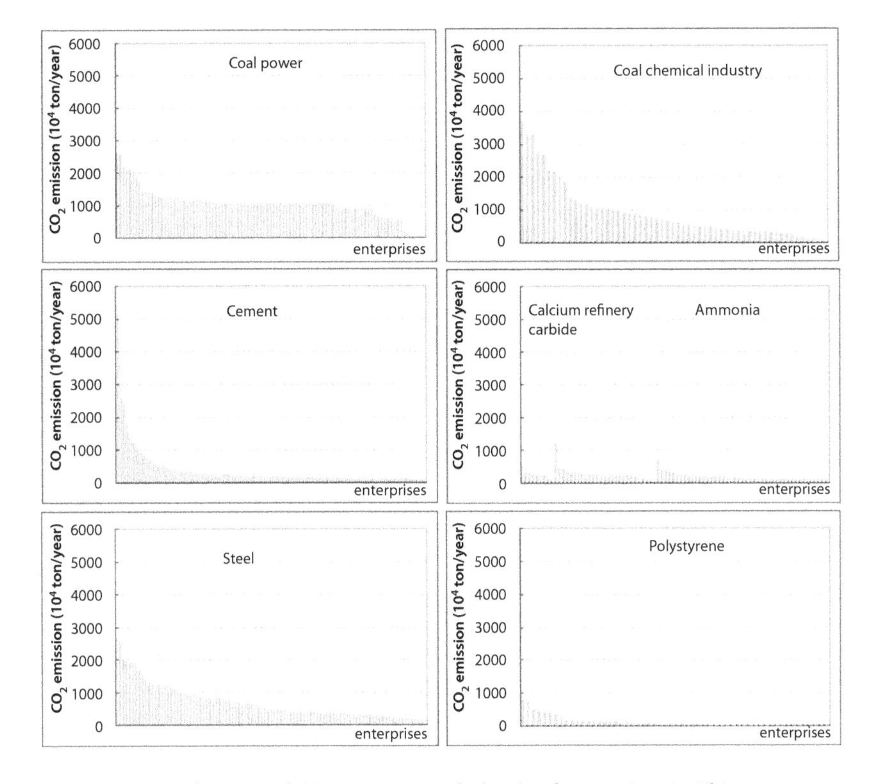

Figure 14.4 Distributions of CO_2 Emissions of 8 kinds of enterprises in China.

Xinjiang Oilfield and Changqing Oilfield, which are surrounded by emission sources of relatively high concentrations, including enterprises of coal chemical, synthetic ammonia and calcium carbide. For Huabei Oilfield, Jidong Oilfield, and Dagang Oilfield, there are more medium concentration sources like polyethylene and low concentration sources like cement and electric power enterprises, while Daqing oilfield and Jilin oilfield in northeastern China are mainly surrounded by emission sources of low concentration, including thermal power plants, refineries and iron and steel enterprises [10].

14.2.5 Characteristic Comparison Between Projects in China and Abroad

Among the world's large-scale comprehensive CCUS projects in operation and implementation (under construction) in recent years, all the emission sources of the items in operation are high concentration industries like natural gas treatment, chemical fertilizer production and syngas. For the

Figure 14.5 Distributions of carbon sources in China.

execution items, CO_2 are captured from power plants and hydrogen production. For planning projects, capture targets extended to steel, cement, kerosene, chemical and other industries. The project scale is 0.4–8.5 million tons per year, mostly more than one million tons per year. The distance of transportation is 0–315 kilometers, and most of which are more than 100 kilometers. As for the sequestration type, 62.5% of the projects in operation and execution are EOR projects, while the proportion of CO_2-EOR items in planning projects is smaller, accounting for 46%. The number of saline layer sequestration projects has increased [11].

Compared with international items, there are less whole-industry-chain CCUS projects in operation and execution in China. Their scales are smaller, while the capture targets are relatively monotonous. Pipeline transportation of long distance is hardly used. CO_2 captured is mainly from the food and chemical industry and saline layer sequestration items are less.

Table 14.1 Characteristic comparison between foreign and chinese projects.

Type	Abroad	China
Whole-Industrial-Chain Project	• More whole-industrial-chain projects	• More single-section projects
Scale of Project	• Most of which are more than 10^6 ton/year.	• The scales of projects in operation and under construction are relatively small (3,000–120,000ton/year) • Scales of the planning projects would reach to 1–5*106 ton/year.
Captured Object	• Emission sources of projects in operation and under construction are from natural gas treatment, fertilizer production and syngas which are high concentration. • For executing projects are from power plants and hydrogen production companies. • For the planning projects, categories of captured objects expanded to steel, cement, kerosene and chemical industry.	• Captured objects of projects in operation and under construction come from power plants and separated natural gas. • For the planning projects are from power plants and coal liquefaction.

(Continued)

Table 14.1 Cont.

Type	**Abroad**	**China**
Transportation	• Most items are greater than 100 km, and the longer one is up to 315 km.	• Pipeline transport only are used in Jilin among all projects in operation and under construction. The distance is 25 km. • In the planning projects, the long distance of pipeline transport could reach 100–200 km.
Type of Sequestration	• The sequestration of projects in operation and under construction is EOR method (62.5). • For the planning items, projects sequestration by saltwater layers increase (54%).	• For projects in operation and under construction, most of which are food and chemical utilization. EOR and saltwater layer sequestration are also adopted in an item separately. • For the planning projects, sequestrations of saltwater layers and abandoned oil reservoirs are considered.

14.3 Industry Patterns & Driving Modes

14.3.1 CCUS Industry Patterns at Home and Aboard

According to the combinations of capture, transportation, utilization and sequestration in the CCUS industry, three industry patterns at home and abroad can be divided. And we can find the following features of each combination:

- CU type: the capture - utilization combination, means that the captured CO_2 is directly used in the chemical, refrigeration and beverages. For instance, the 120,000-ton flue gas capture project in the Shanghai Shidongkou power plant, which is supported by the Huaneng Group, is used for the food industry.
- CTUS type: the capture - transport – utilization - sequestration combination, such as the Enid fertilizer project in Oklahoma, US. The captured gas, the amount of which is 68 million tons per year, transported by land-land pipeline used for CO_2 drive in oil reservoirs.
- CTS type: the capture - transport - sequestration combination, like the Norway's Sleipner project, undertaking in the North Sea, and captured CO_2 will be injected into saline layers.

On the part of the world's large-scale comprehensive projects, the United States, Canada and the Middle East are dominated by the CTUS-EOR industry pattern, while in Europe and Australia-New Zealand, CTS- saline layer and abandoned oilfields are the most. Among the 74 large-scale integrated projects, CTUS (CO_2-EOR) projects exceeded both in the number of items and capture amount in the projects in operation and execution. For planning ones, however, the number of CTS projects (sequestration of saline layers or abandoned oilfields) is in a dominant position.

In China, more attention is paid to the utilization of CO_2 for projects in operation and construction. Therefore, the CU model becomes the main industry pattern, while there are less whole-industry-chain projects with CTS or CTUS patterns. Among planned large-scale projects, the whole-industry-chain and permanent-sequestration items with CTUS or CTS patterns have increased [12].

14.3.2 Driving Modes of CCUS Industry

Basically, public funds, national incentive policies, tax (carbon tax), mandatory emission reduction and carbon trading are five driving modes to the development of CCUS industry at present. The progress of CCUS technology will also bring about a lower cost, and further promote the development of industry [13].

14.3.2.1 *Incentive Policy: Investment and Subsidy from Government*

Incentives include investments and subsidies from government or organizations, tax breaks for mining royalties, CO_2 price guarantees as well as government's guarantees for investment loans and low-interest loans.

Financial support: developed countries have funded 26–36 billion dollars worth of CCUS projects, with most of which being provided to power system. And CCUS industry projects in Canada, Australia and Europe are also eligible for funds. Several other agencies have also funded the CCUS programs.

Tax breaks for royalties: the 2008 Emergency Rescue Act in the US amended the tax credit for investment in advanced coal and coal gasification projects. For advanced coal capture and sequestration projects with emission reduction of more than 65%, the tax credit of 1.25 billon dollars could be increased, up to 30% of its cost, while for coal gasification projects that reduce emissions by more than 75%, the number is 250 million dollars.

CO_2 price guarantee: the Contracts for Difference (CFD), introduced in the draft of the UK Energy Act is a long-term transitional mechanism, used for different arrangements for CCUS, renewable energy and nuclear energy. In order to provide energy, a project always needs to receive a constant price model (also known as "exercise price") effectively through the protocol of CFD. If the exercise price is higher than the wholesale price of power market, the generator will receive funding for the difference; on the contrary, if price in the power market is higher than the exercise price, the generator will compensate for the difference. Exercise price will be set at a level sufficient to support different types of technology.

Government's guarantees for investment loans and low-interest loans are also incentives.

14.3.2.2 *Improvement of Carbon Pricing Mechanism*

The improvement of carbon pricing mechanism includes carbon tax and carbon trading market.

Considering the increase of global carbon dioxide emissions and the pressure on carbon taxation, carbon tax system has become the main policy for developed countries to promote the emission reduction of domestic enterprises. A carbon tax system is established to control greenhouse gas emissions, and on the other hand, carbon tax can provide a price signal for carbon emissions externalities.

The carbon tax system was introduced in Finland in 1990 firstly, followed by other countries like Sweden, Norway, the Netherlands, Denmark, Slovenia, Italy, Germany, Britain and Switzerland. In 2010, the National Development and Reform Commission (NDRC) and the Ministry of Finance jointly issued a special report on the Framework Design of China's carbon tax system and put forward a proposal for the levy of carbon tax.

Carbon trading, or greenhouse gas emissions trading, means a party of the contract can obtain a reduction amount of greenhouse gas emission by paying the other. Carbon trading can be divided into two sorts: one is quota-based trading, where buyers buy emission-reduction quotas under a "cap-and-trade" system. The other is project-based trading, where buyers buy emissions reductions from projects that prove to reduce greenhouse gas emissions like the clean development mechanism (CDM).

At present, many CCUS projects are in the stage of research and demonstration, the main drivers of which are the government's financial support and national incentive policies, as well as the drivers like taxation or other factors. Mandatory emission reduction and carbon trading market might become the main driving factors when the industry moves from the stage of demonstration to commercial operation.

14.3.2.3 The Progress of CCUS Technology: A Lower Cost

The progress of CCUS technology itself can bring about a lower cost, and further promote the development of industry. The development of CCUS technology includes expansions of transport and sequestration, reduced capital costs and technological improvements. Early estimates of the experimental costs for small-scale projects are lower than those observed costs in subsequent large-scale applications. Due to the changes in design and improvements in the early state of commercial production, the cost always increases; but then will decrease with the more mature technology and the gained experience.

Measures to reduce the cost of CO_2 emission reduction include: expanding the scale of transport and sequestration, increasing the utilization; enhancing the financial capacity of CCUS chain, reducing investment risks, enhancing investors' confidence for projects; improving the capture technology and expanding the capture capacity, as well as improving the engineering design and performance of projects.

Table 14.2 Capture costs in each stage of development ($\$/tCO_2$).

Stage of development	PC-Coal	PC-Gas	Oxy	IGCC	Average	Rate of cost reduction
Demonstration (2013-2015)	107	84	105	82	95	
Early commercialization (2020)	79	61	77	70	72	24%
Mature commercialization (2030)	70	51	67	61	62	34%

Capture cost is the main component in cost of emission reduction. The study shows that the capture cost is 24% lower in the early stage of commercialization than in the stage of demonstration, and 34% lower in the stage of mature commercialization.

14.4 Composition & Factors of CO_2 Source Cost

The CO_2 source cost mainly includes capture cost, compression cost and transportation cost. At present, there are three kinds of estimation of CO_2 capture investment in power plants and industrial enterprises: engineering quantity, regression and scale index (scale factor). The method of scale index was used in this article, when the compression and transportation costs were calculated through the method studied by David L. McCollum and Joan M. Ogden at the University of California, Davis, USA [14].

Factors influencing CO_2 source cost include CO_2 flow, emission concentration and transportation distance.

For capture cost, the main factors are emission concentration and flow rate. When the concentration is at a high level, the capture cost is low, and vice versa. At the same concentration, the capture cost decreases with the increase of CO_2 flow rate. The influence degree, however, varies with the concentration. The effect of flow rate is more obvious when CO_2 concentration is low.

As for compression cost, the main factors are CO_2 flow and transportation distance (which is at a limit of 50 km). The trend that CO_2 flow affects the cost appears as: in some flow range, the compression cost decreases with the rise of CO_2 flow. When the flow reaches a certain scale, the compression chains

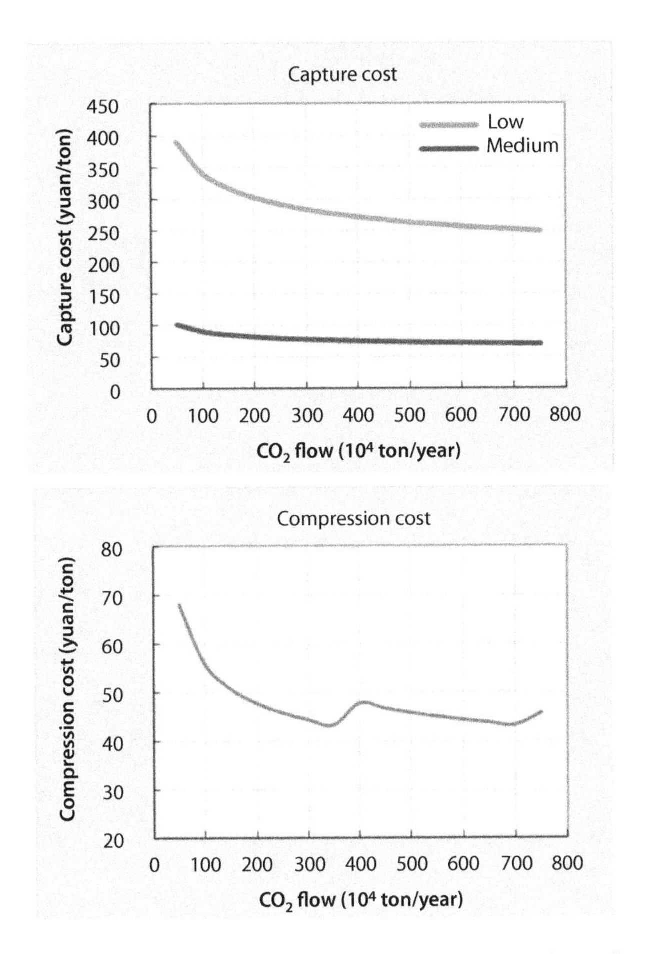

Figure 14.6 Relationship between capture cost, compression cost and CO_2 flow.

need to be increased because of the rise of the compression power, which improve costs of investment and operation, thus causing a jump of the curve.

For transportation cost, the main influence factors are transportation distance and CO_2 flow. Transportation cost increases in power function with the increase of distance and decreases in power function with the increase of CO_2 flow. The longer the distance is, the faster the decreasing speed is.

For CO_2 sources of high concentration, compression cost has a dominant place of 90%, with the cost about 54 ~ 83 yuan per ton; for medium CO_2 concentration the main part is capture cost, accounting for about 60%, and compression costs is accounted for about 35%, while most of the cost at 125 ~ 227 yuan per ton. As for CO_2 source of low concentration, capture cost is accounted for 80%, while the source cost is 300 ~ 400 yuan per ton.

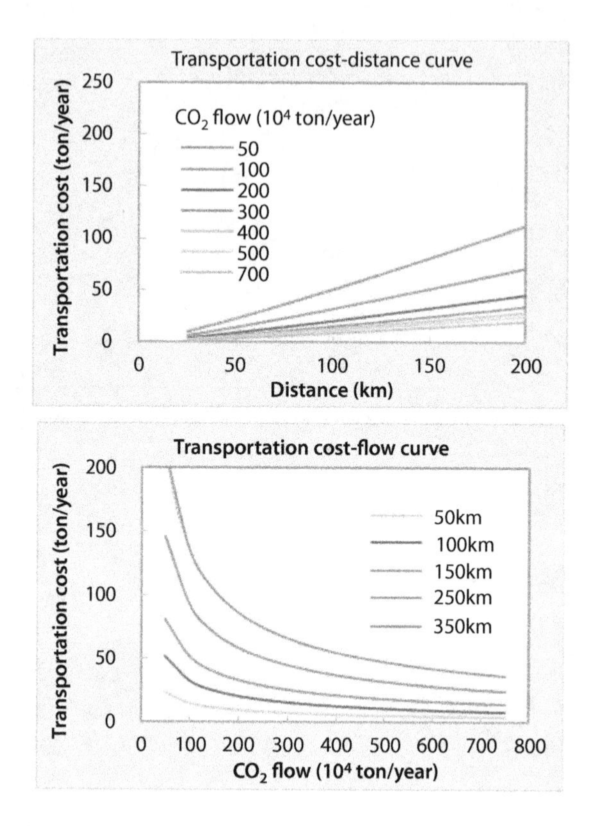

Figure 14.7 Relationship between transportation cost and distance, CO_2 flow.

The bearing capability for CO_2 is lower than the source cost in most oilfields, thus the gap between them needs to seek technology, policies, market and other ways to fill in order to forge ahead and continuously develop. According to the above results, two means could be considered to improve the situation gradually [15, 16].

On the one hand, the CO_2 source could be taken into account for the tendency of the future cost reduction to fill the cost blank. The CO_2 source cost reduction mainly refers to the capture cost of the origin, thus for the supply source which owns origins of low concentration and high capture cost, the effect on increasing the financially feasible projects of oil areas is obvious. If the cost of CO_2 source is reduced by 20–30% or even lower, the number of financially feasible oilfields could be greatly increased.

On the other hand, policy benefits could be considered throughout the links of sequestration in reservoirs, including reducing resource tax or granting certain sequestration subsidies. Then the development scales of CCUS projects of enterprises could be expanded in a great way, especially

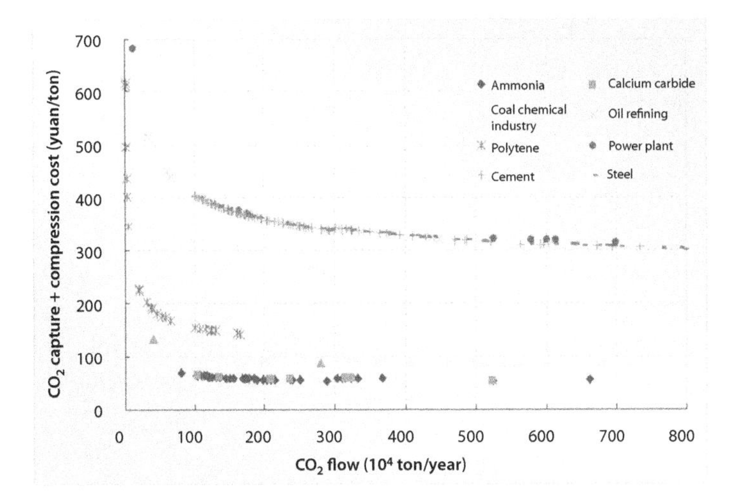

Figure 14.8 Estimated costs of different kinds of emission sources.

at low oil price. For some oil areas, it is necessary to rely on policy support. The CO_2 bearing capacity of each oil field is evaluated with the assumption that resource tax exemption and granting sequestration subsidies respectively.

If the CO_2 source cost is reduced, and the preferential policies such as exemption of resource tax or the granting of sequestration subsidies are adopted, the supply cost could be decreased from sources; meanwhile the capacity cost in sequestration is also able to be improved, so that the number of financially feasible oilfields can be greatly increased.

14.5 Conclusions

1. The types, distributions, scales and features of CCUS projects in China and abroad are compared and analyzed. The target of capture has expanded from power plants and natural gas processing to steel, cement, kerosene, fertilizers, and hydrogen production. CCUS projects are gradually turning from single-section items to whole-industry ones, of which scales become larger and larger with a broad prospect.
2. According to the combination of capture, transportation, utilization and sequestration of CCUS industry, the CCUS industry models at home and aboard can be divided into three types: CU, CTUS, and CTS. The five industry driving

modes include government and public funds, state incentives, taxes (carbon tax), mandatory emission reduction and carbon trading.

3. Cost of CO_2 sources is comprised of three main parts: capture, compression and transportation, all of which are affected by the scale of capture. The cost of capture is also related to the concentration of emission source. For the type of high CO_2 concentration, the expense of compression takes the lead in accounting, and for the low CO_2 concentration type is capture cost. For most oil fields, the tolerance of CO_2 is lower than source cost, and ways like technology, policies and markets are needed to fill the gap.

References

1. CCUS in China: Towards market transformation. *The Climate Group*, 2011.
2. Technology roadmap - carbon capture and sequestration in industrial applications. *International Energy Agency*, 2011.
3. CCUS in China: 18 hot issues. *The Climate Group*, 04, 2011.
4. Cost and performance baseline for fossil energy plants, *DOE/NETL*, 2010.
5. Cost and performance of carbon dioxide capture from power generation, IEA. *Work Report*, 2011.
6. The Global Status of CCS, Global CCS Institute, 2013.
7. Capture, C., Sequestration: assessing the economics, McKinsey & Company, 2008.
8. Raveendran, S.P., The Role of CCS as a mitigation technology and challenges to its commercialization, MIT technology and policy program, 2013.
9. Piessens, K., Laenen, B., Policy support system for carbon capture and sequestration. *Belgian Science Policy*, 2008.
10. KempAlexander, G., Kasim, D.S., The economics of CO2-EOR cluster developments in the UK Central North Sea/Outer Moray Firth. *North Sea Study Occasional Paper*, 123, 2012.
11. Carbon dioxide enhanced oil recovery, NETL, 2010.
12. Technology roadmap-carbon capture and sequestration, IEA, 2013.
13. Carbon capture and sequestration in industry application-technology synthetic working report, UNIDO, 2012.
14. McCollum, D.L., Ogden, J.M., Techno-economic models for carbon dioxide compression, transport, and sequestration & correlations for estimating carbon dioxide density and viscosity, UCD-ITS-RR-06-14, 2006.
15. Yao, W., *Carbon finance: global vision and distribution in China*. China Economic Publishing House, 2010-08-04.

16. De Boer, D., Kater, H., Carbon price research in China. *China Carbon Forum*, 10, 2013.
17. CCS Cost reduction taskforce, the UK carbon capture andsequestration cost reduction task force, 2013.

15

Study on Reasonable Soaking Duration of CO_2 Huff-and-Puff in Tight Oil Reservoirs

Yong Qin

Research Institute of Petroleum Exploration & Development, CNPC, Beijing, China

Abstract

Technology of stimulated reservoir volume fracturing brings high initial oil production in the ultra-low permeability and tight oil reservoir, but rapid decline of oil production, low cumulative oil production and poor economic effect in the depletion production restrict effective development in tight oil reservoirs. CO_2 huff-and-puff has the advantages of less investment, quick response and long stable production time, which is the most realistic technique for enhanced oil recovery in tight oil. The effect of CO_2 huff-and-puff in fractured horizontal wells is greatly influenced by the reservoir conditions and the strategy of injection and production, in which the soaking duration is the key parameter that affects oil displacement. In this paper, the actual reservoir parameters and samples of Block M in Mahu tight sandy conglomerate reservoir in Junggar Basin were used to conduct laboratory tests to study the mass transfer and diffusion rules of CO_2 in dense pores. In view of the phenomenon that the optimal soaking duration exists in the laboratory test, the influence of the relative position of the CO_2 moving front and the pressure drop front of the production on the recovery rate is analyzed. On the reservoir scale, we developed a multi-component model to simulate CO_2 huff-and-puff in fractured horizontal wells, and determined the optimal soaking duration considering the mass transfer and diffusion of CO_2. The results show that the mass transfer and diffusion rate of CO_2 is decreased gradually with time, and the diffusion coefficient is increased with the increase of pressure and temperature. Therefore, CO_2 huff-and-puff can effectively expand the swept volume by mass transfer and diffusion during soaking in the complex network of artificial fractures. We suggested that the correlation between oil recovery and soaking duration is not monotonous; that is, with the increase of soaking duration, the recovery increases first and then

Corresponding author: qinyong2012@petrochina.com.cn

Ying Wu, John J. Carroll and Yongle Hu (eds.) The Three Sisters, (295–310) © 2019 Scrivener Publishing LLC

decreases. The optimal soaking duration is to keep the CO_2 moving front at the same position as the pressure drop front of the production. It is an important technique to increase oil recovery by using CO_2 huff-and-puff for fractured horizontal wells. Determining optimal soaking duration plays an important role in improving oil production efficiency of CO_2 huff-and-puff.

Keywords: CO_2 huff-and-puff, diffusion coefficient, soaking duration, numerical simulation, tight oil reservoir

15.1 Introduction

China's low permeability reservoirs are rich in reserves, and the reasonable and efficient development of low permeability reservoirs will be one of the important ways to produce oil in China. The low permeability reservoir is characterized by low porosity and low permeability, low abundance, significant interlayer heterogeneity, etc., which lead to exposure to many problems in the production process, such as low natural productivity and high water injection pressure, and the oil recovery is less than 30% [1] relying only on natural energy production and water injection. At present, thanks to the horizontal well completion technology and the staged multi-cluster volume fracturing technology, many ultra-low permeability-tight reservoirs at home and abroad have been put into commercial production. From the development trend of overseas EOR technology, gas flooding will be one of the most promising methods to enhance the recovery in low permeability reservoirs in China [2–7], in which CO_2 huff-and-puff stimulation of low permeability reservoirs has the advantages of low investment, quick effect, among others. The factors affecting CO_2 huff-and-puff production efficiency are complex, such as reservoir parameters, operation parameters, geological parameters, etc [8–12], among which the soaking duration [13] is a very important parameter. Therefore, the characteristics of CO_2 huff-and-puff production in low permeability and tight reservoirs are studied in this paper. The influence of diffusion coefficient on the production efficiency is studied by combining the results of the laboratory tests and the numerical simulation method. The reason for the optimal soaking duration is analyzed from the relative position correlation between the CO_2 moving front and the pressure wave propagation front in open production, which will provide an important reference for determining the optimal soaking duration in CO_2 huff-and-puff blocks in the future.

15.2 Mechanism of CO_2 Huff-and-Puff in Developing Low Permeability Reservoirs

CO_2 huff-and-puff mechanisms for the stimulation of low permeability reservoir are multiple, and these mechanisms play an important role in enhancing oil well production and oil recovery. Mainly described as follows [14–18]:

15.2.1 CO_2 Mechanism for Oil Expansion

After being injected into the formation, CO_2 is dissolved in the oil in the formation under the conditions of its temperature and pressure and expands the volume of the oil, and the increased expansion can enable more oil to be produced from the formation, thereby enhancing the oil well production and the oil recovery.

15.2.2 CO_2 Mechanism for Reduction of Oil Viscosity

After being dissolved in the oil, CO_2 makes the oil volume expand, the intermolecular distance increase, the intermolecular force decrease, and the oil viscosity decrease, so that the oil seepage resistance in the formation is reduced, resulting in facilitating the oil production in the formation and enhanced oil recovery.

15.2.3 Acid Plugging Removal

After CO_2 is injected into the formation, it can react with formation water to generate H_2CO_3, which can react with carbonate cements in the rock, so as to dissolve some carbonate cements in the oil layer and improve the permeability of the oil layer.

15.2.4 Dissolved Gas Flooding

After CO_2 injection into the formation, CO_2 dissolved in the oil is separated out along with the reduction of pressure during open production, forming dissolved gas flooding to drive the oil in the formation to the wellbore, thus enhancing the oil well production.

15.2.5 Improved Oil-Water Density Ratio

After CO_2 injection into the formation, the light components of oil in the formation can be extracted, so that the oil density is increased, the density difference between oil and water is reduced, and the oil-water separation in the subsequent waterflooding is decreased, resulting in the enhanced oil recovery.

15.2.6 Improved Oil-Water Fluidity Ratio

When CO_2 is dissolved in oil, it can enable its viscosity to be decreased and its fluidity to be increased, and while CO_2 is dissolved in formation water, the viscosity of formation water will be increased, thereby decreasing the fluidity of water. Finally the reduced water-oil fluidity ratio enlarges the swept volume of the subsequent waterflooding.

15.2.7 Reduced Interfacial Tension

After CO_2 injection into the formation, it will extract and gasify the light hydrocarbon components in the oil, a large number of light hydrocarbons are mixed with CO_2 to greatly reduce the oil-water interfacial tension and the residual oil saturation, so as to enhance the oil recovery.

15.2.8 Formation of Miscible Fluids

When the CO_2 injection pressure is greater than the minimum miscible pressure, CO_2 will be miscible with the oil in the formation, thus reducing the oil-water interfacial tension to zero and significantly enhancing the oil recovery.

15.3 CO_2 Diffusion and Mass Transfer in Dense Pores

The molecular diffusion coefficient of a gas represents its diffusion capacity and is one of the physical properties of a substance. According to Fick's law, the diffusion coefficient is the mass or mole number of a substance diffused vertically through a unit area in the diffusion direction under the condition of a unit concentration gradient for a given time. In the conventional CO_2 huff-and-puff model, the influence of diffusion coefficient is often ignored, but it plays an important role in the process of CO_2 huff-and-puff in tight oil reservoirs [19, 20].

In this paper, the diffusion coefficient of CO_2 in the oil in Block M of the tight sandy conglomerate reservoir in Xinjiang Mahu, is calculated using Songyan Li's method [21]. This method is used to establish Fick diffusion equation to process diffusion and mass transfer, and a comprehensive

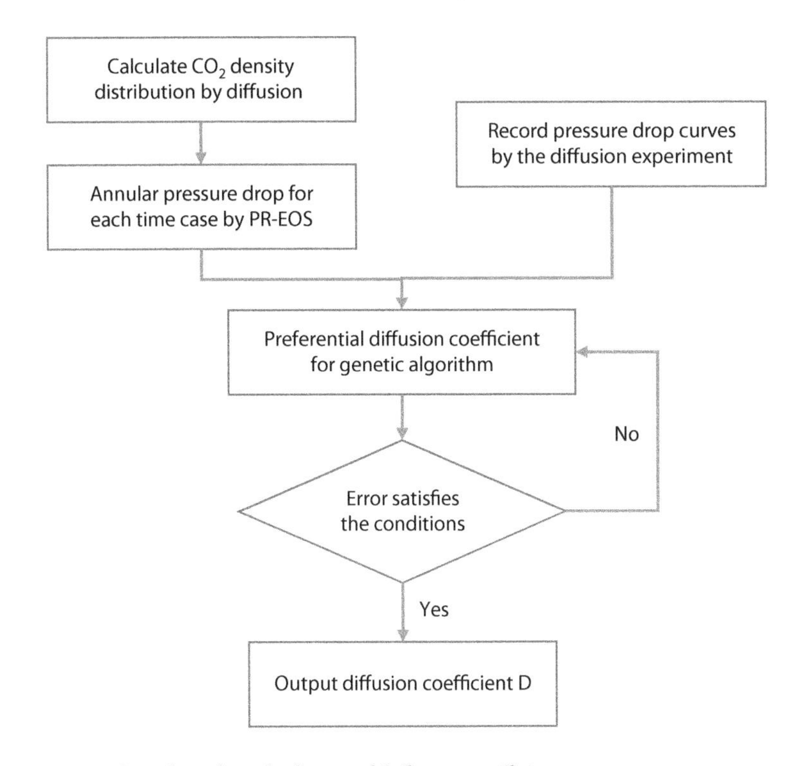

Figure 15.1 Flow chart for calculation of diffusion coefficient.

mathematical model is formed in combination with PR-EOS to process the phase-state change of CO_2 with the oil along the path. The experimental result of a physical model is combined with the calculation result of the mathematical model by using a genetic algorithm, so that the diffusion coefficient of CO_2 in a dense porous medium is obtained (see (Figure 15.1) for the calculation flow). In the method, the PVT pressure drop method is used to indirectly measure the diffusion coefficient of CO_2 in the core, and the core radial diffusion method is used in the experiment, so that the diffusion area is greatly increased, resulting in the increased pressure drop, and the measurement error is reduced; and the experimental equipment can satisfy the HPHT environment, and is more in line with the actual situation of the oil reservoir.

The mass transfer equation of CO_2 is $J=-D\nabla \cdot C$

where: J is CO_2 flow rate, $mol/m^2/s$; D is diffusion coefficient, m^2/s; C is CO_2 concentration, mol/m [3].

The theoretical pressure curve calculated by the mathematical model is fitted with the experimental pressure curve by using the genetic algorithm. The global regression diffusion coefficient is $1.05 \times 10^{-8} m^2/s$, the early stage regression diffusion coefficient is $11.54 \times 10^{-8}m^2/s$, and the later stage regression diffusion coefficient is $0.82 \times 10^{-8}m^2/s$. it can be seen that the two-stage fitting results are obviously better than the global fitting results (Figure 15.2), indicating that the mass transfer rate of CO_2 changes obviously as the diffusion pressure and CO_2 concentration in oil decrease gradually.

It can be seen from (Figure 15.3) by studying the influence of pressure on CO_2 diffusion and mass transfer: with the increase of pressure, the distance between CO_2 molecules is decreased, Van Der Waals' force is boosted and diffusion force is increased. However, as the pressure continues to increase, the influence of the resistance on diffusion coefficient

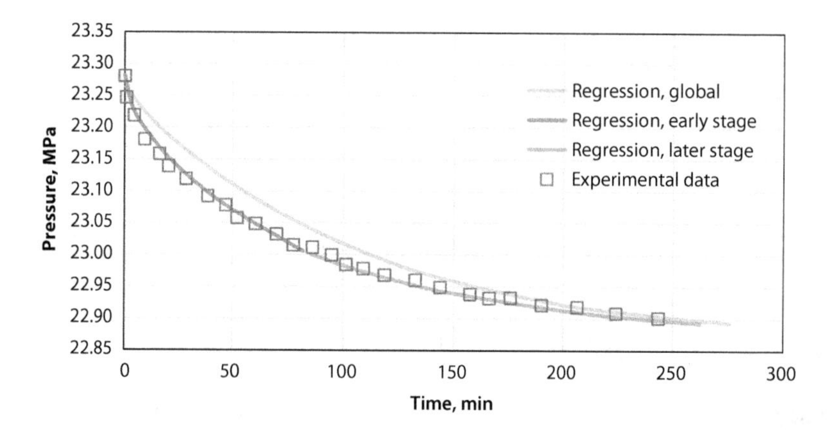

Figure 15.2 Regression curve for diffusion coefficient.

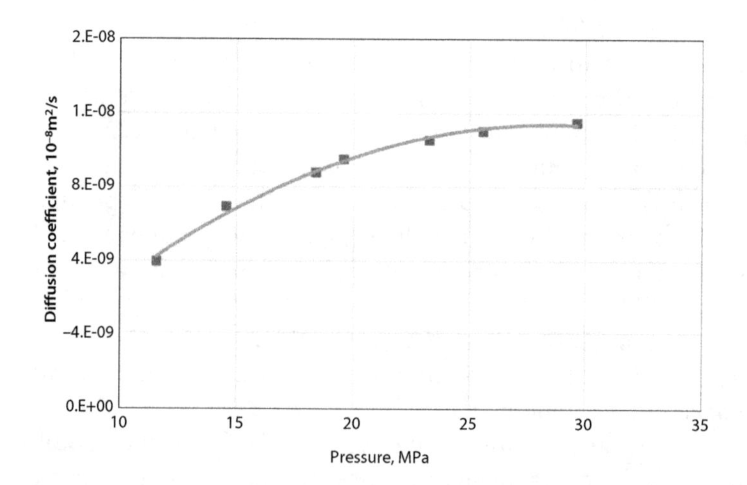

Figure 15.3 Variation curve of diffusion coefficient under different pressures.

is gradually dominated, and the trend of increasing diffusion coefficient is mitigated. In a certain range, with the increase of pressure, the solubility of CO_2 in oil is increased and the resistance of CO_2 diffusion into oil phase is decreased. However, when the pressure reaches a certain value, the solubilization of CO_2 in oil is no longer obvious, which mitigates the increasing trend of CO_2 diffusion coefficient.

A CO_2 diffusion model is established to simulate the diffusion process of a semi-infinite flat plate model;

$$\begin{cases} \frac{\partial \bar{C}}{\partial \tau} = \frac{\partial^2 \bar{C}}{\partial \bar{x}^2} - \bar{C}\frac{\partial \bar{u}}{\partial \bar{x}} - \bar{u}\frac{\partial \bar{C}}{\partial \bar{x}} \\ \bar{C} = 1 \quad \left(\bar{x} = 1,\ \tau > 0\right) \\ \bar{u} = 0,\ \frac{\partial \bar{C}}{\partial \bar{x}} = 0 \left(\bar{x} = 1, \tau > 0\right) \\ \bar{u} = 0,\ \bar{C} = 0 \left(\bar{x} < 1, \tau = 0\right) \\ \bar{u} = 0,\ \bar{C} = 1 \left(\bar{x} < 1, \tau = 0\right) \end{cases} \tag{15.1}$$

$$\begin{cases} \bar{r} = \frac{r}{r_0}\ \bar{C} = \frac{C}{C_0} \\ \tau = \frac{tD}{r_0^2}\ \bar{u} = \frac{ur_0}{D} \end{cases} \tag{15.2}$$

where: $\bar{r}$ is dimensionless radius and r_0 is core radius; $\bar{C}$ is dimensionless concentration, C_0 is CO_2 concentration in oil under experimental conditions; τ is dimensionless time; $\bar{u}$ is dimensionless diffusion velocity, u is CO_2 diffusion velocity, m/s.

The calculated results show that the diffusion coefficient of CO_2 can advance 1.77 m from the fracture to the diffusion front of the matrix after 30 days when the diffusion coefficient is in the order of 10^{-8}. The results show that CO_2 huff-and-puff in fractured tight reservoirs can enlarge swept volume by means of soaking diffusion.

15.4 Production Simulation of CO_2 Huff-and-Puff

15.4.1 Overview of Numerical Model

The selected block is Baikouquan Formation in Xinjiang Mahu sag, which belongs to the fan delta depositional system. Its reservoir thickness of T_1b_3 Formation is 5.0–22.7 m with an average of 12.95 m, which belongs to the

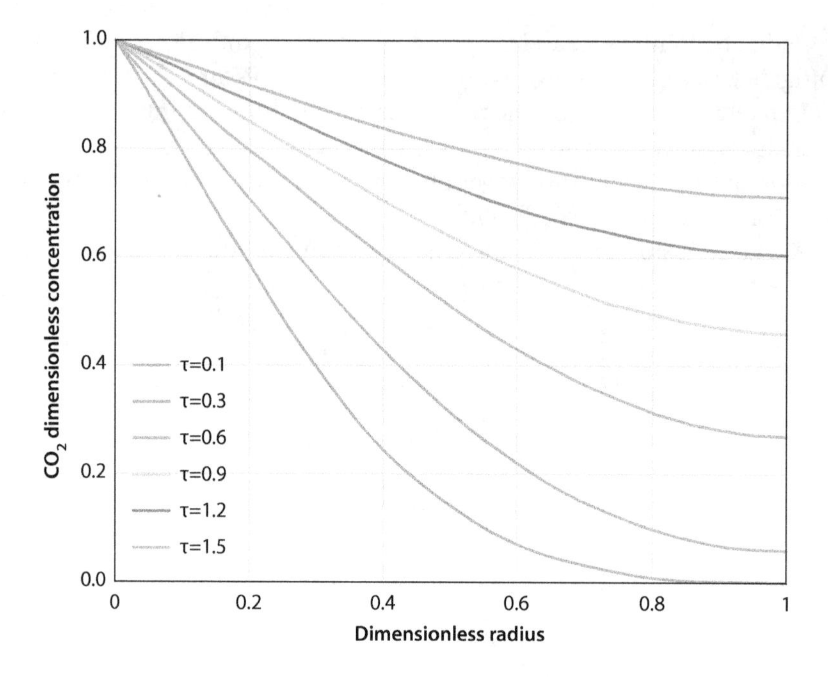

Figure 15.4 Chart of diffusion concentration.

tight sandy conglomerate reservoir. The reservoir is developed using the volume fracturing of the horizontal well in a package manner.

In the process of staged multi-cluster volume fracturing stimulation, there are lots of micro-fractures which are sheared, slid and opened, but not supported, and which are equivalent to natural fractures in modeling and simulated using a dual-porosity and dual-permeability model, that is, each grid in the model has dual properties of matrix and fracture simultaneously. In order to accurately describe the pressure distribution around hydraulic fractures, a log spacing and local encryption method is adopted. The method can not only ensure the accuracy of numerical simulation, but also save simulation time. Considering the actual situation of Mahu oilfield, the model grid number is $100 \times 50 \times 5$, the grid size is Dx = 10 m, Dy = 10 m, Dz = 3 m. The single well model of CO_2 huff-and-puff is shown in (Figure 15.5).

On the basis of satisfying the simulation precision, the original fluid components are combined. After splitting and recombining, the number of the solved equations is reduced and the calculation efficiency is enhanced. The original fluids are divided into seven pseudo-components, namely CO_2, C_1, C_2-C_4, C_5-C_6, C_{7+}, C_{11+} and C_{14+}. See Table 15.1 for related parameters of reservoirs and fluids.

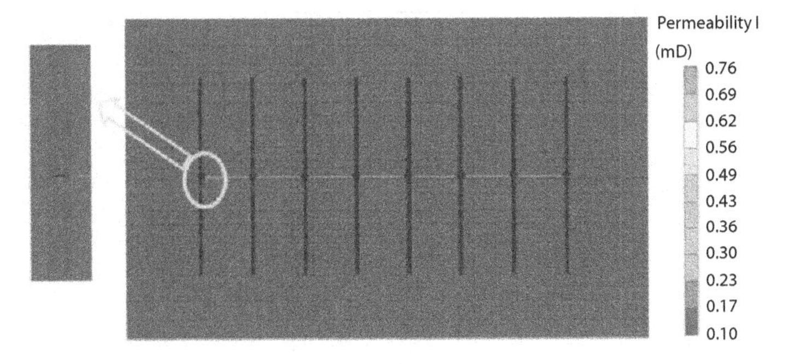

Figure 15.5 The single well model of CO_2 huff-and-puff.

15.4.2 The Process of CO_2 Huff-and-Puff

The field operation for CO_2 huff-and-puff stimulation as an effective method to enhance the oil recovery is generally divided into three stages: the first stage is the use of high pressure to inject CO_2 gas into the bottom-hole of a production well; the production well is then shut down for soaking duration in which CO_2 is brought into full contact with the reservoir oil, which is the process of "huff"; the third stage is the reopening of the production well, during which the CO_2 gas originally injected is produced together with the reservoir oil, which is the process of "puff". In the numerical simulation model, a single well is developed for three years by natural depletion, then CO_2 gas is injected with 30000 m^3/d, the formation regains the in-situ formation pressure by continuous injection, then the well is soaked for one month, and finally the well is reopened for production; if there are multiple rounds of CO_2 huff-and-puff, the above three stages of CO_2 stimulation are repeated.

15.4.3 The Influence of Diffusion Coefficient on Huff-and-Puff

In the conventional CO_2 huff-and-puff model, the influence of diffusion coefficient is often ignored, but CO_2 diffusion and mass transfer plays an important role in enhancing matrix utilization in the process of huff-and-puff in the tight oil reservoir. The results show that the diffusion coefficient affects the final oil recovery by comparing the effects of CO_2 huff-and-puff under different diffusion coefficients. When the diffusion coefficient is increased from $1 \times 10^{-9} m^2/s$ to $1 \times 10^{-7} m^2/s$, the final oil recovery is gradually augmented, as shown in (Figure 15.6). The simulation results show that within the range of experimental data, the larger the CO_2 diffusion coefficient is, the better the CO_2 huff-and-puff effect is.

Table 15.1 Physical parameter sheet of reservoirs and fluids.

Reservoir temperature	Oil density	Reservoir oil viscosity	Porosity	Permeability	In-situ formation pressure	Oil volume factor	Saturation pressure
°C	g/cm^3	mPa·s	%	10^{-3}μm^2	MPa		MPa
76	0.827	0.78	8	0.91	35	1.417	24

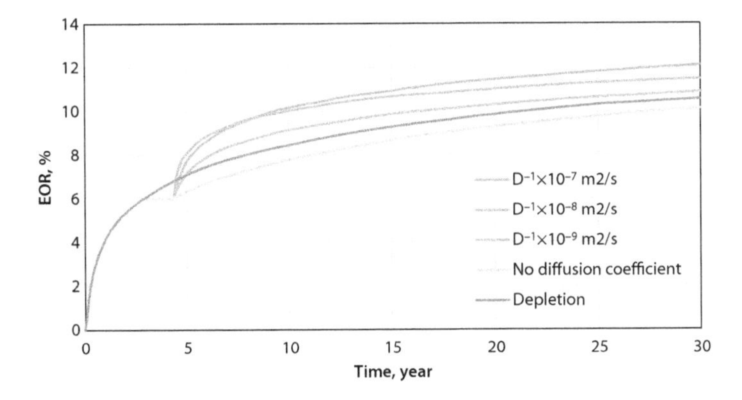

Figure 15.6 Comparison of oil recovery for CO_2 huff-and-puff under different diffusion coefficients.

15.4.4 Optimization of Soaking Duration

CO_2 gas is injected into a single well after three years of natural depletion production with the volume of 30000 m^3/d. The formation regains the in-situ pressure by continuous injection, and the soaking duration of the well is designed to be 20 days, 30 days, 40 days and 50 days respectively. After soaking, the well is opened for production until the formation pressure drops to 75% of the in-situ formation pressure, with 5 rounds of the continued huff-and-puff.

Figure 15.7 is a chart showing the correlation between the oil recovery percent and average oil draining rate with the soaking duration. It can be seen from the chart that when the soaking duration is less than 30 days, the oil recovery percent and average oil draining rate are increased with the increase of the soaking duration; when the soaking duration is more than 30 days, the recovery percent and average oil draining rate are decreased with the increase of the soaking duration, hence the optimal soaking duration is 30 days.

15.4.5 The Influence of Soaking Duration on Huff-and-Puff Results

It can be seen from the curve variation in (Figure 15.7) that the recovery percent is increased first and then decreased with the increase of the soaking duration. The reason for this phenomenon is that there are two fronts while opening production in the horizontal well, namely CO_2 moving front and pressure wave propagation front. Macroscopically, the CO_2 moving front lags behind the pressure wave propagation front when

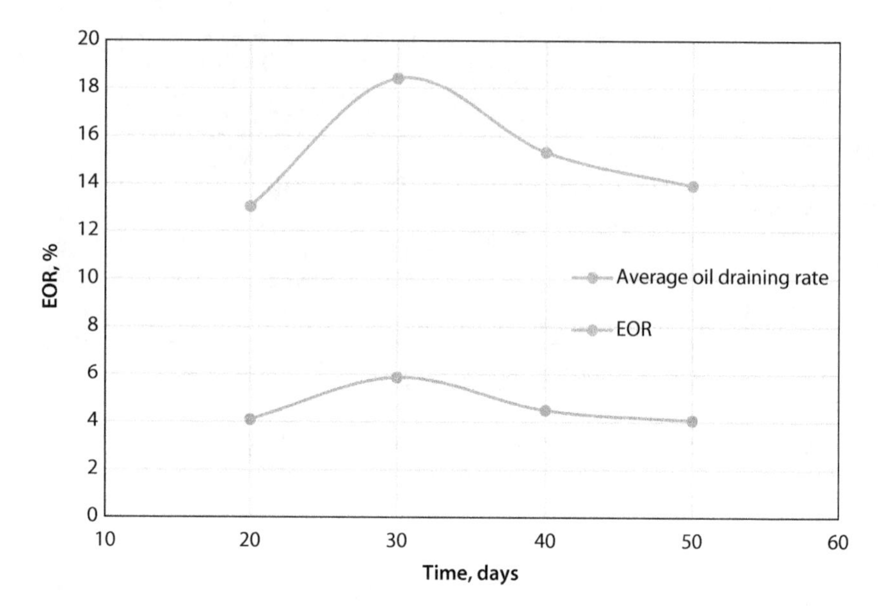

Figure 15.7 The correlation between oil recovery percent and average oil draining rate as a function of soaking duration.

the soaking duration is less than 30 days. Meanwhile, with the increase of soaking duration, the distance between the CO_2 moving front and the pressure wave front will be gradually decreased, CO_2 can act on more oil in the formation, and more oil can be produced when the well is opened. When the soaking duration is 30 days, the CO_2 moving front is ahead of the pressure wave propagation front. At this time, the longer the soaking duration is, the more CO_2 is beyond the pressure wave front, the less CO_2 in the range of pressure drop funnel is, the less functional CO_2 is in open production and the lower the recovery percent is.

It can be intuitively seen from the variation of oil displacement rate for all rounds under different soaking durations in Figure 15.8 that the influence of the relative position correlation between the CO_2 moving front and the pressure wave front is exerted on the huff-and-puff efficiency.

As can be seen from (Figure 15.8), the oil draining rate of CO_2 is decreased with the increased huff-and-puff round after 50 days of soaking, and is the largest in the 1st round and much smaller in the 2nd round than in the 1st round. The main reasons are as follows: on the 50th day of soaking, the moving front of CO_2 is ahead of the pressure wave front; for the 1st round, most of the oil is produced with CO_2 in the range of pressure drop funnel. Since the reservoir permeability is extremely low, it can be considered that there is a very small change of the front of pressure drop funnel

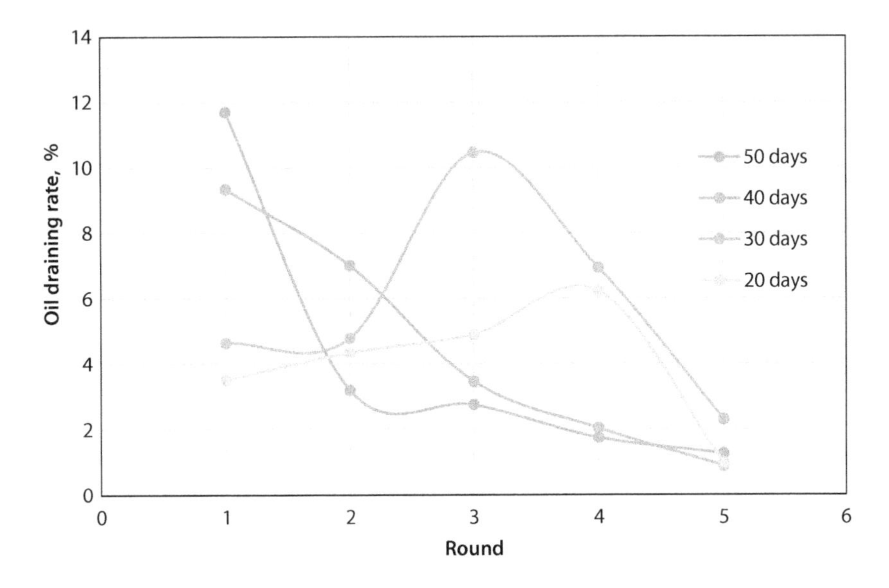

Figure 15.8 The variation of oil displacement rate for all rounds under different soaking durations.

in each huff-and-puff round. In the process of subsequent huff-and-puff stimulation, most of the oil in the range of pressure drop funnel has been recovered, and the moving front of CO_2 has exceeded the front of pressure drop, resulting in the obviously-reduced oil displacement for subsequent rounds.

On the 40th day of soaking, the oil displacement is decreased with the increased rounds, because the CO_2 moving front lags behind the pressure wave front during the first round of huff-and-puff. During the subsequent huff-and-puff operations, the flow resistance of CO_2 is decreased because the oil in the near-wellbore area is produced, which will diffuse to a deeper zone. The CO_2 moving front is close to the front of the pressure drop funnel, and the larger swept area of CO_2 results in the less decrease rate in the oil displacement, but after 3 rounds of huff-and-puff, the CO_2 moving front has exceeded the pressure wave front.

On the 30th day of soaking, the oil displacement rate is lower because the CO_2 moving front for the first round is farther than the front of the pressure drop funnel. With the increased huff-and-puff rounds, the CO_2 moving front is overlapping gradually with the pressure drop front. Meanwhile, the oil production is matched with the fluid supply capacity in the deeper formation, and the recovery percent is the highest after five rounds of huff-and-puff. When the soaking duration is 20 days, the CO_2

swept area is small, the moving front is far away from the pressure drop front all along, and the low CO_2 utilization rate leads to the low average oil displacement and the low final recovery percent, for each round.

15.5 Conclusion

1. CO_2 mass transfer and diffusion rate is characterized by fast first and slow second, CO_2 huff-and-puff stimulation in the complex network of artificial fractures can effectively expand the swept volume by way of soaking diffusion;
2. In the process of CO_2 huff-and-puff stimulation in low permeability reservoirs, there are two fronts: pressure wave propagation front and CO_2 moving front. The two fronts have different positional correlations, which lead to different variation rules of recovery percent and average oil displacement rate;
3. The CO_2 huff-and-puff recovery percent is subject to the variation rule of increase first and then decrease with the augmented soaking duration. The optimal soaking duration in Block M of Mahu oilfield is 30 days, and the optimal production efficiency can be obtained by keeping the CO_2 moving front and the pressure drop front at the same position.

References

1. Yunting, X., Qi, X., Yonggui, G., *et al.*, *Seepage flow mechanism and application in low permeability reservoir.* Bei Jing, Petroleum Industry Press. pp. 25–28, 2006.
2. Huang, Y.F., Huang, G.H., Dong, M.Z., Feng, G.M., *et al.*, Development of an artificial neural network model for predicting minimum miscibility pressure in CO_2 flooding. *Journal of Petroleum Science and Engineering*, 37(1), 83–95, 2003.
3. Erhui, L., Hu Yongle, L.B., *et al.*, Practices of CO_2 EOR in China. *Special Oil and Gas Reservoirs*, 20, 1–7, 2013.
4. Shilun, L., Zhou Shouxin, D.J., Review and prospects for the development of EOR by gas injection at home and abroad. *Oil and Gas Geology and Oil Recovery*, 9(2), 1–5, 2002.

5. Bingsen, Y., Ziqing, Z., Wei, Y., Economic analysis model about the feasibility of CO_2 miscible flooding. *Petroleum Exploration and Development*, 17(6), 95–98, 1990.
6. Hang, Z., Shaoran, R., Yin, Z., *et al.*, Evaluation methods of CO_2 EOR and geo-sequestration potential. *Petroleum Geology and Oilfield Development in Daqing*, 28(3), 116–121, 2009.
7. Pingping, S., Huaiyou, J., Yongwu, C., *et al.*, EOR study of CO_2 injection. *Special Oil and Gas Reservoirs*, 14, 1–4, 2007.
8. Binbin, Z., Guo Ping, L.M., *et al.*, Research on CO_2 stimulation mechanism and numerical simulation. *Petroleum Geology and Oilfield Development in Daqing*, 28(2), 117–120, 2009.
9. Yifei, L., Peiwen, W., Changjiu, T., Lab experiment study on CO_2 in-situ generation and huff-and-puff for producing high pour point oil. *Petroleum Geology and Oilfield Development in Daqing*, 26(3), 111–114, 2007.
10. Xiansong, Z., Xiaoqing, X., Chunbai, H., Optimization of injection and production parameters for steam huff and puff in offshore heavy oilfield. *Special Oil and Gas Reservoirs*, 22(2), 89–93, 2015.
11. Yang Shenglai, L.X., Laboratory evaluation on displacement efficiency of seven schemes of CO_2 injection in heavy oil reservoir.. *Journal of the University of Petroleum, China: Edition of Natural Science*, 25(2), 62–64, 2001.
12. Fusheng, R., Yanping, L., Chunfeng, D., *et al.*, The principle and the character of carbon dioxide huff and puff and its application in low permeability reservoirs. *Fault-Block Oil and Gas Field*, 9(4), 77–79, 2002.
13. Fei, Z., Kaoping, S., Wentao, S., *et al.*, Parameter optimization of single well CO_2 huff and puff for low permeability reservoir. *Special Oil and Gas Reservoirs*, 17(5), 70–72, 2010.
14. Hui, G., Zhang Lihua, L., Study of numerical simulation for recovery of heavy oil reservoir using CO_2 simulation. *Oil and Gas Recovery Technology*, 7(4), 13–15, 2000.
15. Joshi, R., Kelkar, M., Production performance study of West Carney Field, Lincoln County, Oklahoma. *Symposium on Improved Oil Recovery. Society of Petroleum Engineers*, 2004.
16. Xiuling, C., Zhou Zhengping, D.F., Increase production mechanism and application of CO_2 Huff and Puff technology. *Oil Drilling and Production Technology*, 24(4), 45–46, 2002.
17. Ma Kuiqian, H.Q., Jun, Z., *et al.*, Carbon dioxide huff and puff to enhance oil recovery in Nanpu 35-2 Oilfield. *Journal of Chongqing University of Science and Technology*, 13(1), 28–32, 2011.
18. Liangliang, M., CO_2 huff and puff EOR technique for low permeability oil reservoir. *Petroleum Geology and Oilfield Development in Daqing*, 31(4), 144–148, 2012.
19. Aixian, L., Peng, L., Qiang, S., *et al.*, The experimental and modeling studies on the diffusion coefficient of CO_2 in pure water. *Journal of Petrochemical Universities*, 25(6), 5–9, 2012.

20. Biao, G., Jirui, H., Chunlei, Y., *et al.*, Determination of diffusion coefficient for carbon dioxide in the porous media. *Journal of Petrochemical Universities*, 22(4), 38–40, 2009.
21. Li, S., Qiao, C., Zhang, C., Li, Z., Determination of diffusion coefficients of supercritical CO_2 under tight oil reservoir conditions with pressure-decay method. *Journal of CO_2 Utilization*, 24, 430–443, 2018.

16

Potential Evaluation Method of Carbon Dioxide Flooding and Sequestration

Yongle Hu[1], Mingqiang Hao[1], Chao Wang[2], Xinwei Liao[3] and Lina Song[3]

[1]Research Institute of Petroleum Exploration and Development
[2]China University of Geosciences (Beijing)
[3]China University of Petroleum (Beijing)

Abstract
During the process of carbon dioxide (CO_2) injection to enhance oil recovery (EOR), CO_2 geological sequestration is achieved since a large amount of CO_2 is trapped in the formation. When the reservoir pressure is above the minimum miscibility pressure (MMP), CO_2 flooding is the process of miscible flooding. Otherwise, the oil displacement process is immiscible. The EOR mechanism of miscible flooding is different from that of immiscible flooding, and the geological sequestration capacity is different. To evaluate the potential reservoir of CO_2 flooding and sequestration, accurate and reliable evaluation method is the foundation. In view of the above two situations, this paper gives the potential evaluation calculation model of CO_2 flooding and sequestration and the calculation method of CO_2 geological sequestration capacity. Also, the potential evaluation steps are introduced in the form of examples.

Keywords: CO_2, flood, sequestration

16.1 Introduction

Improving energy efficiency, developing renewable energy, CO_2 capturing and storing are three main ways to reduce greenhouse gas emission. Under current economic and technological conditions, the latter is the most effective. There are mainly three options exist for sequestration of CO_2, namely geological sequestration, marine sequestration and vegetation

Ying Wu, John J. Carroll and Yongle Hu (eds.) The Three Sisters, (311–332) © 2019 Scrivener Publishing LLC

sequestration, among which the CO_2 geological sequestration technology is the most mature. The internationally recognized major geologic bodies suitable for CO_2 sequestration are oil and gas reservoirs, saline water formations, coal seams, etc. Oil and gas reservoirs are ideal as sequestration locations because they are known to have a geologic seal that trapped hydrocarbons and they are the most feasible implementation of the current economy. Because the use of CO_2 flooding can not only significantly improve oil and gas recovery, ease pressure on energy demand, but also achieve the purpose of sequestration, it is suitable for the basic national conditions in China and is the ideal way to reduce emissions. Through field pilot test in Chinese oilfields, the good application prospect of CO_2 flooding in low permeability oil field is presented. However, CO_2 flooding is still in the pilot test stage in China. To apply this technology on a large scale, it is necessary to establish a scientific, standardized, widely adaptable and reliable potential evaluation method for CO_2 flooding and sequestration in order to further evaluate the application potential of CO_2 EOR in China, which provides a scientific basis for achieving the medium and long-term development goal of CO_2 flooding development.

16.2 CO_2 Miscible Flooding and Sequestration Potential Evaluation Model and Sequestration Capacity Calculation Method

16.2.1 CO_2 Miscible Flooding and Sequestration Potential Evaluation Model

CO_2 miscible flooding and sequestration potential evaluation model assumptions:

1. Miscibility can be realized by injecting CO_2 under the formation pressure;
2. CO_2 flooding is an isothermal process;
3. Viscous fingering can be described by Koval coefficient;
4. When the injection mode is alternating between water and gas, the water and CO_2 are injected alternately in a certain proportion;
5. There is no free gas.

According to the mass conservation equation, we can get

$$\frac{\partial C_i}{\partial t_D} + \frac{\partial F_i}{\partial X_D} = 0 \tag{16.1}$$

where subscript $i = 1$ is the water component; $i = 2$ is the crude oil components; $i = 3$ is the injected CO_2 gas component; $X_D = \frac{X}{L}$ is the dimensionless distance; $t_D = \int_0^t q \frac{dt}{dV_p}$ is the dimensionless time in the pores; C_i is the total concentration of component i; F_i is the total flow of component i:

$$C_i = C_{i1}S_1 + C_{i2}S_2 \tag{16.2}$$

$$F_i = C_{i1}f_1 + C_{i2}f_2 \tag{16.3}$$

where C_{ij} is the concentration of the component i in the j phase; subscript $j = 1$ is the water phase, $j = 2$ is the oil phase; S_j is the j phase saturation; f_j is the j phase fractional flow:

$$f_j = \frac{Q_j}{Q} \tag{16.4}$$

where Q_j is the j phase flow; Q is the total flow of oil and water.

Partial differential equation (16.1) can also be expressed

$$\frac{\partial C_i}{\partial t_D} + \left(\frac{\partial F_i}{\partial C_i}\right)_{X_D} \frac{\partial C_i}{\partial X_D} = 0 \tag{16.5}$$

$\left(\frac{\partial F_i}{\partial C_i}\right)$ defines a concentration rate. According to the compatibility condition, the concentration rate of all components at the same concentration is equal, that is,

$$V_{ci} = \left(\frac{\partial F_i}{\partial C_i}\right)\lambda, \; i = 1, 2, 3 \tag{16.6}$$

The equation (16.6) can be expressed as the eigenvalue problem, and the two characteristic rates can be solved by using the fractional flow theory.

$$\lambda\pm = 0.5\left\{F_{22} + F_{33} \pm \left[\left(F_{22} - F_{33}\right)^2 + 4F_{32}.F_{23}\right]^{1/2}\right\} \tag{16.7}$$

in which, $F_{23} = \left(\frac{\partial F_2}{\partial C_3}\right)_{C_2}$.

Rate $\lambda\pm$ defines two similar component lines (directions): fast line and slow line. The fast line must give the initial conditions (in the reservoir), and the slow line needs to explicit injection conditions. Firstly, this model calculates the two-phase evaporation and the fractional flow along the component line and obtains the fast and slow line. Then, find their intersection as is shown in Figure 16.1. (The component line changes from the fast line to the slow line).

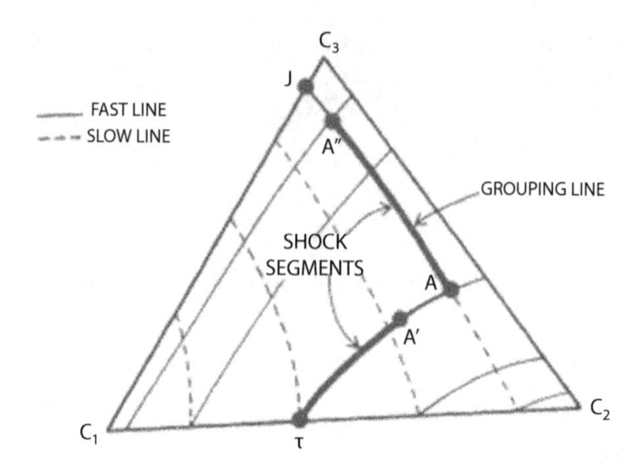

Figure 16.1 Component diagram.

In the model, the breakthrough of injected CO_2 gas and the crude oil recovery need to be calculated by the modified fractional flow theory, which includes the effects of viscous fingering, areal sweep efficiency, vertical heterogeneity and gravity separation. It is still solved by the characteristic line method.

16.2.2 CO_2 Miscible Flooding and Sequestration Capacity Calculation Method

1. Oil production and oil recovery calculation

Dimensionless time is necessary to calculate oil production and oil recovery. When calculating dimensionless time in the model, Claridge's concept of "intrusion area" and "non-intrusive area" are cited, as Figure 16.2 shows. After the areal sweep efficiency is calculated, the concept of "intrusive area" and "non-intrusive area" is used to calculate the dimensionless time.

The following formula is the dimensionless time for the one-dimensional model regardless of the non-intrusive area.

$$T_{D1} = \frac{vol_{injected(gas+water)}}{vol_{intrusive\ area}} \tag{16.8}$$

in which, $vol_{injected(gas+water)}$ is the total volume of injected water and CO_2 gas, m³; $vol_{intrusive\ area}$ is the intrusive volume of the injected solvent, m³. So,

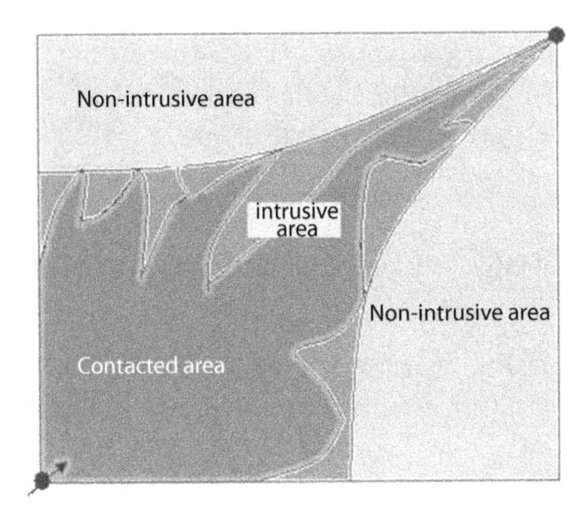

Figure 16.2 Map of contact and intrusion in 1/4 area of the five-spot pattern.

$$T_{D1} = \left(1.0 + R_{WAG}\right) \times C_3 \tag{16.9}$$

where R_{WAG} is volume ratio of injected water and CO_2 gas, it can be represented by the following formula.

$$R_{WAG} = \frac{vol_{injected\ water}}{vol_{injected\ gas}} \tag{16.10}$$

where $vol_{injected\ gas}$ is the volume of injected CO_2 gas, m^3; $vol_{injected\ water}$ is the volume of injected water m^3. If consider non-invasive area

$$TI_D = T_{D1} \times E_A \tag{16.11}$$

where TI_D is *real* dimensionless time; E_A is areal sweep efficiency.

$$TI_D = \frac{\Delta t \times Q \times 365.0}{V_p} \tag{16.12}$$

In the formula, Δt is the time step in the model calculation, a; Q is total injection rate, m^3/d; $V_p$is unit volume of well pattern, m^3:

$$V_p = A \times \phi \times h \tag{16.13}$$

In the formula, A is pattern area, m^2; ϕ is porosity, decimal; h is reservoir thickness, m.

In the two-dimensional model, introduce the second dimensionless time T_{D2}

$$T_{D2} = \frac{vol_{injected\ gas}}{vol_{pattern\ injected\ gas}} = \frac{vol_{injected\ gas}}{vol_{injected(water+gas)}} \times \frac{vol_{injected(water+gas)}}{vol_{well\ pattern}} \times \frac{vol_{well\ pattern}}{vol_{pattern\ injected\ gas}} \tag{16.14}$$

In the formula, $vol_{pattern\ injected\ gas}$ is the volume of CO_2 gas in the well pattern, m^3; $vol_{well\ pattern}$ is the volume of the well pattern, m^3.

So, after injected gas breaks through

$$T_{D2} = \frac{TI_D}{[(1.0 + R_{WAG}) \times C_3(I)]} \tag{16.15}$$

When the front edge breaks through, formula (16.15)

$$T_{D2} = E_A,\ C_3(I) = C_3(1) \tag{16.16}$$

the dimensionless time needs to be computed iteratively. Then, the output of oil, water and CO_2 after breakthrough are calculated by the following formula.

$$Q_O(I) = \frac{Q}{B_O} \times \left[\left(1.0 - F_{STL}\right) \times F_2\left(T_{D1}\right) + F_{STL} \times F_{OINIT}\right] \tag{16.17}$$

$$Q_{CO_2}(I) = \frac{Q}{B_{CO_2}} \times \left[\left(1.0 - F_{STL}\right) \times F_3\left(T_{D1}\right) + F_{STL} \times F_3(I)\right] \tag{16.18}$$

$$Q_W(I) = \frac{Q}{B_W} \times \left[\left(1.0 - F_{STL}\right) \times F_1\left(T_{D1}\right) + F_{STL} \times \left(1.0 - F_{OINIT} - F_3(I)\right)\right] \tag{16.19}$$

In the formula, B_O is the volume coefficient of oil, decimal; B_{CO_2} is the volume coefficient of CO_2, decimal; B_W is the volume coefficient of water, decimal; F_1 is water fractional flow, decimal; F_2 is oil fractional flow, decimal; F_3 is CO_2 fractional flow, decimal; F_{OINIT} is oil initial fractional flow.

$$F_1\left(T_{D1}\right) = 1.0 - F_2\left(T_{D1}\right) - F_3\left(T_{D1}\right) \tag{16.20}$$

$$F_{STL} = \frac{dE_A}{dT_{D2}} \tag{16.21}$$

In the formula, $(1.0 - F_{STL})$ represents the contribution of the intrusive area, F_{STL} represents the contribution of the non-intrusive area.

Before the breakthrough, production of oil, water and CO_2 is

$$Q_O(I) = F_{OINIT} \times \frac{Q}{B_O} \tag{16.22}$$

$$Q_{CO_2}(I) = 0.0 \tag{16.23}$$

$$Q_W(I) = \frac{Q}{B_W} \times \left[1.0 - F_{ONIT} - F_3(I)\right] \tag{16.24}$$

Therefore, we can get the stage oil recovery.

$$E_R = \frac{Q_{OT}}{N} \tag{16.25}$$

In the formula, N_{OOIP} is original oil in place, 10^6 t; Q_{OT} is the calculated oil recovery from miscible flooding with CO_2 injection, 10^6 t.

2. CO_2 sequestration capacity and sequestration coefficient calculation

In CO_2 miscible flooding model, the amount of CO_2 sequestration capacity can be calculated according to the following formula by obtaining the amount of cumulative injection and cumulative production of CO_2.

$$M_{CO_2} = Q_{CO_2IN} - Q_{CO_2O} \tag{16.26}$$

In the formula, M_{CO_2} is CO_2 sequestration capacity, 10^6 t; $Q_{CO_2IN}$is the cumulative amount of CO_2 injection , 10^6 t; Q_{CO_2O} is the cumulative amount of CO_2 production , 10^6 t.

Therefore, Stage of sequestration coefficient R_{SCO2} is

$$R_{SCO_2} = \frac{M_{CO_2}}{N_{OOIP}} \tag{16.27}$$

16.2.3 CO_2 Miscible Flooding and Sequestration Potential Evaluation Model and Sequestration Capacity Calculation Method

CO_2 immiscible flooding and sequestration potential evaluation model assumptions:

1. Miscibility cannot be realized by injecting CO_2 under the formation pressure;
2. CO_2 flooding is isothermal process;
3. Viscous fingering can be described by Koval coefficient;
4. When the injection mode is alternating between water and gas, the water and CO_2 are injected alternately in a certain proportion;
5. There is no free gas.

According to the mass conservation equation, we can obtain the same fractional flow theory equation as the CO_2 miscible flooding model. However, in CO_2 immiscible flooding, there is an oil-gas interface between CO_2 and oil due to the immiscible state between CO_2 and oil. It means that there is a separate gas phase slug. Therefore, the fractional flow theory applied in the model must consider the gas phase as a separate phase.

The total concentration C_i and total flow F_i of the component i need to be obtained by the following calculation:

$$C_i = C_{i1}S_j + C_{i2}S_2 + C_{i3}S_3, \; i = 1, 2, 3 \tag{16.28}$$

$$F_i = C_{i1}f_j + C_{i2}f_2 + C_{i3}f_3,\ i = 1, 2, 3 \tag{16.29}$$

In the formula, subscript $j = 1$ is water phase, $j = 2$ is oil phase, and $j = 3$ is gas phase; S_j is j phase saturation; $f_j$is j phase fractional flow. According to the assumptions, oil and water cannot evaporate into the gas phase, so, $C_{33} = 1, C_{13} = C_{23} = 0$.

As the same as CO_2 miscible flooding prediction model, in the model of CO_2 immiscible flooding, the breakthrough of which are oil and injection gas, and the oil recovery are also calculated by the modified fractional flow theory. The modified fractional flow theory includes the effects of viscous fingering, areal sweep efficiency, vertical heterogeneity and gravity separation. Also, it is still solved by the characteristic method. The calculation methods of oil, gas, water three-phase production and CO_2 sequestration capacity are basically similar to the CO_2 miscible flooding model, expect the difference in the description methods of influencing factors which won't be described in detail here.

16.3 Potential Evaluation Model and Calculation Method of CO_2 Sequestration

The potential of CO_2 geological sequestration can be divided into three levels: theoretical sequestration capacity, effective sequestration capacity and practical sequestration capacity.

16.3.1 Calculation Method of CO_2 Theoretical Sequestration Capacity

Theoretical sequestration capacity shows that CO_2 can be fully filled in reservoir pore space. CO_2 can also dissolve the sequestration capacity with maximum saturation in formation fluid, so it's an acceptable physical limit in the geological system.

Considering the problem of dissolved trap mechanism, according to the material balance method, the calculation model of CO_2 theoretical sequestration capacity in the reservoir is

$$\begin{aligned} M_{CO_2t} = \rho_{CO_2r} \times \Big[& E_R \times A \times h \times \varphi \times \left(1 - S_{wi}\right)/10^6 - V_{iw} + V_{pw} \\ & + C_{ws} \times \left(A \times h \times \varphi \times S_{wi}/10^6 + V_{iw} - V_{pw}\right) \\ & + C_{os} \times \left(1 - E_R\right) \times A \times h \times \phi \times \left(1 - S_{wi}\right)/10^6 \Big] \end{aligned} \tag{16.30}$$

In the formula, M_{CO_2t} is the theoretical sequestration capacity of CO_2 in reservoir, 10^6 t; ρ_{CO_2r} is the density of CO_2 in reservoir conditions, kg/m^3; E_R is oil recovery, dimensionless; B_O is the volume coefficient of oil, m^3/ m^3; A is reservoir area, km^2; h is the thickness of the reservoir, m; ϕ is reservoir porosity, dimensionless; S_{wi} is bound water saturation, dimensionless; V_{iw} is the amount of water injected into the reservoir, 10^9 m^3; V_{pw} is the amount of water produced from the reservoir, 10^9 m^3; C_{ws} is the solubility coefficient of CO_2 in water, m^3/m^3; C_{os} is the solubility coefficient of CO_2 in oil, m^3/m^3.

In the formula (16.30), the density ρ_{CO2} of CO_2 in reservoir conditions can be solved by the density map of state equation; The original oil in place N_{OOIP} can be obtained by the evaluation of petroleum resources or the reserve database; The surface area A and effective thickness h of reservoir in China can be determined according to 《Standards of the geology and mineral industry of the People's Republic of China DZ/T 0217— 2005 oil and gas reserves calculation specifications》, that is to use the contour area balance method or the well point control area method or the uniform area method. The effective porosity of reservoir rock can be directly analyzed by the core analysis data. It can also be determined by logging interpretation, and the relative error of porosity which is obtained by logging interpretation and rock analysis is not more than ±8%. Fractured Porous Reservoir must determine the matrix porosity and fracture, vug porosity or use the effective thickness section volume balance method. The water saturation S_w can be calculated by Locked-up function or J-function. Recovery E_R can be calculated by CO_2 flooding and sequestration potential evaluation model, depending on the reservoir type, drive type, reservoir characteristics, fluid property, development way, well pattern and so on. We can also select the empirical formula method, analogue method and numerical simulation method. The total volume of injected water V_{iw} and produced water V_{pw} can be obtained through the production records. The solubility coefficient of CO_2 in water C_{ws} and in oil C_{os} can be obtained according to experimental results or empirical formula.

16.3.2 Calculation Method of CO_2 Effective Sequestration Capacity

Effective sequestration capacity shows that the reservoir properties (including permeability, porosity and heterogeneity, etc.), reservoir sealing, sequestration depth of reservoir, reservoir pressure system and pore volume are considered for affecting the sequestration capacity from the technical level (including geological and engineering factors).

The following formula can be used to calculate the effective sequestration capacity of CO_2 in the reservoir when realizing CO_2 geological sequestration in the process of CO_2 EOR.

$$M_{CO_2e} = R_{SCO_2} . N_{OOIP} \tag{16.31}$$

In the formula, M_{CO_2e} is the effective sequestration capacity of CO_2 in the reservoir, 10^6 t; N_{OOIP} is original oil in place, 10^6 t; R_{SCO_2} is the sequestration coefficient of CO_2 in the reservoir.

The sequestration coefficient of CO_2 R_{SCO_2} is affected by the factors like the areal permeability, ratio of vertical and horizontal permeability $\frac{K_v}{K_h}$, areal heterogeneity, sedimentary rhythm and thickness of the reservoir, viscosity and density of the oil, CO_2 injection method, and miscible conditions of CO_2 flooding. We can calculate it through CO_2 flooding and sequestration potential evaluation model which are mentioned above. In addition, the empirical formula method, analogy method and numerical simulation method can be selected, too.

Eqn. (16.32) is an empirical calculation model for CO_2 sequestration coefficient under continuous CO_2 immiscible flooding

$$\begin{aligned} R_{SCO_2} &= 1.69 - 7.11 \times 10^{-2}A - 2.16 \times 10^{-3}B - 8.05 \times 10^{-4}C \\ &-3.93D - 1.09E + 3.66 \times 10^{-2}F + 1.05 \times 10^{-2}B \times D \\ &+2.23D \times F - 1.91B^2 + 2.91D^2 \end{aligned} \tag{16.32}$$

Eqn. (16.33) is an empirical calculation model for CO_2 sequestration coefficient under continuous CO_2 miscible flooding.

$$\begin{aligned} R_{sco_2} &= -0.55 + 2.37E - 0.17A \times D + 9.9 \times 10^{-2}A \times F + 0.13D \times F \\ &-0.32B \times E - 0.18(A - 0.81)^2 - 1.48(D - 0.59)^2 \\ &-1.2 \times 10^{-5}C^2 - 0.93(B - 0.58)^2 - 1.38(E - 1.3)^2 \end{aligned} \tag{16.33}$$

Eqn. (16.34) is an empirical calculation model for CO_2 sequestration coefficient under water and CO_2 alternate immiscible flooding

$$\begin{aligned} R_{sco_2} &= 0.18 - 3.42 \times 10^{-3}A - 2.56 \times 10^{-3}C - 0.12E + 1.15 \times 10^{-3}B \\ &-6.58 \times 10^{-4}A \times E - 7.72 \times 10^{-3}B \times C - 5.72 \times 10^{-3}C \times E \\ &-0.19B \times F + 0.39B \times E - 4.05 \times 10^{-2}E \times F \end{aligned} \tag{16.34}$$

Eqn. (16.35) is an empirical calculation model for CO_2 sequestration coefficient under water and CO_2 alternate miscible flooding

$$R_{sco_2} = 1.41 - 0.27A \times D + 3.1 \times 10^{-3}A \times C + 0.19A \times B + 0.21D \times B$$
$$-0.33F \times B - 0.45(A - 0.78)^2 - 0.79(D - 0.35)^2 \qquad (16.35)$$
$$-0.76(B - 0.54)^2 - 2.3 \times 10^{-5}C - 0.66(E - 1.0)^2$$

In all of the above, A is the aspect ratio of permeability, $A = k_{xy}/K_z$; B is the heterogeneity coefficient of reservoir; C is gas to oil mobility ratio, $C = \frac{k_{rg}^{o}\mu_o}{k_{ro}^{o}\mu_g}$; D is water saturation; E is the ratio of difference between injection pressure and MMP to MMP.; F is the ratio of difference between production pressure and MMP to MMP.

16.3.3 Calculation Method of Carbon Dioxide Practical Sequestration Capacity

The practical sequestration capacity takes technology, laws, policies, infrastructure and economy into consideration, changing with technology, policies, regulations and economic conditions. Similar to the proved reserves in the evaluation of energy and mining resources, the practical sequestration capacity needs to be calculated according to the actual situation.

16.4 An Example of CO_2 Flooding and Sequestration Potential Evaluation

Based on the theoretical method of CO_2 flooding and sequestration and the evaluation software, the actual data of typical well group in a block of A oilfield is taken as an example to illustrate the potential evaluation process of CO_2 flooding and sequestration.

16.4.1 Data Collection and Sorting

For an evaluation reservoir, the information that needs to be collected includes fluid parameters, reservoir parameters, the relative permeability parameter (including oil-water relative permeability curve and data, oil-gas relative permeability curve and data) and other data, as shown in Table 16.1.

Table 16.1 Parameter data table.

Category	Variable name	Value	Unit
Fluid Parameters	Oil Viscosity	1.403	mPa.s
	Water Viscosity	1.0	mPa.s
	Oil Phase Volume Coefficient	1.279	m^3/m^3
	Dissolved Gas and Oil Ratio	97.5	m^3/m^3
	Crude Oil Density	0.7295	Kg/m^3
	Formation Water Salinity	35506.0	ppm
	CO_2 Relative Density	0.7	-

(Continued)

Table 16.1 Cont.

Category	Variable name	Value	Unit
Reservoir Parameters	Reservoir Temperature	80.4	°C
	Reservoir Pressure	19.1	MPa
	Minimum Miscibility Pressure	18.937	MPa
	Areal Heterogeneity Coefficient	0.7	-
	Mean Permeability	0.58	mD
	Total Thickness	10.5	m
	Porosity	9.4	%
	Small Layer Numbers	3	层
	Oil Saturation	62.0	%
	Gas Saturation	0.0	%
	Water Saturation	38.0	%
	Horizontal and Vertical Permeability Ratio	0.1	-

(Continued)

Table 16.1 Cont.

Category	Variable name	Value	Unit
Other Parameters	Well Array Spacing	480	m
	Well Spacing	150	m
	Type of Well Pattern	diamond-shape inverted nine-spot	addressable well pattern mode
	Reserves	81022.5	10^3 kg

16.4.2 Calculate the Minimum Miscibility Pressure and Judge Whether or Not to Be Miscible

The minimum miscibility pressure (MMP) of CO_2 flooding is calculated according to the following formula, and judge whether the evaluation block can be miscible phase.

$$MMP = \left[-329.558 + \left(7.727 \times WM \times 1.005^{T}\right) - \left(4.377 \times MW\right)\right] /145 \quad (16.36)$$

For Changqing, Daqing and Jilin oilfield, molecular weight (MW) can be calculated with the following formula

$$MW = \left(\frac{8864.9}{G}\right)^{\frac{1}{1.012}} \quad (16.37)$$

For Xinjiang, Tuha oilfield, MW can be calculated with the following formula

$$MW = \left(\frac{12880}{G}\right)^{\frac{1}{1.012}} \quad (16.38)$$

$$G = \frac{141.5}{\gamma_0} - 131.5 \quad (16.39)$$

In the formula, MMP is the minimum miscibility pressure of CO_2 flooding, MPa; T is formation temperature, °F; γ_0 is crude oil ground density, kg/m^3.

The ground density of oil collected in this block is 0.831 kg/m^3, and the formation temperature is 177.5°F. Therefore, based on the equation above, MMP is 18.937 MPa The average formation pressure of the block is 19.1 MPa, so CO_2 miscible flooding can be realized in this block.

16.4.3 Choose Suitable Reservoirs for CO_2 Flooding and Sequestration

According to the screening criteria as shown in Table 16.2, screen reservoirs suitable for CO_2 flooding and sequestration, determine the miscible flooding reservoir and immiscible flooding reservoirs, and we can have a suitability evaluation of CO_2 geological sequestration at the same time. The evaluation of reservoir data collected in this block indicates that this block is suitable for CO_2 miscible flooding and geological sequestration.

16.4.4 Determine CO_2 EOR and Geological Sequestration Coefficient

According to the theory and method introduced in this paper, the evaluation model is applied to calculate the potential of oil recovery and

Table 16.2 Reservoir screening criteria for CO_2 flooding and sequestration.

Screening Program		Miscible Flooding	Immiscible Flooding	Marginal Reservoir	Corresponding Factors
Oil properties	Crude Oil Gravity (OAPI)	>25	>11	>11	Miscible Ability
	Crude Oil Viscosity (mPa.s)	<10	<600	-	Miscibility characteristics and Injection Ability
	Crude Oil Composition	High Content ofC2~C10			Miscible Ability

(Continued)

Table 16.2 Cont.

Reservoir Characteristic	Reservoir Depth (m)	900～3000		>900	>900	Miscible Ability
	Mean Permeability (md)	Irrespective				Injection Ability
	reservoir temperature (oC)	<90				Miscible Ability
	oil saturation (%)	>30		>30	-	EOR Potential
	variation coefficient	<0.75		<0.75	-	Sweep Efficiency
	Vertical and Horizontal Permeability Ratio	<0.1		<0.1		Buoyancy Effect
	Kh (m3)	>10-13～10-14		>10-13～10-14		Floodability
	So.φ	>0.05		>0.05	-	Sequestration Ability
	Reservoir pressure (MPa)	Original Injection	Water Flooding Post injection	-	-	Miscible Conditions
		Pi>MMP	Pcurrent>MMP			
Cap rock Characteristic	Cap Rock Sealing	No Development of Cap Rock Fractures				Security
	Escaped Quantity	★				Security

(Continued)

Table 16.2 Cont.

Economic Factors	CO2 Cost	★	★	★	Economical Operation
	Transportation Cost	★	★	★	Economical Operation
	Ground cost	★	★	★	Economical Operation

the geological sequestration coefficient with CO_2 flooding. The results show that water flooding recovery is 20.63%, CO_2 flooding recovery is 33.8%, and the CO_2 sequestration coefficient is 0.299. We can see that CO_2 flooding recovery is 13.17% higher than water flooding recovery.

16.4.5 Determine the Oil Increment and CO_2 Sequestration Capacity

According to the theory and method introduced in this paper, we can determine the different calculation methods of CO_2 sequestration capacity, the oil increment and CO_2 sequestration capacity. The calculation results are: the oil increment is 10670.6 t, CO_2 theoretical sequestration capacity is 96902.9 t, and the effective sequestration capacity is 24225.7 t.

16.4.6 Analyze Evaluation Results

Organize and analyze the evaluation results of CO_2 flooding and sequestration potential, then draw the distribution diagram of the evaluation results of CO_2 flooding and sequestration potential.

16.5 Conclusions

1. Based on fractional flow theory model, the article modifies the model considering the miscible and immiscible phase, CO_2 sweep volume, diffusion dispersion and other factors and establishes the sequestration potential evaluation model and calculation methods of CO_2 miscible flooding and immiscible flooding.
2. In three levels, namely theoretical sequestration capacity, effective sequestration capacity and practical sequestration capacity, this article establishes potential evaluation models of CO_2 geological sequestration.
3. An example is presented to introduce the methods and steps of the evaluation of CO_2 flooding and sequestration potential.

References

1. World Resources Institute., *World Resources 1996-1997*. London, Oxford University Press. pp. 326–327, 1996.

2. Burke, P.A., Synopsis: Recent progress in understanding of Co_2 corrosion. *Advances in Co_2 Corrosion.* Houston, National Association of Corrosion Engineers, 1984.
3. Gale, J., Geological sequestration of Co_2: What's known, where are the gaps, and what more needs to be done. *Greenhouse Gas Control Technologies.* 1. Amsterdam, Elsevier. pp. 207–212, 2003.
4. Gaspar, A.T.F.S., Lima, G.A.C., Suslick, S.B., "Co_2 capture and sequestration in mature oil reservoir: Physical description, EOR and economic valuation of a case of a brazilian mature field". *SPE 94181, presented at the Europec/EAGE Annual Conference, Madrid, Spain*, 2005.
5. Holt, T.J., Lindeberg, E.G.B., Taber, J.J.. *"Technologies and possibilities for larger-scale Co_2 separation and underground sequestration", SPE 63103, presented at the 2000 SPE Annual Technical Conference and Exhibition, Dallas, Texas,* 2000.
6. Orr, F.M. Jr.2004"Sequestration of Carbon Dioxide in Geologic Formations".paper SPE 88842, Distinguished Author Series.
7. Klins, M.A., *Carbon Dioxide Flooding Mechanism and Engineering Design.* Petroleum Industry Press, 1989.
8. Koval, E.J., A method for predicting the performance of unstable miscible flooding. *Soc. Pet. Eng. J.*, 145–154, 1963.
9. Paul, G.W., Lake, L.W., Gould, T.L.1984"A Simplified Predictive Model for Miscible Flooding,".paper SPE 13238.
10. Hirasaki, G.J., Application of the theory of multicomponent, multiphase Flooding to three-component, two-phase surfactant flooding. *Soc. Pet. Eng. J.*, 21(02), 191–204, 1981.
11. Development and verification of simplified prediction models for enhanced oil recovery application-CO_2 (Miscible Flood) predictive model, U. S. *Department of Energy Report.*
12. Dykstra, H., Parsons, R.L., *The Prediction of Oil Recovery by Waterflood," Secondary Recovery of Oil in the United States.* 2nd ed. pp. 160–174, 1950.
13. Tielong, C., *Tertiary Oil Recovery.* Petroleum Industry Press, 2000.
14. Jishui, J., Jishui, S., Improve oil recovery technology. *Petroleum Industry Press,* 1999.
15. Dicharry, R.M., Perryman, T.L., Ronquille, J.D., Evaluation and design of a Co_2 miscible flood project-SACROC unit, kelly-snyder field. *Journal of Petroleum Technology*, 25(11), 1309–1318, 1973.
16. Langston, M.V., Hoadley, S.F., Young, D.N., "Definitive Co_2 flooding response in the SACROC unit". *SPE/DOE 17321, presented at the Enhanced Oil Recovery Symposium, Tulsa, OK,* 1988.
17. Hawkins, J.T., Benvegnu, A.J., Wingate, T.P., McKamie, J.D., Pickard, C.D., Altum, J.T., SACROC unit Co_2 flood: Multidisciplinary team improves reservoir management and decreases operating costs. *SPE Reservoir Engineering,* 11(03), 141–148, 1996.

18. Holm, L.W., O'Brien, L.J., 1986. "Factors to consider when designing a Co_2 flood".SPE 14105, presented at the SPE 1986 International Meeting on Petroleum Engineering,Beijing, China;March 17-20.
19. Bears, D.A., Wied, R.F., Martin, A.D., Doyle, R.P., Paradis CO_2 flood gathering, injection, and production systems. *Journal of Petroleum Technology*, 36(08), 1312–1320, 1984.
20. Bachu, S., Shaw, J., Evaluation of the CO sequestration capacity in alberta's oil and gas reservoirs at depletion and the effect of underlying aquifers. *Journal of Canadian Petroleum Technology*, 42(9), 51–61, 2003.
21. Li Shilun, Z.Z., Xinquan, R., *et al.*, *Gas Injection Technology Improves the Oil Recovery*. Chengdu, Sichuan Science and Technology Press, 2001.
22. Xiangwei, L., Peili, L., The mechanism and research status of the oil recovery improved by Co_2 immiscible flooding. *Petroleum Geology and Engineering*, 21(2), 2007.
23. American, F.I., Stokka, S., Miscible flooding develop oilfield. *Petroleum Industry Press*, 6, 1989.
24. Li Xiangyuan, L.X., Standard study of carbon dioxide selecting well in low permeability thin oil reservoirs. *Oil and Gas Geology and Recovery*, 8(5), 66–68, 2001.
25. Juan, W., Ping, G., Maolin, Z., Yanmei, X., Peng, X., Study on influencing factors of hydrocarbon gas injection. *Journal of Southwest Petroleum University*, 29(4), 69–70, 2007.
26. Pingping, S., Xinwei, L., *Carbon Dioxide Geological Sequestration and Oil Recovery Improvment Technology*. Beijing, Petroleum Industry Press, 2009.
27. Yongzhi, Y., Study on Co_2 geological sequestration and oil recovery improvment evaluation system: [Doctoral Dissertation]. *China University of Petroleum (Beijing)*, 2007.

17

Emergency Response Planning for Acid Gas Injection Wells

Ray Mireault

Independent Consultant, Calgary, Alberta, Canada

Abstract

Safety planning for hazardous operations has two overriding objectives: reduce the likelihood of incurring a failure and reduce the consequences of failure. In the spirit of pro-active planning, this paper presents the thermodynamics associated with an uncontrolled flow from an acid gas injection well and the health, safety and environmental implications of a wellhead exit gas temperature in the order of -50 °C.

At such a cold exit temperature, a denser-than-air fluid has no loft and could quickly settle back to earth. With only minimal mixing with the atmosphere the result would be a cold, stable and expanding "blanket" of 2-phase acid gas at the wellsite and surrounding area. Since acid gas injection wells are normally long-life wells that inject significant volumes of acid gas over their lifetime, a loss of well control implies the potential for a large release volume as a base-case scenario.

The industry has no experience conducting emergency well control operations in an atmosphere that consists entirely of H_2S and CO_2. How would the industry deal with a scenario that:

- Precludes lighting the escaping H_2S to reduce the risks to on-site personnel and the public, as is done with a sour gas well release?
- Requires wearing bulky "space" suits and breathing apparatus if personnel attempt to regain control at the wellhead, as is done with conventional well control operations?

Corresponding author: ray.mireault.peng@gmail.com

Ying Wu, John J. Carroll and Yongle Hu (eds.) The Three Sisters, (333–346) © 2019 Scrivener Publishing LLC

- Prevents operation of the field equipment internal combustion engines within the acid gas "blanket"?

The paper first considers the feasibility of pumping oxygen and heat into the blanket in order to ignite and dissipate it and thereafter undertake conventional well control operations. It then considers the practicalities of remote well control operations for medium and high release rate scenarios and identifies areas for further study and discussion.

17.1 Introduction

It was a winter morning like any other when Tom left home for his shift at the gas plant. After three days of snow the clouds had cleared and he could see the stars in the pre-dawn sky. He enjoyed the silence in the crisp air as he checked the truck and then started the drive. "At least there's no wind today," he said to himself. "If the plows have been out, the roads should be clear".

The plant was located several kilometers off the main highway from town and Tom drove carefully as the access road wound its way through forested foothills topography. If you ran into trouble here you were stuck until another plant employee or contractor found you. Tom wasn't interested in being the topic of a plant safety incident discussion.

However, as the truck started down the section of road that led into the river valley, the headlights illuminated what appeared to be a bank of fog draped across the road. "Funny," thought Tom, "It's cold enough I wouldn't have guessed that we would have fog today".

Months later, the emergency response teams found Tom and all the other personnel who were either at the plant or on the access road that morning. At-site personnel had perished first, so they were unable to warn those approaching the plant through the morning hours. Investigators concluded that personnel had quickly lost consciousness and succumbed when they encountered a ground level "cloud" that was created when the plant's acid gas disposal well "blew out". Weeks of effort to regain well control and reduce H_2S and CO_2 air concentrations to safe levels were hampered by the stability and toxicity of the release cloud, which killed any creatures and some plants in its path.

Could the foregoing be the script for a Hollywood disaster movie? Perhaps, but the purpose of this paper is to raise industry awareness *a priori* so that unlike the *Deepwater Horizon*, any future movie is not based on an actual series of events.

17.2 Hydrocarbon Well Blowout Control Practices

The industry has established well control practices to effectively deal with a blowout from oil and/or gas wells because they take advantage of hydrocarbon fluid properties. In particular, natural gas:

1. Is lighter or less dense than air. The specific gravity of natural gas typically ranges from 0.60 to 0.70. Methane has a specific gravity of 0.55.
2. Is gaseous even at winter temperatures. The boiling point of methane at atmospheric pressure is −161.5 °C.
3. Is highly flammable. The lower and upper explosive limits for methane are 5 and 15% by volume in air (https://www.mathesongas.com/pdfs/products/Lower-(LEL)-&-Upper-(UEL)-Explosive-Limits-.pdf)

Thus, in a blowout, escaping gas rises into the upper levels of the atmosphere solely due to the density difference with air. The rising plume mixes with air to create a flammable mixture that often ignites from either an errant spark/heat source or deliberately in the case of sour gas blowouts, converting highly toxic H_2S into less dangerous SO_2. The heat generated simultaneously increases air flow to the wellhead and increases plume loft, which further mixes, dilutes and disperses the combustion products over a very large area. By continually drawing fresh air to the well at ground level, the air convection pattern allows well control operations to access the wellsite.

17.3 Acid Gas Blowout Thermodynamics

In contrast, H_2S and CO_2:

1. Are denser than air. The specific gravity for H_2S and CO_2 is 1.18 and 1.52 respectively.
2. May be a liquid or solid during a blowout. The boiling point for CO_2 is −78.5 °C; for H_2S, −60 °C.
3. Are non-flammable in the case of CO_2 while H_2S may not ignite and/or sustain a flame under blowout conditions because:

a. The H_2S concentration in the escaping plume is too low or too high. The H_2S lower and upper explosive limits are 4 to 44%.
b. The H_2S vaporization rate is too low to sustain a flame, particularly in a very cold mixture with a significant CO_2 concentration. The auto-ignition temperature of H_2S is 232 °C and its boiling point is −60 °C (https://en.wikipedia.org/wiki/Hydrogen_sulfide).

From thermodynamics, we know that acid gas will undergo a Joules-Thompson isenthalpic cooling effect as it travels from the bottom of the wellbore to surface and escapes to atmosphere. Flash calculations using the Peng-Robinson equation of state demonstrate that the acid gas escape temperature can be as low as −68 to −86 °C, depending on acid gas composition (Table 17.1). These low temperatures can occur over a wide range of reservoir temperatures as pressures approach 45 to 80% of the hydrostatic gradient. The combination of CO_2 content and cold gas temperature brings into question whether any of the compositions could be ignited.

The significance is that immediate ignition of a sour gas blowout is done to protect the public and on-site personnel from toxic H_2S and is the foundation for the industry's sour gas well control procedures. An H_2S concentration of 100 ppm is considered immediately dangerous to life and health. But an additional concern for acid gas is that high CO_2 concentrations can cause asphyxiation. According to one safety data sheet, a 10% CO_2 concentration can cause unconsciousness in 1 min or less (http://www.generalair.com/pdf/Safety%20Topics/Carbon%20Dioxide%20Asphyxiation.pdf).

17.4 Acid Gas Wellbore Dynamics

Wellbore modeling was undertaken to assess whether the escaping acid gas temperature could increase significantly during a blowout via wellbore heat transfer. Wellbores make poor heat exchangers and the short residence time in a blowout scenario further limits heat transfer capability. Nonetheless, modeling was undertaken using IHSMarkit VirtuWell™ software combined with the VMGThermo™ fluid property library, which uses the Peng-Robinson equation of state for acid gas properties.

Six of the thermodynamic cases were arbitrarily selected for modeling (see Table 17.2). An 88.9 mm production tubing size was assumed for each case. The stabilized blowout rate and acid gas escape temperature were determined from the model, assuming isenthalpic expansion from the bottom of the wellbore to surface.

Table 17.1 Acid gas escape temperature at surface.

Depth	Conditions at Bottom of Wellbore			Acid Gas Escape Temperature at Ambient Pressure °C		
m	% of Hydrostatic Gradient	kPaa	°C	78/20/2 %H_2S/ CO_2/C_1	49/49/2 %H_2S/ CO_2/C_1	20/78/2 %H_2S/ CO_2/C_1
1000	51%	5000	50	-67.2	-16	-8.1
1000	82%	8000	50	-70.2	-77.5	-68.6
1500	48%	7000	68	-66.8	-20.7	-7.8
1500	82%	12000	68	-69.2	-77.3	-86.5
2000	46%	9000	85	-66	-20.5	-3.4
2000	82%	16000	85	-68.5	-76.5	-86.4
2500	45%	11000	100	-66.7	-17.6	2.4
2500	82%	20000	100	-68	-75.9	-83.2
3000	48%	14000	120	-57.4	-8.8	13.5
3000	82%	24000	120	-67.5	-74.9	-53.2

Table 17.2 Cases selected for wellbore modelling.

Depth	Conditions at Bottom of Wellbore			Blowout Rate		Acid Gas Escape Temperature at Ambient Pressure °C		
m	% of Hydrostatic Gradient	kPaa	°C	$10^3 m^3/d$	m^3/sec	78/20/2 %H_2S/CO_2/C_1	49/49/2 %H_2S/CO_2/C_1	20/78/2 %H_2S/CO_2/C_1
1000	51%	5000	50	238	2.75	-67.2	-16	-8.1
1000	82%	8000	50	455	5.27	-70.2	-77.5	-68.6
1500	48%	7000	68			-66.8	-20.7	-7.8
1500	82%	12000	68			-69.2	-77.3	-86.5
2000	46%	9000	85	386	4.47	-66	-20.5	-3.4
2000	82%	16000	85	520	6.02	-68.5	-76.5	-86.4
2500	45%	11000	100			-66.7	-17.6	2.4
2500	82%	20000	100			-68	-75.9	-83.2
3000	48%	14000	120	318	3.68	-57.4	-8.8	13.5
3000	82%	24000	120	571	6.61	-67.5	-74.9	-53.2

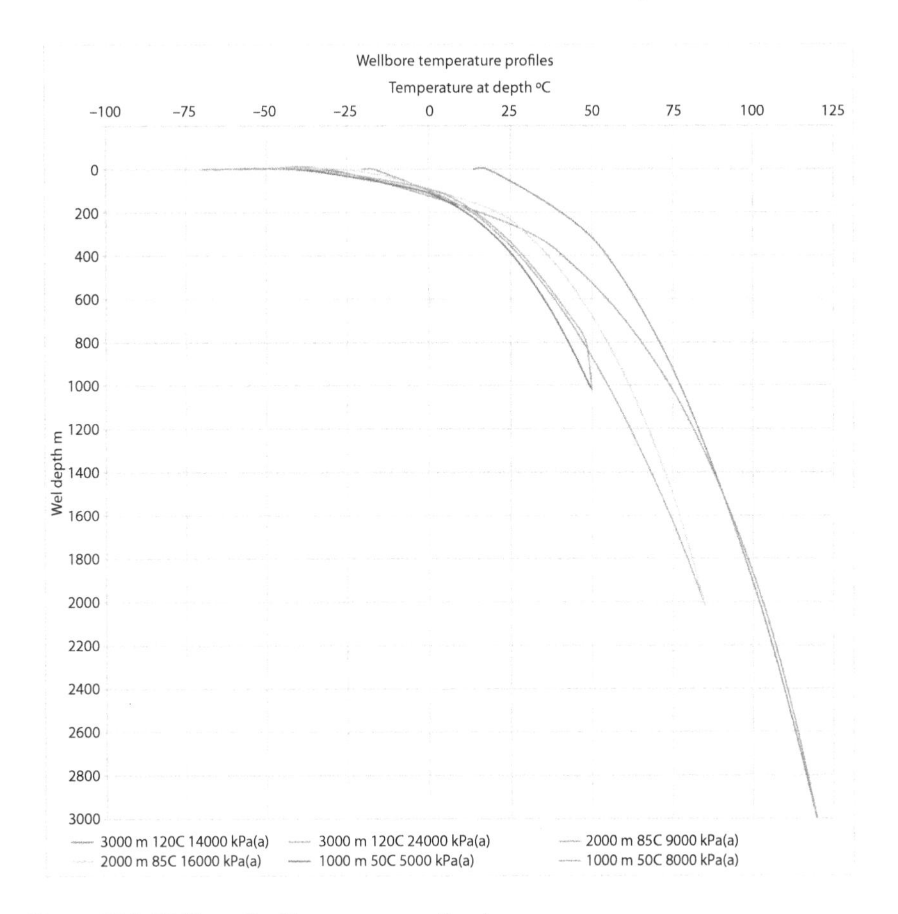

Figure 17.1 Wellbore Gas Temperature vs. Depth.

A profile of fluid temperature vs. well depth (Figure 17.1) illustrates that in all cases, the majority of cooling occurs within 150 to 300 m of the wellhead. Since 300 m of 88.9 mm tubing has a capacity of about 1.14 m^3 (0.0038 m^3/lineal m) the gas residence time in the upper 300 m of tubing is fractions of a second. With little opportunity for heat exchange, the exit gas temperature will be approximately as predicted from isenthalpic fluid expansion.

Additional Implications From the Modeling Include:.

- All other things being equal, increasing bottomhole pressure increases the blowout release rate and, in some cases, further reduces exit gas temperature by a disproportionate amount.

- A higher release rate means a bigger surface volume in a shorter period of time.

17.5 Acid Gas Plume Behaviour

The author's knowledge of plume modeling predictions for acid gas blowouts comes from his interaction with plume modeling companies in studies done for acid gas disposal well applications. With gas exit temperatures of −50 to −80 °C, simulations consistently indicated the plume would rise into the air a short distance and then fall back to earth, which is as expected for a heavier than air material. Industry safety training cautions that H_2S and CO_2 collect in low-lying areas.

A short travel time also implies little mixing between the escaping gas and air. Thus, we have the spectre of a damaged wellhead at the center of a large, toxic, stable and growing ground-level blanket of acid gas that cannot be ignited, as is done in conventional well control. How quickly the blanket envelopes the surrounding area depends on the release rate, while the thickness of the blanket and its direction of travel is influenced by air temperature, wind conditions and topography.

Note that there is no "safe" topography. Mountainous terrain may tend to create a very thick blanket and direct its expansion along the length of a valley. Treed terrain would likely trap the blanket and reduce or delay any dilution and dissipation from the wind and sun. Open prairie or an offshore blowout might reduce the depth of the cloud but allow it to spread more quickly over a larger area. Under any scenario, the consequences of a blowout could be severe.

17.6 Analogue Performance

An informal presentation [1] on a blowout from a CO_2 injection well that was part of an EOR scheme at a temperate latitude in Eastern Europe confirms that the blanket cloud of gas scenario is credible and also provides some insight on the procedure used to regain control of a limited volume release.

The presentation photos showed a stable blanket of CO_2 that reached from ground level to the tops of the deciduous trees (perhaps 10 to 12 m) in hilly, rolling farmland. After evacuating area residents and waiting till the well depleted, the exhaust from a jet aircraft engine was used to heat "cut" a path 100 + metres through the CO_2 blanket to the then dormant wellhead. Workers equipped with breathing apparatus on air lines, similar to underwater diving bell equipment, could then see well enough to go in and secure the wellhead. Pumps located outside the cloud periphery pumped fresh air through each worker's air hose. The presentation did not explain if the rig was used to secure the well and if so, how the rig engines were able to operate. Presumably, the remainder of the CO_2 blanket eventually warmed up and dissipated.

Another documented blowout incident [2] also experienced extreme cooling of the escaping effluent. Lynch *et al.*, reported frozen chokelines and casing valves and "softball size chunks of solid CO_2 spewing hundreds of feet into the air" out of surface fissures that resulted from an underground blowout while drilling a CO_2 reservoir in the Sheep Mountain Unit, Huerfano County, Colorado.

17.7 Acid Gas Well Control Procedures

Based on the foregoing, it should be obvious that industry well control procedures developed for hydrocarbon blowouts cannot be expected to work on an acid gas blowout. Nor can the "evacuate and wait" procedure used in the limited volume CO_2 blowout analogue be directly applied to an acid gas release, given the toxicity of H_2S.

From first principles, there are two ways to regain control of a well. Either:

1. Close the BOP or wellhead valves.
2. Pump a column of fluid into the wellbore with sufficient hydrostatic head to overcome bottomhole (reservoir) pressure.

Operations can be conducted at the wellsite or remotely, via a relief well. Normally, conventional well blowout control operations are initiated at the wellsite because a successful wellsite operation can more quickly bring a blowout back under control. Sometimes relief well drilling commences in concert with the wellsite operation but it is a backup plan in the event that the wellsite efforts prove unsuccessful. Although a wellsite operation may take days to weeks, it is preferable to the longer time frame required to drill a relief well and then conduct a remote kill operation.

But how do we conduct operations at a wellsite enveloped by a toxic atmosphere with limited visibility? How do we safely work on wellheads and piping at temperatures that turn steel brittle as glass? Further, how do we protect site personnel and the public from a ground level cloud if it can't be ignited? To illustrate, if a 3–6 m^3/sec release rate creates a cloud that is 4 m high and 800 m wide, it will advance at a rate of 80 to 160 m/day. At some point, the affected area becomes too large to evacuate.

For low-rate/limited volume scenarios perhaps it is possible to mix enough heat, air and fuel into the cloud to sustain a flame within the mixing zone? Could exhaust from a jet aircraft engine mix sufficient air and heat into at least a portion of the cloud to enable combustion? Maybe supplemental fuel (natural gas or propane) could be piped into the cloud to create a flammable mixture and reduce H_2S and CO_2 concentrations to safe levels? Once the cloud is under control, presumably current well control practices could be used to access and regain control at the wellhead.

The approach might prove viable for relatively low release rates and/or limited release volumes from a well and may also be applicable to situations of pipeline rupture and plant surface piping and equipment failure. However, significant study is required before truck mounted jet engines become a standard part of the safety equipment inventory. Areas for study likely include:

- The range of acid gas compositions, gas temperatures and emission rates for which ignition is both feasible and practical.
- Deployment procedures to ignite the cloud using jet engine exhaust or an alternate means of mixing and providing heat, such as piping hydrocarbons into the cloud.
- Escape plume and burning cloud behaviour under different temperatures, equipment failure configurations, topography and atmospheric conditions.
- Suitable safety equipment, clothing and breathing apparatus for such an environment.
- Identification and development of critical mechanical components and equipment that could be exposed to −60 to −80 °C temperatures.

However, at higher release rates that exceed logistical capabilities and for gas clouds that cannot be burned to gain access to the wellsite, well control will have to be achieved remotely. Under such circumstances, predrilling a relief well from a safe location may be the only option to protect public safety and the environment. One advantage of this approach is that an extremely swift response, in the order of hours, is possible if sufficient kill fluid is stockpiled at the relief wellsite and set to discharge into the disposal wellbore in the event that bottomhole pressure in the injection well drops below the parameters under regular injection operations. Areas of study include establishing the:

- Circumstances and criteria that warrant pre-drilling the relief well.
- Criteria to determine relief wellsite setback distance from the disposal wellhead.
- Requirements and procedures for drilling, completion and testing of the relief well to confirm that it is fit-for-purpose.
- Composition of the kill fluid and the required stockpile volume. Drilling mud requires constant mixing to prevent settling of solids in the storage tanks. Brine water might be an alternative but the maximum pressure in the disposal zone would have to be less than the hydrostatic gradient.
- Design philosophy for the kill system monitoring and control. Should it be a fail-safe system that activates with or without human action? What is the trigger criteria? Should the system be equipped with manual overrides and if so, how would they function?
- Responsibility for ongoing operation, maintenance, monitoring and testing of the relief well system? What arrangement best ensures the ongoing reliability of the system?

In summary, while it is hoped that a blowout of an acid gas disposal well never becomes a reality, the track record of even the nuclear industry demonstrates that people make mistakes and accidents happen. It behooves the oil industry to actively plan and prepare for an acid gas blowout because the potential consequences are too severe to ignore.

References

1. Informal presentation made during the SPE International Conference on CO_2 Capture. *Storage and Utilization held in New Orleans, Louisiana, USA*, 10–12, 2010.
2. Lynch, R.D., McBride, E.J., Perkins, T.K., Wiley, M.E., Dynamic Kill of an Uncontrolled CO_2 Well. *p. 1268. Journal of Petroleum Technology*, 37(07), 1267–1276, 1985.

Appendix

Phase envelopes for the three acid gas compositions in the paper – 78/20/2, 49/49/2 % and 20/78/2 % $H_2S/CO_2/C_1$ and the 10 bottom of wellbore conditions are presented as Figure. 17.2, Figure. 17.3 and Figure. 17.4.

Bottom-of-wellbore conditions that fall below the isenthalpic expansion line tend to transition directly from dense phase to gas and consequently have a warmer surface escape temperature, as shown in Table 17.1 and illustrated by the JT points presented in Figure 17.3 and Figure 17.4.

Bottom-of-wellbore conditions for each gas composition that are above or on the isenthalpic expansion line shown in the Figures have a similar acid gas escape temperature. For 78% H_2S it is about −68 °C, for 49% H_2S, −76 °C and for 20% H_2S, 86 °C (see Table 17.1).

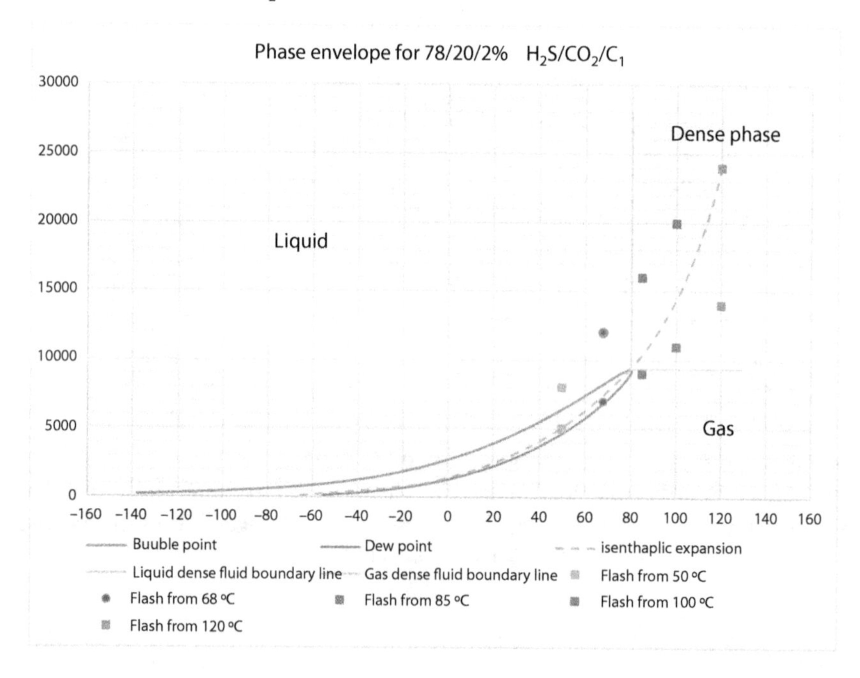

Figure 17.2 78/20/2% $H_2S/CO_2/C_1$ Phase Envelope with Isenthalpic Expansion Bottomhole Conditions.

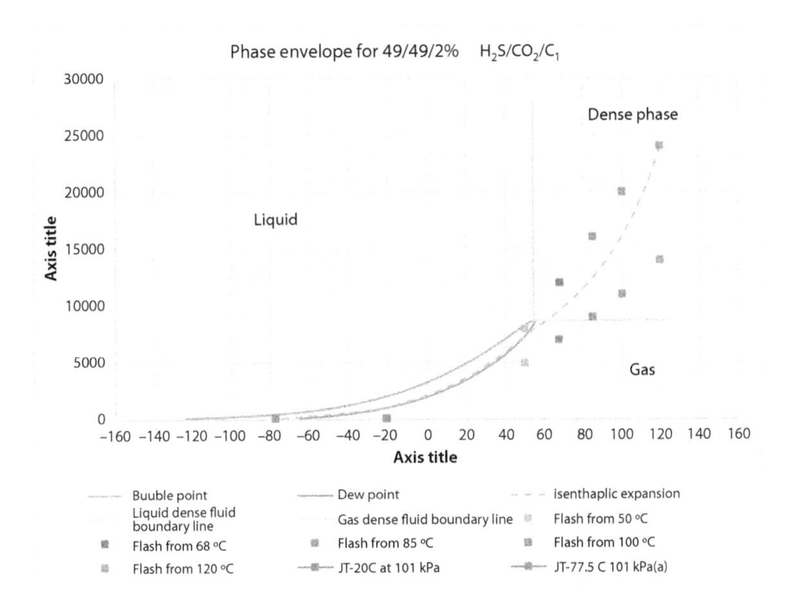

Figure 17.3 49/49/2% $H_2S/CO_2/C_1$ Phase Envelope with Isenthalpic Expansion Bottomhole Conditions.

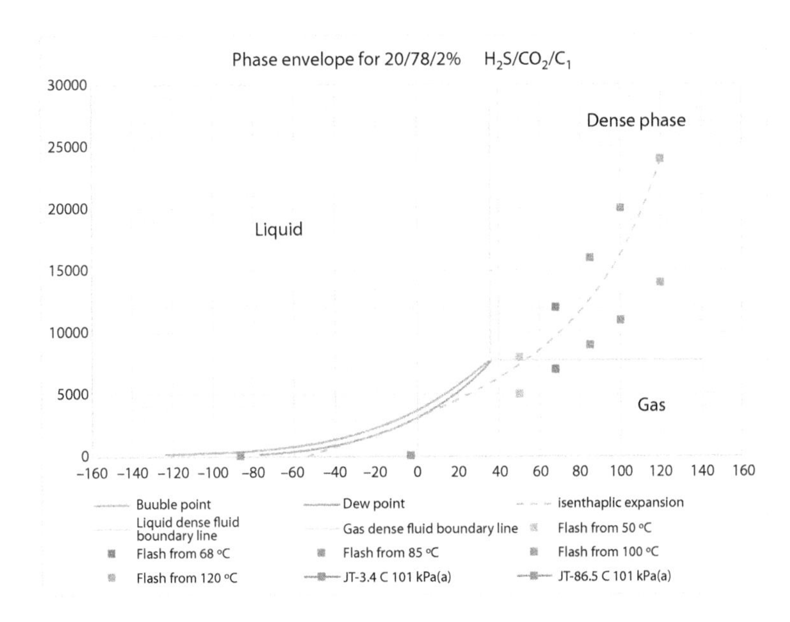

Figure 17.4 20/78/2% $H_2S/CO_2/C_1$ Phase Envelope with Isenthalpic Expansion Bottomhole Conditions.

Index

CPSIA information can be obtained
at www.ICGtesting.com
Printed in the USA
BVHW040712160619
551070BV00009B/64/P

9 781119 510062